Fuel Flexible Energy

Related titles

Combined Cycle Systems for Near-Zero Emission Power Generation
(ISBN: 978-0-85709-013-3)

Biomass Combustion Science, Technology and Engineering
(ISBN: 978-0-85709-131-4)

Advances in Clean Hydrocarbon Fuel Processing
(ISBN: 978-1-84569-727-3)

Woodhead Publishing Series in Energy: Number 91

Fuel Flexible Energy Generation

Solid, Liquid and Gaseous Fuels

Edited by

John Oakey

AMSTERDAM • BOSTON • CAMBRIDGE • HEIDELBERG
LONDON • NEW YORK • OXFORD • PARIS • SAN DIEGO
SAN FRANCISCO • SINGAPORE • SYDNEY • TOKYO
Woodhead Publishing is an imprint of Elsevier

Woodhead Publishing is an imprint of Elsevier
80 High Street, Sawston, Cambridge, CB22 3HJ, UK
225 Wyman Street, Waltham, MA 02451, USA
Langford Lane, Kidlington, OX5 1GB, UK

Copyright © 2016 Elsevier Ltd. All rights reserved.

No part of this publication may be reproduced or transmitted in any form or by any means, electronic or mechanical, including photocopying, recording, or any information storage and retrieval system, without permission in writing from the publisher. Details on how to seek permission, further information about the Publisher's permissions policies and our arrangements with organizations such as the Copyright Clearance Center and the Copyright Licensing Agency, can be found at our website: www.elsevier.com/permissions.

This book and the individual contributions contained in it are protected under copyright by the Publisher (other than as may be noted herein).

Notices
Knowledge and best practice in this field are constantly changing. As new research and experience broaden our understanding, changes in research methods, professional practices, or medical treatment may become necessary.

Practitioners and researchers must always rely on their own experience and knowledge in evaluating and using any information, methods, compounds, or experiments described herein. In using such information or methods they should be mindful of their own safety and the safety of others, including parties for whom they have a professional responsibility.

To the fullest extent of the law, neither the Publisher nor the authors, contributors, or editors, assume any liability for any injury and/or damage to persons or property as a matter of products liability, negligence or otherwise, or from any use or operation of any methods, products, instructions, or ideas contained in the material herein.

ISBN: 978-1-78242-378-2 (print)
ISBN: 978-1-78242-399-7 (online)

British Library Cataloguing-in-Publication Data
A catalogue record for this book is available from the British Library

Library of Congress Control Number: 2015951098

For information on all Woodhead Publishing publications
visit our website at http://store.elsevier.com/

Contents

List of contributors	ix
Woodhead Publishing Series in Energy	xi

Part One Introduction and fuel types — 1

1 Introduction to fuel flexible energy — 3
James G. Speight

1.1	Introduction	3
1.2	Conventional energy sources	5
1.3	Unconventional energy sources	11
1.4	Fischer–Tropsch process	15
1.5	Electrical energy	16
1.6	Fuel flexibility	19
1.7	Conclusions	20
	References	22

2 Solid fuel types for energy generation: coal and fossil carbon-derivative solid fuels — 29
P. Grammelis, N. Margaritis, E. Karampinis

2.1	Introduction	29
2.2	Fossil fuel feedstocks	29
2.3	Fuels derived from coal	44
2.4	Coal supply chain main characteristics	48
2.5	Future trends	53
2.6	Summary	55
	Sources of further information	56
	References	56

3 Biomass and agricultural residues for energy generation — 59
Eija Alakangas

3.1	Introduction	59
3.2	Biomass resources and supply chains	64
3.3	Biomass properties and measurement of properties	70
3.4	Future trends	90
	Symbols and abbreviations	92
	Terminology	92
	References	95

Part Two Fuel preparation, handling and transport 97

4 Biomass fuel transport and handling 99
Michael S.A. Bradley
  4.1 Introduction 99
  4.2 The challenges of biomass handling 102
  4.3 Sources and types of biomass, and classifications according to handling properties 108
  4.4 Other considerations for compatibility of different fuels with a handling system 113
  4.5 Conclusions 118
       References 120

5 Fuel pre-processing, pre-treatment and storage for co-firing of biomass and coal 121
Michiel C. Carbo, Pedro M.R. Abelha, Mariusz K. Cieplik, Carlos Mourão, Jaap H.A. Kiel
  5.1 Handling and storage of biomass at coal-fired power plants 121
  5.2 Biomass pre-treatment technologies 125
  5.3 Industrial-scale experience with pre-treated biomass 126
  5.4 Biological degradation 126
  5.5 Pneumatic conveying 127
  5.6 Mechanical durability and storage 132
  5.7 Explosivity 133
  5.8 Conclusions and future trends 141
       Nomenclature 141
       Acknowledgement 142
       References 142

6 Production of syngas, synfuel, bio-oils, and biogas from coal, biomass, and opportunity fuels 145
James G. Speight
  6.1 Introduction 145
  6.2 Gasification 145
  6.3 Biogas 158
  6.4 Other methods for producing synthesis gas 160
  6.5 Syngas conversion to products 161
  6.6 Current status and future trends 168
       References 170

Part Three Combustion and conversion technologies 175

7 Technology options for large-scale solid-fuel combustion 177
Markus Hurskainen, Pasi Vainikka
  7.1 Introduction 177
  7.2 Combustion technologies for solid fuels 177

	7.3	Summary	197
		References	198
8	**Plant integrity in solid fuel-flexible power generation**		201
	Nigel J. Simms		
	8.1	Introduction	201
	8.2	Potential solid fuels	202
	8.3	Power plant types, component operating environments and fuel options	212
	8.4	Degradation mechanisms and modelling	217
	8.5	Flexible fuel use	230
	8.6	Quantification of damage and protective measures	232
	8.7	Future trends	234
		Sources of further information	236
		References	236
9	**Fuel flexible gas production: biomass, coal and bio-solid wastes**		241
	Shusheng Pang		
	9.1	Introduction	241
	9.2	Characteristics of biomass, coal and bio-solid wastes	243
	9.3	Co-gasification of biomass and coal, and co-gasification of biomass and bio-solid wastes	246
	9.4	Co-pyrolysis of blended solid fuels	262
	9.5	Concluding remarks	266
		References	266
10	**Technology options and plant design issues for fuel-flexible gas turbines**		271
	Jenny Larfeldt		
	10.1	Introduction	271
	10.2	Gas turbines in plants	271
	10.3	Fuel-flexible gas turbines	273
	10.4	Gaseous fuels for gas turbine operation	275
	10.5	Gas turbine combustion-related challenges for gaseous fuel flexibility	277
	10.6	Other fuel flexibility impacts on the gas turbine	280
	10.7	Fuel-flexible gas turbine installation	281
	10.8	Gas turbine with external heating integrated in plants	283
	10.9	CO_2 capture in gas-turbine integrated plants	287
	10.10	Other integrated cycles	288
		References	290
11	**Fuel flexibility with dual-fuel engines**		293
	Jacob Klimstra		
	11.1	Introduction	293
	11.2	The four-stroke spark-ignited gas engine	294

11.3	The diesel engine	**295**
11.4	Fuel specifications	**296**
11.5	Systems for creating fuel flexibility	**299**
11.6	Plant performance	**302**
11.7	Conclusions	**304**
	References	**304**

Index **305**

List of contributors

Pedro M.R. Abelha   Energy Research Centre of the Netherlands (ECN), LE, Petten, The Netherlands

Eija Alakangas   VTT Technical Research Centre of Finland Ltd, Jyväskylä, Finland

Michael S.A. Bradley   University of Greenwich, Chatham, Kent, UK

Michiel C. Carbo   Energy Research Centre of the Netherlands (ECN), Petten, The Netherlands

Mariusz K. Cieplik   Energy Research Centre of the Netherlands (ECN), Petten, The Netherlands

Panagiotis Grammelis   Centre for Research & Technology Hellas/Chemical Process & Energy Resources Institute (CERTH/CPERI), Ptolemaida, Greece

Markus Hurskainen   VTT Technical Research Centre of Finland Ltd, Jyväskylä, Finland

Emmanouil Karampinis   Centre for Research & Technology Hellas/Chemical Process & Energy Resources Institute (CERTH/CPERI), Ptolemaida, Greece

Jaap H.A. Kiel   Energy Research Centre of the Netherlands (ECN), Petten, The Netherlands

Jacob Klimstra   Jacob Klimstra Consultancy, Broeksterwald, The Netherlands

Jenny Larfeldt   Siemens Industrial Turbomachinery AB, Finspong, Sweden

Nikolaos Margaritis   Centre for Research & Technology Hellas/Chemical Process & Energy Resources Institute (CERTH/CPERI), Ptolemaida, Greece

Carlos Mourão   Energy Research Centre of the Netherlands (ECN), Petten, The Netherlands

Shusheng Pang   University of Canterbury, Christchurch, New Zealand

Nigel J. Simms   Centre for Power Engineering, Cranfield University, Cranfield, Bedfordshire, UK

James G. Speight   CD&W Inc., Laramie, WY, USA

Pasi Vainikka   VTT Technical Research Centre of Finland Ltd, Jyväskylä, Finland

Woodhead Publishing Series in Energy

1. Generating power at high efficiency: Combined cycle technology for sustainable energy production
   Eric Jeffs
2. Advanced separation techniques for nuclear fuel reprocessing and radioactive waste treatment
   Edited by Kenneth L. Nash and Gregg J. Lumetta
3. Bioalcohol production: Biochemical conversion of lignocellulosic biomass
   Edited by Keith W. Waldron
4. Understanding and mitigating ageing in nuclear power plants: Materials and operational aspects of plant life management (PLiM)
   Edited by Philip G. Tipping
5. Advanced power plant materials, design and technology
   Edited by Dermot Roddy
6. Stand-alone and hybrid wind energy systems: Technology, energy storage and applications
   Edited by John K. Kaldellis
7. Biodiesel science and technology: From soil to oil
   Jan C. J. Bart, Natale Palmeri and Stefano Cavallaro
8. Developments and innovation in carbon dioxide (CO_2) capture and storage technology Volume 1: Carbon dioxide (CO_2) capture, transport and industrial applications
   Edited by M. Mercedes Maroto-Valer
9. Geological repository systems for safe disposal of spent nuclear fuels and radioactive waste
   Edited by Joonhong Ahn and Michael J. Apted
10. Wind energy systems: Optimising design and construction for safe and reliable operation
    Edited by John D. Sørensen and Jens N. Sørensen
11. Solid oxide fuel cell technology: Principles, performance and operations
    Kevin Huang and John Bannister Goodenough
12. Handbook of advanced radioactive waste conditioning technologies
    Edited by Michael I. Ojovan
13. Membranes for clean and renewable power applications
    Edited by Annarosa Gugliuzza and Angelo Basile
14. Materials for energy efficiency and thermal comfort in buildings
    Edited by Matthew R. Hall
15. Handbook of biofuels production: Processes and technologies
    Edited by Rafael Luque, Juan Campelo and James Clark
16. Developments and innovation in carbon dioxide (CO_2) capture and storage technology Volume 2: Carbon dioxide (CO_2) storage and utilisation
    Edited by M. Mercedes Maroto-Valer

17 Oxy-fuel combustion for power generation and carbon dioxide (CO_2) capture
Edited by Ligang Zheng
18 Small and micro combined heat and power (CHP) systems: Advanced design, performance, materials and applications
Edited by Robert Beith
19 Advances in clean hydrocarbon fuel processing: Science and technology
Edited by M. Rashid Khan
20 Modern gas turbine systems: High efficiency, low emission, fuel flexible power generation
Edited by Peter Jansohn
21 Concentrating solar power technology: Principles, developments and applications
Edited by Keith Lovegrove and Wes Stein
22 Nuclear corrosion science and engineering
Edited by Damien Féron
23 Power plant life management and performance improvement
Edited by John E. Oakey
24 Electrical drives for direct drive renewable energy systems
Edited by Markus Mueller and Henk Polinder
25 Advanced membrane science and technology for sustainable energy and environmental applications
Edited by Angelo Basile and Suzana Pereira Nunes
26 Irradiation embrittlement of reactor pressure vessels (RPVs) in nuclear power plants
Edited by Naoki Soneda
27 High temperature superconductors (HTS) for energy applications
Edited by Ziad Melhem
28 Infrastructure and methodologies for the justification of nuclear power programmes
Edited by Agustín Alonso
29 Waste to energy conversion technology
Edited by Naomi B. Klinghoffer and Marco J. Castaldi
30 Polymer electrolyte membrane and direct methanol fuel cell technology Volume 1: Fundamentals and performance of low temperature fuel cells
Edited by Christoph Hartnig and Christina Roth
31 Polymer electrolyte membrane and direct methanol fuel cell technology Volume 2: *In situ* characterization techniques for low temperature fuel cells
Edited by Christoph Hartnig and Christina Roth
32 Combined cycle systems for near-zero emission power generation
Edited by Ashok D. Rao
33 Modern earth buildings: Materials, engineering, construction and applications
Edited by Matthew R. Hall, Rick Lindsay and Meror Krayenhoff
34 Metropolitan sustainability: Understanding and improving the urban environment
Edited by Frank Zeman
35 Functional materials for sustainable energy applications
Edited by John A. Kilner, Stephen J. Skinner, Stuart J. C. Irvine and Peter P. Edwards
36 Nuclear decommissioning: Planning, execution and international experience
Edited by Michele Laraia
37 Nuclear fuel cycle science and engineering
Edited by Ian Crossland
38 Electricity transmission, distribution and storage systems
Edited by Ziad Melhem

39	**Advances in biodiesel production: Processes and technologies**
	Edited by Rafael Luque and Juan A. Melero
40	**Biomass combustion science, technology and engineering**
	Edited by Lasse Rosendahl
41	**Ultra-supercritical coal power plants: Materials, technologies and optimisation**
	Edited by Dongke Zhang
42	**Radionuclide behaviour in the natural environment: Science, implications and lessons for the nuclear industry**
	Edited by Christophe Poinssot and Horst Geckeis
43	**Calcium and chemical looping technology for power generation and carbon dioxide (CO_2) capture: Solid oxygen- and CO_2-carriers**
	Paul Fennell and E. J. Anthony
44	**Materials' ageing and degradation in light water reactors: Mechanisms, and management**
	Edited by K. L. Murty
45	**Structural alloys for power plants: Operational challenges and high-temperature materials**
	Edited by Amir Shirzadi and Susan Jackson
46	**Biolubricants: Science and technology**
	Jan C. J. Bart, Emanuele Gucciardi and Stefano Cavallaro
47	**Advances in wind turbine blade design and materials**
	Edited by Povl Brøndsted and Rogier P. L. Nijssen
48	**Radioactive waste management and contaminated site clean-up: Processes, technologies and international experience**
	Edited by William E. Lee, Michael I. Ojovan, Carol M. Jantzen
49	**Probabilistic safety assessment for optimum nuclear power plant life management (PLiM): Theory and application of reliability analysis methods for major power plant components**
	Gennadij V. Arkadov, Alexander F. Getman and Andrei N. Rodionov
50	**The coal handbook: Towards cleaner production Volume 1: Coal production**
	Edited by Dave Osborne
51	**The coal handbook: Towards cleaner production Volume 2: Coal utilisation**
	Edited by Dave Osborne
52	**The biogas handbook: Science, production and applications**
	Edited by Arthur Wellinger, Jerry Murphy and David Baxter
53	**Advances in biorefineries: Biomass and waste supply chain exploitation**
	Edited by Keith Waldron
54	**Geological storage of carbon dioxide (CO_2): Geoscience, technologies, environmental aspects and legal frameworks**
	Edited by Jon Gluyas and Simon Mathias
55	**Handbook of membrane reactors Volume 1: Fundamental materials science, design and optimisation**
	Edited by Angelo Basile
56	**Handbook of membrane reactors Volume 2: Reactor types and industrial applications**
	Edited by Angelo Basile
57	**Alternative fuels and advanced vehicle technologies for improved environmental performance: Towards zero carbon transportation**
	Edited by Richard Folkson

58 **Handbook of microalgal bioprocess engineering**
   Christopher Lan and Bei Wang
59 **Fluidized bed technologies for near-zero emission combustion and gasification**
   Edited by Fabrizio Scala
60 **Managing nuclear projects: A comprehensive management resource**
   Edited by Jas Devgun
61 **Handbook of Process Integration (PI): Minimisation of energy and water use, waste and emissions**
   Edited by Jiří J. Klemeš
62 **Coal power plant materials and life assessment**
   Edited by Ahmed Shibli
63 **Advances in hydrogen production, storage and distribution**
   Edited by Ahmed Basile and Adolfo Iulianelli
64 **Handbook of small modular nuclear reactors**
   Edited by Mario D. Carelli and Dan T. Ingersoll
65 **Superconductors in the power grid: Materials and applications**
   Edited by Christopher Rey
66 **Advances in thermal energy storage systems: Methods and applications**
   Edited by Luisa F. Cabeza
67 **Advances in batteries for medium and large-scale energy storage**
   Edited by Chris Menictas, Maria Skyllas-Kazacos and Tuti Mariana Lim
68 **Palladium membrane technology for hydrogen production, carbon capture and other applications**
   Edited by Aggelos Doukelis, Kyriakos Panopoulos, Antonios Koumanakos and Emmanouil Kakaras
69 **Gasification for synthetic fuel production: Fundamentals, processes and applications**
   Edited by Rafael Luque and James G. Speight
70 **Renewable heating and cooling: Technologies and applications**
   Edited by Gerhard Stryi-Hipp
71 **Environmental remediation and restoration of contaminated nuclear and NORM sites**
   Edited by Leo van Velzen
72 **Eco-friendly innovation in electricity networks**
   Edited by Jean-Luc Bessede
73 **The 2011 Fukushima nuclear power plant accident: How and why it happened**
   Yotaro Hatamura, Seiji Abe, Masao Fuchigami and Naoto Kasahara. Translated by Kenji Iino
74 **Lignocellulose biorefinery engineering: Principles and applications**
   Hongzhang Chen
75 **Advances in membrane technologies for water treatment: Materials, processes and applications**
   Edited by Angelo Basile, Alfredo Cassano and Navin Rastogi
76 **Membrane reactors for energy applications and basic chemical production**
   Edited by Angelo Basile, Luisa Di Paola, Faisal Hai and Vincenzo Piemonte
77 **Pervaporation, vapour permeation and membrane distillation: Principles and applications**
   Edited by Angelo Basile, Alberto Figoli and Mohamed Khayet
78 **Safe and secure transport and storage of radioactive materials**
   Edited by Ken Sorenson

79	**Reprocessing and recycling of spent nuclear fuel** *Edited by Robin Taylor*
80	**Advances in battery technologies for electric vehicles** *Edited by Bruno Scrosati, Jürgen Garche and Werner Tillmetz*
81	**Rechargeable lithium batteries: From fundamentals to applications** *Edited by Alejandro A. Franco*
82	**Calcium and chemical looping technology for power generation and carbon dioxide (CO_2) capture** *Edited by Paul Fennell and Ben Anthony*
83	**Compendium of hydrogen energy volume 1: Hydrogen production and purificiation** *Edited by Velu Subramani, Angelo Basile and T. Nejat Veziroglu*
84	**Compendium of hydrogen energy volume 2: Hydrogen storage, transmission, transportation and infrastructure** *Edited by Ram Gupta, Angelo Basile and T. Nejat Veziroglu*
85	**Compendium of hydrogen energy volume 3: Hydrogen energy conversion** *Edited by Frano Barbir, Angelo Basile and T. Nejat Veziroglu*
86	**Compendium of hydrogen energy volume 4: Hydrogen use, safety and the hydrogen economy** *Edited by Michael Ball, Angelo Basile and T. Nejat Veziroglu*
87	**Advanced district heating and cooling (DHC) systems** *Edited by Robin Wiltshire*
88	**Microbial electrochemical and fuel cells: Fundamentals and Applications** *Edited by Keith Scott and Eileen Hao Yu*
89	**Renewable heating and cooling: Technologies and applications** *Edited by Gerhard Stryi-Hipp*
90	**Small modular reactors: Nuclear power fad or future?** *Edited by Daniel T. Ingersoll*
91	**Fuel flexible energy generation: Solid, liquid and gaseous fuels** *Edited by John Oakey*

Part One

Introduction and fuel types

Introduction to fuel flexible energy

James G. Speight
CD&W Inc., Laramie, WY, USA

1.1 Introduction

Fuels for domestic and industrial use are changing with the passage of time and have continued to do so since whale oil was first used as illuminating oil followed by kerosene as the illuminant. From that time, with the onset of the modern petroleum industry in 1856, petroleum and coal have been the dominant fuel sources and are now joined by natural gas, shale gas, oil from shale, tar sand bitumen, and oil shale (from which shale oil is produced by thermal decomposition of the kerogen) in the shale—these are the so-called *conventional fuel sources* and which will be the dominant fuel sources for the next several decades (Speight and Ozum, 2002; Hsu and Robinson, 2006; Gary et al., 2007; Speight, 2007, 2008, 2009; Bower, 2009; Wihbey, 2009; Crane et al., 2010; Levant, 2010; Speight, 2011a,b,c, 2012, 2013a,b,c, 2014a,b). All of these sources are fossil fuel sources, are nonrenewable, and cannot be replaced without invoking the concept of geological time and, therefore, sustainability of current fuel sources is open to debate (Crane et al., 2010; Zatzman, 2012).

To clarify and avoid the confusion regarding recent terminology, the term *oil from shale* is a petroleum-type oil than *can be recovered in its natural state* from shale formations, whereas *shale oil* is a petroleum substitute oil which *does not exist in a natural state* in shale and is produced by thermal decomposition of the organic material (kerogen) in oil shale. Shale oil is produced from either mined or unmined (in situ) shale (Speight and Ozum, 2002; Hsu and Robinson, 2006; Gary et al., 2007; Speight, 2013a, 2014a).

Nonconventional energy sources (also called *alternative energy sources*) are any sources or substances that can be used to produce fuels, other than conventional fuels. These are sources that are continuously replenished by natural processes, including biomass, hydropower, nuclear power, solar energy, and tidal energy. Examples of nonconventional fuels include biodiesel, bioalcohols (methanol, ethanol, butanol produced from biological sources), hydrogen, and fuels from other nonconventional (nonfossil fuel) sources (Speight, 2011a,b).

Furthermore, a biofuel is any as solid, liquid, or gaseous fuel consisting of, or derived from, biomass. Biomass can also be used directly for heating or power—known as *biomass fuel*. Biofuel can be produced from any carbon source that can be replenished rapidly, for example, plants. Many different plants and plant-derived materials are used for biofuel manufacture and the various technologies applied to produce biofuels hold great promise (Speight, 2008; Giampietro and Mayumi, 2009; EREC, 2010; Langeveld et al., 2010; Speight, 2011a,b).

On the other hand, there are fuels known as flexible fuels that are typically mixture of fuels such as gasoline and ethanol. Thus, a flexible-fuel vehicle (FFV) is, for example, an automobile that can alternate between two or more sources of fuel such as gasoline and ethanol mixtures. Flexible-fuel vehicles are already in production by automobile manufacturers and are engineered to run on blends of gasoline and ethanol in any percentage up to 85%. For example, E85 is a liquid fuel that is 85% v/v ethanol and 15% v/v gasoline—the mixture can be seasonally adjusted for variations in the weather and may, at times, be less than 85% v/v ethanol. To be considered an alternative fuel vehicle (for tax incentives), the automobile or truck must be able to operate on up to 85% v/v ethanol. However, to use any ethanol blend in accordance with the specifications provided by the manufacturer, each manufacturer will, more than likely, have individual specifications. Generally, all gasoline-fueled vehicles are FFVs insofar as they are able to operate on gasoline and ethanol blends up to 10% v/v ethanol—in fact, most gasoline sold in the United States has approximately that amount of ethanol to meet clean air or emissions regulations.

Ethanol is the most common alternate fuel that is used in FFVs and, as the use of ethanol and the appearance of ethanol-fueled became available during the late 1990s, the common use of the term *flexible-fuel vehicle* became synonymous with the use of ethanol as a vehicle fuel (Ryan and Turton, 2007). In the United States and many other countries, FFVs are often referred to as *E85 vehicles* or *flex vehicles* (also *flex cars flexi-fuel vehicles*). In addition, the term flexible-fuel vehicles is sometimes used to include other alternative fuel vehicles that can run with compressed natural gas (CNG), liquefied petroleum gas (LPG), or hydrogen. However, such vehicles are typically *bi-fuel vehicles* and not FFVs, because the alternate (nongasoline) fuel is stored in a separate tank, and the engine runs on one fuel at a time—the bi-fuel vehicles have the capability to switch back and forth between gasoline and the other fuel. On the other hand, FFVs are based on a dual-fuel system that supplies both fuels into the combustion chamber at the same time in measured proportions.

An extension of the flexible-fuel concept is the *multifuel vehicle* that is capable of operating with more than two fuels, such as a CNG−ethanol−gasoline-fueled vehicle. The term multifuel is applied to any type of engine, boiler, or other fuel-burning device that is designed to burn multiple types of fuels. A common application of multifuel technology is in military settings, in which the normally used diesel or gas turbine fuel might not be available during combat operations for vehicles or other fuel-burning units. However, the growing need to establish fuel sources other than petroleum for transportation and other nontransportation uses has led to the development of multifuel technology for nonmilitary use.

It is the purpose of this chapter to present an overview of the production and uses of conventional fuels, which are in constant demand. For the purposes of this chapter, *petroleum products* and *fuels* are those bulk fractions that are derived from petroleum and have commercial value as a bulk product (Speight, 2014a). In the strictest sense, petrochemicals are also petroleum products but they are individual chemicals that are used as the basic building blocks of the chemical industry.

1.2 Conventional energy sources

By definition, conventional energy sources include oil, gas, and coal (i.e., the fossil fuels) and the fuels derived therefrom have provided a measure (but not a guaranteed measure) of energy security in the past 100 years, which now may be on the wane even though projections indicate that the refining industry as is currently recognized may last another 50 years—subject, of course, to the politics of the various nations involved in oil production and use (Bower, 2009; Speight, 2011c; Hamilton, 2013; Khoshnaw, 2013; Luciani, 2013).

1.2.1 Petroleum

Petroleum (also called *crude oil*) also includes crude oil, natural gas, and heavy oil (a type of petroleum)—tar sand bitumen (called *oil sand bitumen* in Canada) is not included because, based on the definition by the Congress of the United States, it is not a type of petroleum (Speight, 2009, 2014a; Levant, 2010). Both crude oil and natural gas are predominantly mixtures of hydrocarbons with low amounts of heteroatoms (nitrogen oxygen, sulfur nickel, vanadium)—heavy oil and tar sand bitumen contain substantially more heteroatoms. Under conditions of standard temperature and pressure at the surface, the lower molecular weight hydrocarbons methane, ethane, propane, and butane, occur as gases, whereas the higher molecular weight constituents are in the form of liquids, semisolids, and/or solids.

Petroleum and natural gas that have not been refined have low value and are not generally used as such but are transported (typically by pipeline, but also by ocean tanker) to a *refinery*. At the refinery, the different hydrocarbon constituents are separated into various components from which saleable products are derived that may be suitable for use as fuel gases, liquid fuels, lubricants, wax, asphalt, and as feedstock for petrochemicals. The complexity of petroleum is reflected in the variations in distribution of the various fractions—the actual proportions of low-boiling fractions (0−205 °C, 32−400 °F), medium-boiling fractions (205−345 °C, 400−650 °F), and high-boiling fractions (345 °C, >650 °F) vary significantly from one crude oil to another.

A petroleum refinery is a group of manufacturing plants (unit processes) (Figure 1.1) that is used to separate petroleum into fractions and the subsequent treating of these fractions to yield marketable products, particularly fuels (Speight and Ozum, 2002; Hsu and Robinson, 2006; Gary et al., 2007; Speight, 2013a, 2014a). Refinery configuration is not stable and will vary from refinery to refinery, depending on the crude oil feedstocks. In times past, a refinery usually accepted a single crude oil, but the modern refinery is more likely to accept several crude oils as a blend—in addition to conventional crude oils, some blends contain one or more heavy oils whereas others also contain small amounts of tar sand bitumen.

In general, crude oil—the term here is used in a general sense to mean refinery feedstock, whatever the composition—once refined, yields three groups of products that are produced when it is separated into boiling-range fractions (*distillation cuts*)

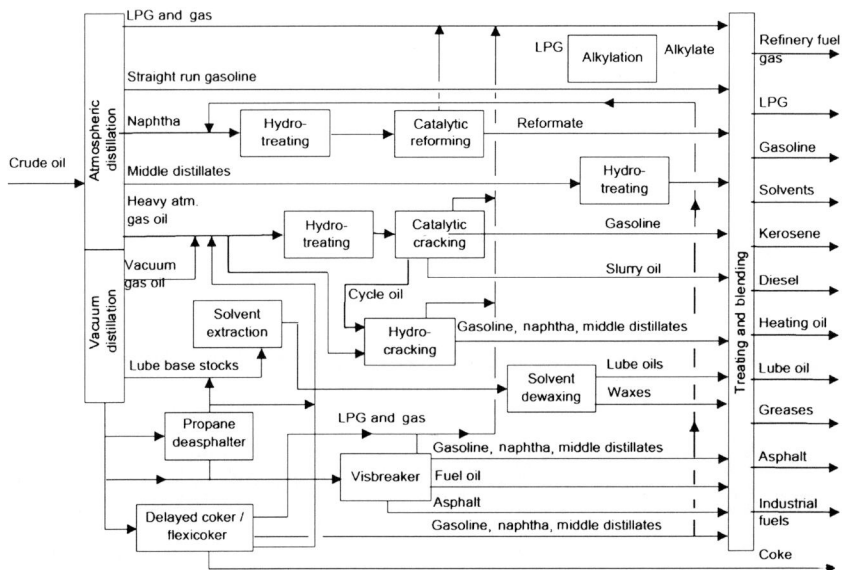

Figure 1.1 Schematic overview of a refinery.

(Table 1.1; Speight and Ozum, 2002; Hsu and Robinson, 2006; Gary et al., 2007; Speight, 2014a). The yields and quality of refined petroleum products produced by any given oil refinery depends on the mixture of crude oil used as feedstock and the configuration of the refinery facilities. Light sweet (low-sulfur) crude oil is generally more expensive and has inherently greater yields of higher value, low-boiling products such naphtha and kerosene from which gasoline, aviation gasoline, jet fuel, and diesel fuel are produced. Heavy sour (high-sulfur) crude oil is generally less expensive and produces greater yields of lower value higher boiling products that must be converted into lower boiling products within the refinery system (Speight and Ozum, 2002; Hsu and Robinson, 2006; Gary et al., 2007; Speight, 2014a).

1.2.2 Natural gas

Natural gas, which is predominantly methane, occurs in underground reservoirs separately or in association with crude oil (Mokhatab et al., 2006; Speight, 2007, 2014a). The principal constituent of natural gas is methane (CH_4)—other constituents are paraffinic hydrocarbons such as ethane (CH_3CH_3), propane ($CH_3CH_2CH_3$), and the butanes ($CH_3CH_2CH_2CH_3$ and/or $(CH_3)_3CH$) (Table 1.2; Mokhatab et al., 2006; Speight, 2008, 2014a). Many natural gases contain nitrogen (N_2) as well as carbon dioxide (CO_2) and hydrogen sulfide (H_2S). Trace quantities of argon, hydrogen, and helium may also be present. Generally, hydrocarbons having a higher molecular weight than methane are removed from natural gas prior to its use as a fuel;

Table 1.1 **Crude petroleum is a mixture of compounds that can be separated into different generic boiling fractions—further refining of these fractions produces saleable products**

Fraction	Boiling range[a] °C	Boiling range[a] °F	Designation (arbitrarily based on boiling range)
Light naphtha	0–150	30–300	Light distillate
Heavy naphtha	150–205	300–400	Middle distillate
Kerosene	205–260	400–500	Middle distillate
Light gas oil	260–345	500–650	Middle distillate
Heavy gas oil	345–500	650–930	Heavy distillate
Lubricating oil	>400	>750	Heavy distillate
Vacuum gas oil	425–600	800–1100	Heavy distillate
Residuum	>500	>930	Residuum

[a] For convenience, boiling ranges are converted to the nearest 5°.

at this stage of natural gas refining, carbon dioxide, hydrogen sulfide, and any other n-hydrocarbons are removed.

Liquefied petroleum gas (LPG) is the term applied to certain specific hydrocarbons and their mixtures, which exist in the gaseous state under ambient conditions of temperature and pressure but can be converted to the liquid state under conditions of moderate pressure at ambient temperature. These are the low-boiling hydrocarbons

Table 1.2 **Range of composition of natural gas**

Methane	CH_4	70–90%
Ethane	C_2H_6	0–20%
Propane	C_3H_8	
Butane	C_4H_{10}	
Pentane and higher hydrocarbons	C_5H_{12}	0–10%
Carbon dioxide	CO_2	0–8%
Oxygen	O_2	0–0.2%
Nitrogen	N_2	0–5%
Hydrogen sulfide, carbonyl sulfide	H_2S, COS	0–5%
Rare gases: argon, helium, neon, xenon	Ar, He, Ne, Xe	Trace

constituents of the paraffin series: ethane (CH_3CH_3, which is used as a petrochemical feedstock), propane ($CH_3CH_2CH_3$), butane ($CH_3CH_2CH_2CH_3$), and *iso*-butane ($CH_3CH(CH_3)CH_3$). The most common commercial products are propane, butane, or some mixture of the two and are generally extracted from natural gas or crude petroleum. However, there are specifications for LPG (ASTM D1835) that depend upon the required volatility.

1.2.3 Coal

Coal (the term is used generically throughout the book to include all types of coal) is a brown-to-black *organic sedimentary rock* of biochemical origin, which is combustible and occurs in rock strata (coal beds, coal seams). Coal is composed primarily of carbon with variable proportions of hydrogen, nitrogen, oxygen, and sulfur. In the United States, deposits of coal, sandstone, shale, and limestone are often found together in sequences hundreds of feet thick that were laid down predominantly during the Mississippian (approximately 360–325 million years ago) and the Pennsylvanian periods (approximately 325–300 million years ago) due to the significant sequences found in those states (i.e., Mississippi and Pennsylvania).

Production of coal is by both underground and open-pit mining. Surface, large-scale coal operations are a relatively recent development, commencing as late as the 1970s. Underground mining of coal seams presents many of the same problems as mining of other bedded mineral deposits, together with some problems unique to coal. Furthermore, through a process known as in situ *gasification* (Speight, 2013a), coal beds can be converted to gaseous products underground. To do this, the coal is ignited, air and steam are pumped into the burning seam, and the resulting gases (typically carbon monoxide, hydrogen, carbon dioxide, and hydrocarbons) are pumped to the surface for cleaning and use.

Coal is the world's most abundant and widely distributed fossil fuel and possibly the least understood in terms of its importance to the world economy. The United States has a vast supply of coal, with almost 30% of world reserves and more than 1600 billion tons (1600×10^9 tons) as remaining coal resources. The United States is also the world's second largest coal producer after China and annually produces more than twice as much coal as India, the third largest producer. Coal is a major contributor to the energy needs of the world—approximately five billion tons (5×10^9 tons) of coal are mined on a worldwide basis and, of the total coal mined, approximately 80% ($>4 \times 10^9$ tons) are required annually to generate electricity (Speight, 2013b).

There are different coal types and each coal has a different quality—peat, the lowest member of the series, is not considered coal and is more closely related to biomass than to coal. Coal is classified into four main types: (1) lignite, (2) subbituminous coal, (3) bituminous coal, and (4) anthracite:

Lignite (brown coal) is the least mature of the coal types and provides the least yield of energy; it is often crumbly, relatively moist, and powdery. It is the lowest rank of coal, with a heating value of 4000−8300 Btu per pound. Most lignite mined in the United States comes from Texas. Lignite is mainly used to produce electricity.

Subbituminous coal is poorly indurated and brownish in color, but more like bituminous coal than lignite. It typically contains less heating value (8300−13,000 Btu per pound) and more moisture than bituminous coal.

Bituminous coal was formed by added heat and pressure on lignite and is the black, soft, slick rock and the most common coal used around the world. Made of many tiny layers, bituminous coal looks smooth and sometimes shiny. It is the most abundant type of coal found in the United States and has two to three times the heating value of lignite. Bituminous coal contains 11,000−15,500 Btu per pound. Bituminous coal is used to generate electricity and is an important fuel for the steel and iron industries.

Anthracite is usually considered the highest grade of coal and is actually considered metamorphic. Compared to other coal types, anthracite is much harder, has a glassy luster, and is denser and blacker with few impurities. It is largely used for heating domestically as it burns with little smoke. It is deep black and looks almost metallic due to its glossy surface. Like bituminous coal, anthracite coal is a big energy producer, containing nearly 15,000 Btu per pound.

Coal is often sold by *grade*—the grade of a coal establishes its economic value for a specific end use. *Coal grade* refers to the amount of mineral matter that is present in the coal and is a measure of coal quality. Sulfur content, ash fusion temperature (i.e., the temperature at which measurement the ash melts and fuses), and quantity of trace elements in coal are also used as means of grading coal. Although formal classification systems have not been developed using coal grade as the means of defining the class of coal, grade is important to the coal user.

Finally, *steam coal*, which is not a specific rank of coal, is a grade of coal that falls between bituminous coal and anthracite, once widely used as a fuel for steam locomotives. In this specialized use, it is sometimes known as *sea-coal* in the United States. Small steam coal (*dry small steam nuts*, DSSN) was used as a fuel for domestic water heating. In addition, the material known as *jet* is the gem variety of coal. Jet is generally derived from anthracite and lacks a crystalline structure, so it is considered a mineraloid. Mineraloids are often mistaken for minerals and are sometimes classified as minerals, but lack the necessary crystalline structure to be truly classified as a mineral. Jet is, being one of the products of an organic process, remains removed from full mineral status.

Fuel companies convert coal into gaseous or liquid products. Coal-based gaseous fuels are produced through the process of *gasification* (Speight, 2012, 2013a,b). In the gasification process, coal is heated in the presence of steam and oxygen to produce *synthesis gas*, a mixture of carbon monoxide, hydrogen, and methane used directly as fuel or refined into cleaner-burning gas.

Liquefaction processes convert coal into a liquid fuel that has a composition similar to that of crude petroleum (Chapters 18 and 19) (Speight, 2008, 2014a). In general, coal is liquefied by breaking hydrocarbon molecules into smaller molecules. Coal contains more carbon than hydrogen, so hydrogen must be added—directly as hydrogen gas or indirectly as water—to bond with the carbon chain fragments.

1.2.4 Shale gas

Shale formations (composed mainly of clay-size mineral grains) are the most abundant sedimentary rocks in the crust of the Earth—organic shale formations are source rocks as well as the reservoir basement and cap rocks that trap oil and gas (Speight, 2014a). Shale is a fissile (referring to the ability of the shale to split into thin sheets along bedding), terrigenous (referring to the origin of the sediment) sedimentary rock in which particles are mostly of silt and clay size (Blatt and Tracy, 2000).

Shale exists in two general varieties, based on organic content: (1) dark or (2) light. Dark-colored or black shale formations are organic rich, whereas the lighter-colored shale formations are organic lean. Organic-rich shale formations were deposited under conditions of little or no oxygen in the water, which preserved the organic material from decay. The organic matter was mostly plant debris that had accumulated with the sediment. The presence of organic debris in black shale formations makes the formations candidates for oil and gas generation. If the organic material is preserved and properly heated after burial oil, natural gas might be produced. The Marcellus Shale, Appalachian Shale, Haynesville Shale, and Eagle Ford Shale, as well as the Barnett Shale, the Fayetteville Shale, and other gas-producing rocks, are all dark gray or black shale formations that yield natural gas (Speight, 2013c).

Shale formations are ubiquitous in sedimentary basins and, as a result, the main organic-rich shale formations have already been identified in most regions of the world. The depths vary from near surface to several 1000 feet underground, whereas the thickness varies from tens of feet to several hundred feet. However, each shale formation has different geological characteristics that affect the way gas can be produced, the technologies needed, and the economics of production (Scouten, 1990; Speight, 2012).

The amount of natural gas liquids (NGLs—hydrocarbons such as propane, butane, pentane, hexane, heptane, and even octane) commonly associated with natural gas production present in the gas can also vary considerably, with important implications for the economics of production. Although most dry gas sources in the United States are uneconomic at low natural gas prices, sources with significant liquid content can be produced for the value of the liquids only (the market value of NGLs is correlated with oil prices, rather than gas prices), making gas an essentially free byproduct.

The Barnett Shale of Texas was the first major natural gas field developed in a shale reservoir rock. Producing gas from the Barnett Shale was a challenge because the pore spaces in shale are sufficient small that gas has difficulty moving through the shale and into the well. It was necessary to fracture the shale to liberate the gas from the pore spaces and allow that gas to flow to the well.

Natural gas production from shale gas reservoirs (using hydraulic fracturing, hydrofracking methods) is now proven feasible from numerous operations in various shale gas reservoirs in North America. However, maximization of reservoir producibility can only be achieved by a thorough understanding of the occurrence and properties of the shale gas resources as well as the producibility of the gas from the reservoir (Kundert and Mullen, 2009). Although distinct in focus, these needs to demonstrate the importance of the thorough characterization of a shale gas reservoir as well as an understanding of how earth materials deform over various time scales and how that affects the current state of stress in the crust (Speight, 2012, 2014a).

Furthermore, although shale gas resources represent a significant portion of current and future production, all shale gas is not constant in composition, and gas processing requirements for shale gas can vary from area to area (Schettler et al., 1989; Bullin and Krouskop, 2008; Wihbey, 2009; Weiland and Hatcher, 2012).

1.2.5 Oil from shale

Tight shale formations, which are impermeable rock and nonporous sandstone or limestone formations and exist (typically) at depths greater than 10,000 feet below the surface, also contain natural gas and petroleum. Although the viability of a typical sandstone reservoir (containing petroleum and/or natural gas) is determined by porosity and permeability, tight shale formations have very little porosity and permeability. In some cases, the oil and gas can be found in small, isolated zones within short distances of each other, but due to the density of the rock formation, are inaccessible via the same vertical well.

1.3 Unconventional energy sources

For the purpose of this text, unconventional fuel sources are those sources of fuels other than the more readily accessible fuel sources, and these include tar sand bitumen, oil shale, and biomass. All three have received some acceptance but have not been fully developed to meet their maximum potential. The use of these fuel sources is likely to be fully assimilated into the fuel production scenarios within the foreseeable future. Moreover, in the 50-plus-year time period these fuels along with other natural fuel sources will be an important part of the fuel-generating scenarios on a worldwide basis (Gudmestad et al., 2010; Speight, 2011c).

1.3.1 Tar sand bitumen

The expression *tar sand* is commonly used in the petroleum industry to describe sandstone reservoirs that are impregnated with a heavy, viscous black crude organic material (Speight, 2009, 2014a). However, the term *tar sand* is actually a misnomer; more correctly, the name *tar* is usually applied to the heavy product remaining after the destructive distillation of coal or other organic matter (Speight, 2009, 2013a, 2014a). Because it is incorrect to refer to native bituminous materials as *tar* or *pitch*, alternative names, such as *bituminous sand* or *oil sand*, are gradually finding usage, with the former name (bituminous sands) more technically correct. The term *oil sand* is also used in the same way as the term *tar sand*, and these terms are often used interchangeably.

Tar sand bitumen—a major source of synthetic crude oil in Canada—has increased in popularity over the last four to five decades. The term *bitumen* (also, on occasion, referred to as *native asphalt*, and *extra heavy oil*) includes a wide variety of reddish brown to black materials of semisolid, viscous to brittle character that can exist in nature with no mineral impurity or with mineral matter contents that exceed 50% by weight. Bitumen has been known for millennia and is frequently found filling pores and crevices of sandstone, limestone, or argillaceous sediments, in which case the organic and associated mineral matrix is often known as *rock asphalt* (Speight, 2014a).

Bitumen from the Canadian tars sand deposits is a high-boiling material with that is immobile in the deposit and contains little, if any, material boiling below 350 °C (660 °F); the boiling range is approximately the same as the boiling range of an atmospheric residuum. In addition, *tar sands* have been defined in the United States (FE-76-4) as:

> ...*the several rock types that contain an extremely viscous hydrocarbon which is not recoverable in its natural state by conventional oil well production methods including currently used enhanced recovery techniques. The hydrocarbon-bearing rocks are variously known as bitumen-rocks oil, impregnated rocks, oil sands, and rock asphalt.*

In summary, bitumen found in tar sand deposits is an extremely viscous material that is *immobile under reservoir conditions* and cannot be recovered through a well by the application of secondary or enhanced recovery techniques.

Physical properties such as API gravity, elemental analysis, and composition, fall short of giving an adequate definition of tar sand and tar sand bitumen. The properties of the bulk deposit and, most of all, the necessary recovery methods, form the basis of the definition of these materials. Only then is it possible to classify petroleum, heavy oil, and tar sand bitumen (Speight, 2014a).

1.3.2 Oil shale

Oil shale represents a large and mostly untapped hydrocarbon resource. Like tar sand (*oil sand* in Canada) and coal, oil shale is considered unconventional because oil cannot be produced directly from the resource by sinking a well and pumping. Oil has to be produced thermally from the shale. The organic material contained in the shale is called *kerogen*, a solid material intimately bound within the mineral matrix (Scouten, 1990; Lee, 1996; Ots, 2007; Speight, 2007, 2008, 2012, 2013a).

In fact, the term *oil shale* describes an organic-rich rock from which only small amounts of the carbonaceous material can be removed by extraction (with common petroleum-based solvents) but which produces variable quantities of distillate (*shale oil*) when raised to temperatures in excess of 350 °C (660 °F). Thus, oil shale is assessed by the ability of the mineral to produce shale oil in terms of gallons per ton (g/t) by means of a test method (Fischer Assay) in which the oil shale is heated to 500 °C (930 °F) to produce distillate.

Oil shale is distributed widely throughout the world with known deposits in every continent. Oil shale ranging from Cambrian to Tertiary age occurs in many parts of the world (Scouten, 1990; Lee, 1996; Speight, 2008, 2012). Deposits range from small occurrences of little or no economic value to those of enormous size that occupy thousands of square miles and contain several billions of barrels of potentially producible shale oil. However, petroleum-based crude oil is cheaper to produce today than shale oil but the increasing costs of petroleum-based products present opportunities for supplying some of the fossil energy needs of the world in the future though development of oil shale deposits (Bartis et al., 2005; Andrews, 2006).

The organic matter (generally called *kerogen*) in oil shale is a complex moisture and is derived from the carbon-containing remains of algae, spores, pollen, plant cuticle, and corky fragments of herbaceous and woody plants, plant resins, plant waxes, and other cellular remains of lacustrine, marine, and land plants (Durand, 1980; Scouten, 1990; Hunt, 1996; Dyni, 2003, 2006). These materials are composed chiefly of carbon, hydrogen, oxygen, nitrogen, and sulfur. Generally, the organic matter is unstructured and is best described as amorphous (*bituminite*), the origin of which has not been conclusively identified but is theorized to be a mixture of degraded algal or bacterial remains. Other carbon-containing materials such as phosphate and carbonate minerals may also be present, which, although of organic origin, are excluded from the definition of organic matter in oil shale and are considered to be part of the mineral matrix of the oil shale.

The thermal decomposition of kerogen produces three products: (1) gases, (2) oil, and (3) a carbonaceous (high-carbon) deposit remaining in the rock on (the surface or in the pores) as char—a similar coke-like residue. Water is also produced but is separated from the oil. The relative proportions of gas, oil, and char vary with the pyrolysis temperature and to some extent with the organic content of the raw shale. All three products are contaminated with nonhydrocarbon compounds, and the amounts of the contaminants also vary with the pyrolysis temperature and the character of the oil shale (Scouten, 1990; Brendow, 2003, 2009; Dyni, 2003, 2006).

The method commonly used in the United States for assessing the quality of oil shale in terms of gas yield and oil yield, is the *modified Fischer assay* test method (Scouten, 1990; Speight, 2012). Some laboratories have further modified the Fischer assay method to better evaluate different types of oil shale and different methods of oil shale processing. The standard Fischer assay test method (ASTM D3904, withdrawn in 1996 but still used in many laboratories) consists of heating a 100-g sample crushed to −8 mesh (2.38-mm) screen in a small aluminum retort to 500 °C (930 °F) at a rate of 12 °C (21.6 °F) per minute and held at that temperature for 40 min. The distilled vapors of gas, oil, and water are passed through a condenser cooled with ice water into a graduated centrifuge tube. The oil and water are then separated by centrifuging. The quantities reported are the weight percentages of shale oil (and its specific gravity), water, shale residue, and (by difference) gas plus losses.

1.3.3 Biomass

There are several potential fuels that can be produced from biomass—the fuels of interest in the present context are (1) bio-oil, also known as pyrolysis oil, and/or bio-crude and (2) hydrogen.

Biomass is a renewable resource, the utilization for which has received great attention due to environmental considerations and the increasing demands of energy world-wide (Speight, 2008; Giampietro and Mayumi, 2009; EREC, 2010; Seifried and Witzel, 2010; Speight, 2011a,d). Because the widespread use of fossil fuels within the current energy infrastructure is considered the largest source of anthropogenic emissions, which, with the evolution of the climate of the Earth, can lead to global climate change (Zanganeh and Shafeen, 2007; Pittock, 2009; FitzRoy and Papyrakis, 2010; Sorokhtin et al., 2011), many countries have become interest in biomass as a fuel source to expand energy production (Karekezi et al., 2004). In fact, biomass accounts for 35% of primary energy consumption in developing countries, raising the world total to 14% of primary energy consumption (Hoogwijk et al., 2005; Demirbaş, 2006).

Bio-oil is a multicomponent mixture produced by thermal decompositon of biomass and is removed as distillate from the reaction zone (Czernik and Bridgwater, 2004; Zhang et al., 2007). However, bio-oil may have a similar composition to that of the original biomass, which is different to the composition of petroleum-derived fuels (Brammer et al., 2006; Pütün et al., 2006; Maher and Bressler, 2007).

Hydrogen, which can also be produced from biomass, is the lightest element and is a colorless, odorless, tasteless, and nontoxic gas found in the air at concentrations of about 100 ppm (0.01%) (Suban et al., 2001). It is the most abundant element in the universe, making up 75% of normal matter by mass and over 90% by number of atoms (Mariolakos et al., 2007). Hydrogen has been recognized as a promising, green, and ideal energy carrier of the future due to its high energy yield and clean, efficient, renewable, sustainable, and recyclable nature (Mohan et al., 2008; Seifried and Witzel, 2010). Hydrogen can be used as a transportation fuel, whereas neither nuclear energy nor solar energy can be used directly.

All primary energy sources can be used in the hydrogen-producing process. Currently, the primary route for hydrogen production is the conversion of natural gas and other light hydrocarbons (Wiltowski et al., 2008). The production of hydrogen from fossil fuels causes the coproduction of carbon dioxide, which is assumed mainly responsible for the so-called *greenhouse effect* and needs planned options for sequestration rather than being allowed a free emission into the atmosphere (Resini et al., 2006; Wu et al., 2011). These processes use nonrenewable energy sources to produce hydrogen and are not sustainable—renewable energy sources and technologies for hydrogen production will be necessary during coming decades (Seifried and Witzel, 2010).

Hydrogen can be produced from biomass by pyrolysis, gasification, steam gasification, steam-reforming of bio-oils, and enzymatic decomposition of sugars. The yield of hydrogen that can be produced from biomass is relatively low, 16—18% w/w based on dry biomass weight (Demirbaş, 2001). In the pyrolysis and gasification processes, the water—gas shift reaction is used to convert the reformed gas into hydrogen, and pressure-swing adsorption is used to purify the product (Demirbaş, 2008a). In general, the gasification temperature is higher than that of pyrolysis and the yield of hydrogen from gasification is higher than that from the pyrolysis (Maschio et al., 1994; Dupont et al., 2007; Balat, 2009).

1.4 Fischer–Tropsch process

The Fischer–Tropsch process is a means by which liquid fuels can be produced from a variety of carbonaceous feedstocks, which include residua, coal, biomass, and any carbonaceous waste (typically semisolid or solid waste) from any of the conventional and unconventional fuel sources (Speight, 2008, 2013a; Chadeesingh, 2011; Speight, 2013a,b; Bahadori, 2014; Speight, 2014a,b).

In the first stage of the process, the carbonaceous feedstock is sent to a gasifier (the type of which is dependent upon the nature of the feedstock) to produce *synthesis gas* (*syngas*, a mixture of carbon monoxide and hydrogen). The synthesis gas is then converted to a mixture of hydrocarbons of different chain length such as the hydrocarbon constituents of gasoline and diesel oil as well as wax, olefins, and alcohols:

Paraffin synthesis
$nCO + (2n + 1)H_2 \leftrightarrow C_nH_{2n+2} + nH_2O$

Olefin synthesis
$nCO + 2nH_2 \leftrightarrow C_nH_{2n} + nH_2O$

Alcohol synthesis
$nCO + 2nH_2 \leftrightarrow C_nH_{2n+1}OH + (n - 1)H_2O$

In the above equations, n is the average length of the hydrocarbon chain and m is the number of hydrogen atoms per carbon. All reactions are exothermic and the product is a mixture of different hydrocarbons in which paraffin and olefins are the main parts (Stelmachowski and Nowicki, 2003).

The variants of the Fischer–Tropsch process use catalysts based mainly on iron (Fe), cobalt (Co), ruthenium (Ru), and potassium (K), depending upon the desired product distribution. The process parameters (temperature and pressure) also influence the product distribution (Overend, 2004; Chadeesingh, 2011). The use of iron-based catalysts is often preferred because of their high activity as well as their participation in the water–gas shift reaction (Rao et al., 1992; Jothimurugesan et al., 2000). The design of the gasifier, which is integrated with the Fischer–Tropsch reactor, must be aimed at achieving a high yield of hydrocarbon products. It is important to avoid methane formation as much as possible in the gasifier and convert all carbon in the feedstock biomass to carbon monoxide—carbon dioxide production also occurs (Balat, 2006).

The process is particularly suitable for the production of high-quality diesel, because the products are mainly straight-chain paraffins that possess a high cetane number, which indicates a cleaner burning of the diesel with reduced emissions. Physical properties of the Fischer–Tropsch diesel fuel are very similar to petroleum-based no. 2 diesel fuel, and if correctly processed, contain no aromatics or sulfur compounds.

1.5 Electrical energy

In addition to liquid fuels, a major source of energy comes from the production of electric power, typically from the combustion of coal, petroleum, or natural gas, as well as other carbonaceous feedstocks, and is a mature and well-established technology in the industrialized countries of the world (Speight, 2013a,b, 2014a,b). Furthermore, coal-generated power, an established electricity source that provides vast quantities of inexpensive, reliable power, has become more important as supplies of oil and natural gas diminish. In addition, known coal reserves are expected to last for centuries at current rates of use (Speight, 2013a, 2014b).

1.5.1 Power plant operations

In the process of using a carbonaceous feedstock to generate electricity, the chemical energy of the feedstock is converted to thermal energy which is then used to generate high-pressure steam that passes through a turbine to generate electrical power.

In the power plant, high-temperature, high-pressure steam is generated in the boiler and then enters the steam turbine. At the other end of the steam turbine is the condenser, which is maintained at a low temperature and pressure. Steam rushing from the high-pressure boiler to the low-pressure condenser drives the turbine blades, which powers the electric generator. Steam expands as it works; hence, the turbine is wider at the exit end of the steam. The theoretical thermal efficiency of the unit is dependent on the high pressure and temperature in the boiler and the low temperature and pressure in the condenser.

Most plants built in the 1980s and early 1990s produce about 500 MW (500×10^6 W) of power, whereas many of the modern plants produce about 1000 MW. Also the efficiency (ratio of electrical energy produced to energy released by the coal burned) of conventional coal-fired plants is increased from under 35% to close to 45%. Furthermore, power plants for electricity generation are defined by functional type: (1) base load, (2) peak load, and (3) combined cycle—each has advantages and disadvantages.

Base-load power plants have the lowest operating cost and generate power most in any given year. Base-load power plants are also subdivided into four types: (1) highly efficient combined-cycle plants fueled by natural gas, (2) nuclear power plants, (3) steam power plants fueled primarily by coal, and (4) hydropower plants. Coal and nuclear power plants are the primary types of base-load power plants used in the midwestern United States.

Peak-load power plants are relatively simple cycle gas turbines that have the highest operating cost but are the cheapest to build. They are operated infrequently, are used to meet peak electricity demands in period of high use, and are primarily fueled with natural gas or oil.

In a *combined-cycle power plant*, the feedstock is first combusted in a combustion turbine, using the heated exhaust gases to generate electricity. The exhaust gases are used to heat water in a boiler, creating steam to drive a second turbine. Apart from combustion, synthesis gas can be directly used as a fuel for power generation. Alternatively, hydrogen (another important fuel) can be separated from the product gases and used as a fuel in an open- or combined-cycle process.

1.5.2 Other fuels

Gasification of fossil fuels, biomass materials, and wastes has been used for many years to convert organic solids and liquids into useful gaseous, liquid, and cleaner solid fuels (Speight, 2011a).

1.5.2.1 Biomass

Coal gasification is an established technology (Hotchkiss, 2003; Speight, 2013a,b), whereas biomass gasification has been the focus of research in recent years to estimate efficiency and performance of the gasification process using various types of biomass. These include such things as sugarcane residue (Gabra et al., 2001), rice hulls (Boateng et al., 1992), pine sawdust (Lv et al., 2004), almond shells (Rapagnà and Latif, 1997; Rapagnà et al., 2000), wheat straw (Ergudenler and Ghaly, 1993), food waste (Ko et al., 2001), and wood biomass (Pakdel and Roy, 1991; Bhattacharaya et al., 1999; Chen et al., 1992; Hanaoka et al., 2005). Recently, there has been significant research interest in cogasification of various biomass and coal mixtures such as coal and cedar wood chips (Kumabe et al., 2007), coal and sawdust (Vélez et al., 2009), coal and pine chips (Pan et al., 2000), coal and silver birch wood (Collot et al., 1999), and coal and birch wood (Brage et al., 2000). Cogasification of coal and biomass has some synergy—the process not only produces a low-carbon footprint on the environment, but also improves the H_2/CO ratio in the produced gas that is required for liquid-fuel synthesis (Sjöström et al., 1999; Kumabe et al., 2007). In addition, inorganic matter present in biomass catalyzes the gasification of coal. However, cogasification processes require custom fittings and optimized processes for the coal and region-specific wood residues.

Although cogasification of coal and biomass is advantageous from a chemical point of view, some practical problems have been associated the process on upstream, gasification, and downstream processes. On the upstream side, the particle size of the coal and biomass is required to be uniform for optimum gasification. In addition, moisture content and pretreatment (torrefaction) are very important during upstream processing.

Finally, the presence of mineral matter in the coal–biomass feedstock is not appropriate for fluidized-bed gasification. The low-melting ash obtained from woody biomass leads to agglomeration that causes defluidization of the ash, in turn causing sintering, deposition, and corrosion of the gasifier construction metal bed (Vélez et al., 2009). Biomass containing alkali oxides and salts produces yields higher than 5% w/w ash but causes clinkering/slagging problems (McKendry, 2002). It is imperative to be aware of the melting of biomass ash, its chemistry within the gasification bed (no bed, silica/sand, or calcium bed), and the fate of alkali metals when using fluidized-bed gasifiers.

Coal producers, biomass fuel producers, and to a lesser extent waste companies are enthusiastic about supplying cogasification power plants and realize the benefits of cogasification with alternative fuels. The benefits of a cogasification technology involving coal and biomass include use of a reliable coal supply with gate-fee waste and biomass, which allow the economies of scale from a larger plant than could be supplied just with waste and biomass. In addition, the technology offers a future option for refineries for hydrogen production and fuel development. In fact, oil refineries and petrochemical plants are opportunities for gasifiers when hydrogen is particularly valuable (Speight, 2011b).

1.5.2.2 Waste

Waste may be municipal solid waste that has had minimal presorting, or refuse-derived fuel (RDF), which has had significant pretreatment. Other more specific wastes—but excluding hazardous waste—and possibly including petroleum residua and petroleum coke, already provide niche opportunities for coutilization (Bahadori, 2014; Speight, 2013a, 2014a,b).

Coutilization of waste and biomass with coal may provide economies of scale that help achieve the policy objectives identified above at an affordable cost (Bower, 2009; Hamilton, 2103). In some countries, governments propose cogasification processes as being *well suited for community-sized developments*, suggesting that waste should be dealt with in smaller plants serving towns and cities, rather than moved to large, central plants (satisfying the so-called *proximity principle*).

Combining biomass, refuse, and coal overcomes the potential unreliability of biomass, the potential longer-term changes in refuse, and the size limitation of a power plant using only waste and/or biomass. It also allows benefit from a premium electricity price for electricity from biomass and the gate fee associated with waste. If the power plant is gasification based, rather than direct combustion, further benefits may be available. These include a premium price for the electricity from waste, the range of technologies available for the gas to electricity part of the process, gas cleaning prior to the main combustion stage instead of after combustion, and public image, which is currently generally better for gasification than for combustion. These considerations lead to the current study of cogasification of wastes/biomass with coal (Speight, 2008).

Use of waste materials as cogasification feedstocks may attract significant disposal credits. Cleaner biomass materials are renewable fuels and may attract premium prices for the electricity generated. Availability of sufficient fuel locally for an economic plant size is often a major issue, as is the reliability of the fuel supply. Use of more-predictably available coal alongside these fuels overcomes some of these difficulties and risks. Coal could be regarded as the sustainable energy source that keeps the plant running when the fuels producing the better revenue streams are not available in sufficient quantities.

Furthermore, the disposal of municipal and industrial wastes has become an important problem because the traditional means of disposal, landfill, has become environmentally much less acceptable than previously. New, much stricter regulation of these disposal methods will make the economics of waste processing for resource recovery much more favorable. One method of processing waste streams is to convert the energy value of the combustible waste into a fuel; coprocessing such waste with coal is also an option (Speight, 2008, 2013a, 2014b).

1.6 Fuel flexibility

1.6.1 Environmental issues

As the oil and gas industry faces the challenges of complying with environmental legislation and addressing the global need for energy conservation, the efficiency of the prime movers used to deliver electrical power or mechanical drive has become important. The oil and gas sector requires prime movers to deliver electrical power and drive mechanical operations. These prime movers are typically either combustion engines or combustion turbines. As oil and gas become more difficult to recover and operators attempt to extract more from existing wells, the demand for investments in power generation continues to increase. Innovations in combustion engine technology have made the meeting of these challenges far easier than before.

The ability to run on a wide range of fuels is seeing combustion engines play a major role in the drive to reduce flaring—a practice that is coming increasingly under the spotlight due to environmental concerns and the need for energy conservation. With the clear benefits of better reliability, greater fuel flexibility, and lower operating costs, the oil and gas industry can now focus on using the more efficient and environmentally sound solutions. Furthermore, in the coming decades, the fuel supply will be strongly diversified: new gases, liquid, and solid fuels derived from biomass, residues, and other sustainable sources will coexist with fossil supplies, such as liquefied natural gas, gasoline, and diesel fuel.

1.6.2 Trends and technological challenges

It is evident that in the transition from a fossil fuel-based to a renewable energy-based society, the whole energy sector is facing a paradigm change regarding conversion and distribution of energy (Seifried and Witzel, 2010). The main reasons for this change are the diminishing easily accessible oil reserves increasing the extraction cost, and the ever-increasing demand by emerging economies. Therefore, in a future facing shortage on the principal energy source for the transportation sector, there is a need for finding a suitable and sustainable alternative energy source and fuel substitutes (Speight, 2008; Giampietro and Mayumi, 2009; EREC, 2010; Langeveld et al., 2010; Speight, 2011a,b).

Research programs need to focus on the clean and efficient utilization of the range of fuels in energy supply for the production of heat and power. New gaseous, liquid, and solid fuels have different combustion properties than those traditionally supplied, and insight into the influence of the physical properties and chemical composition of the fuel on the combustion process and emissions in different applications is required to maximize the interchangeable use of these fuels. In addition, the flexibility of combustion systems toward large variations in fuel composition will be studied and will lead to completely new systems with new sensing and control methods.

In the case of biomass, the first objective is biomass resource mapping, in which feasible feedstocks are identified based on factors such as abundance, availability, accessibility, and feedstock composition. In addition, waste streams like sewage sludge and agricultural slurry in addition to energy crops will be investigated. The focus on such issues will require a multidisciplinary approach. Therefore, a strong cooperation between the liquid fuels industries, the combustion industry, and biomass utilization industry with input not only from chemical-oriented disciplines but also from engineering-oriented disciplines is required.

The combustion processes must be stable and clean, and more scientific insight is required in the impact of the different chemical and physical properties of future fuels such as highly viscous, fibrous minerals containing low-energy density, multicomponents. Although the modeling of combustion processes and detailed experimental probing of combustion processes has improved in recent decades, the challenges are to cope with new chemical pathways because of the new chemical composition of the fuels. Finally, fuel properties and gas composition at flame temperatures may challenge material selection.

1.7 Conclusions

Currently, the world is facing three critical problems: (1) variable fuel prices that are often to the high side of the price range, (2) climatic changes, and (3) air pollution. Current oil and gas reserves are expected to last several decades at the current rates of consumption and transport fuels are almost totally dependent on fossil fuel-based fuels, particularly petroleum-based fuels such as gasoline, diesel fuel, LPG, and CNG (Li et al., 2009; Speight, 2011a). In addition to the finite nature of fossil-fuel resources, increasing concerns regarding environmental impact, especially related to greenhouse gas emissions, and health and safety considerations are forcing the search for new energy sources and alternative ways to power the various modes of transportation (Speight, 2011d).

Biofuels such as bioethanol and biodiesel are possible replacements for fossil fuels and have the potential to make a significant contribution in reducing the dependency on fossil-fuel imports, especially in the transport sector (Giampietro and Mayumi, 2009; EREC, 2010; Langeveld et al., 2010). Another advantage of biofuels is their contribution to climate protection: as biofuels are usually considered carbon dioxide neutral, their use helps to reduce greenhouse gas emissions at the time the climate of the Earth is evolving (Bunse et al., 2006; Zanganeh and Shafeen, 2007; Pittock, 2009; FitzRoy and Papyrakis, 2010; Sorokhtin et al., 2011). The biofuel industry has the potential to offer a source of income and large new markets for rural and small farmers and is of great interest (EREC, 2010; Langeveld et al., 2010). The growing international demand for biofuel is of particular interest to developing countries seeking opportunities for economic growth and trade. Developing countries have a comparative advantage for biofuel production because of greater availability of land, favorable climatic conditions for agriculture and lower labor costs. However, there may be other socioeconomic and environmental implications affecting the potential for developing countries to benefit from the increased global demand for biofuel (Giampietro and Mayumi, 2009; EREC, 2010; Langeveld et al., 2010). Large-scale production of biofuels offers opportunity for certain developing countries to reduce their dependence on oil imports. In industrialized countries, there is a growing trend toward employing modern technologies and efficient bioenergy conversion using a range of biofuels (Speight, 2011a).

Bioethanol is currently added to gasoline, but can be used pure. Using bioethanol-blended fuel for automobiles can significantly reduce petroleum use and greenhouse gas emissions. Gasoline and bioethanol mixtures are called gasohol. E10, sometimes called gasohol, is a fuel mixture of 10% bioethanol and 90% gasoline that can be used in the internal combustion engines of most modern vehicles. Bioethanol can be used as a 5% blend with petroleum under the EU quality standard EN 228. This blend requires no engine modification and is covered by vehicle warranties. With engine modification, bioethanol can be used at higher levels, for example, E85 (85% bioethanol). Adding bioethanol to gasoline increases the oxygen content of the fuel, improving the combustion of gasoline and reducing the exhaust emissions normally attributed to imperfect combustion in motor vehicles, such as CO and unburned hydrocarbons (Malça and Freire, 2006).

Biodiesel can be used as pure fuel or blended at any level with petroleum-based diesel for use by diesel engines. The most common biodiesel blends are B2 (2% biodiesel and 98% petroleum diesel), B5 (5% biodiesel and 95% petroleum diesel), and B20 (20% biodiesel and 80% petroleum diesel). Biodiesel can also be used in its pure form (B100), but may require certain engine modifications to avoid maintenance and performance problems. Biodiesel is widely used in many European countries as a blend of 5% biodiesel and 95% petroleum diesel (B5). In the United States, B20 (20% biodiesel and 80% petroleum diesel) is the most commonly used biodiesel blend. In addition, because biodiesel is largely made from vegetable oils, it reduces life-cycle GHG emissions by as much as 78% (Ban-Weiss et al., 2007). Biodiesel and its blends can be run in diesel engines without any significant modifications to the engines, and reduce engine emissions of hydrocarbons (HC), carbon monoxide (CO), sulfur dioxide (SO_2), and particulate matter (PM) relative to petroleum diesel. Biodiesel blends of up to 20% reduce the emissions of HC, CO, SO_2, and particulates, as well as improve the engine performance (Sastry et al., 2006). Emissions of nitrogen oxides (NO_x) increase with the concentration of biodiesel in the fuel. In October 2002, the U.S. Environmental Protection Agency assessed the impact of biodiesel fuel on emissions and published a draft report summarizing the results. The study reported that emissions from soybean-based B20 fuel compared to petroleum diesel have 10.1% less PM, 21.1% less HC, and 11% less CO. These are offset by a 2% increase in NO_x emissions.

Biofuels are attracting growing interest around the world, with some governments announcing commitments to biofuel programs as a way to both reduce greenhouse gas emissions and dependence on petroleum-based fuels. The United States, Brazil, and several EU member states have the largest programs promoting biofuels in the world (EREC, 2010; Langeveld et al., 2010; Speight, 2011a). The recent commitment by the United States government to increase bioenergy threefold in 10 years has added impetus to the search for viable biofuels. In South America, Brazil continued policies that mandate at least 22% bioethanol on motor fuels and encourage the use of vehicles that use hydrous bioethanol ((96 bioethanol + 4 water)/100) to replace gasoline (Stevens et al., 2004).

The *food versus fuel dilemma* relates to the risk of diverting farmland or crops for liquid biofuels production in detriment of the food supply on a global scale. There is disagreement about (1) the significance of this effect, (2) the cause, (3) the impact, and (4) the means to resolve the aforementioned issues (Speight and Singh, 2014). Biofuel production has increased in recent years. Some commodities such as corn, sugar cane, and vegetable oil can be used either as food, feed, or feedstock for biofuels. For example, vegetable oils have recently become more attractive because of their environmental benefits and the fact that they are made from renewable resources (Demirbaş, 2008b). Vegetable oils are a renewable and potentially inexhaustible source of energy with energy content close to diesel fuel (Seifried and Witzel, 2010). On the other hand, extensive use of vegetable oils may cause other significant problems such as starvation in developing countries.

References

Andrews, A., 2006. Oil Shale: History, Incentives, and Policy. Specialist, Industrial Engineering and Infrastructure Policy Resources, Science, and Industry Division. Congressional Research Service, The Library of Congress, Washington, DC.

ASTM D1835, 2015. Standard Specification for Liquefied Petroleum (LP) Gases. Annual Book of Standards. ASTM International, West Conshohocken, Pennsylvania.

ASTM D3904, 2015. Test Method for Oil from Oil Shale (Resource Evaluation by the Fischer Assay Procedure) (Withdrawn 1996). ASTM International, West Conshohocken, Pennsylvania.

Bahadori, A., 2014. Waste Management in the Chemical and Petroleum Industries. John Wiley & Sons Inc., Chichester, West Sussex, UK.

Balat, M., 2006. Sustainable transportation fuels from biomass materials. Energy Education Science and Technology 17, 83–103.

Balat, M., 2009. Gasification of biomass to produce gaseous products. Energy Sources Part A 31, 516–526.

Ban-Weiss, G.A., Chen, J.Y., Buchholz, B.A., Dibble, R.W., 2007. A numerical investigation into the anomalous slight NO_x increase when burning biodiesel: a new (old) theory. Fuel Processing Technology 88, 659–667.

Bartis, J.T., LaTourrette, T., Dixon, L., 2005. Oil Shale Development in the United States: Prospects and Policy Issues. Prepared for the National Energy Technology of the United States Department of Energy. Rand Corporation, Santa Monica, CA.

Bhattacharya, S., Md Mizanur Rahman Siddique, A.H., Pham, H.-L., 1999. A study in wood gasification on low tar production. Energy 24, 285–296.

Blatt, H., Tracy, R.J., 2000. Petrology: Igneous, Sedimentary, and Metamorphic. W.H. Freeman and Company, New York.

Boateng, A.A., Walawender, W.P., Fan, L.T., Chee, C.S., 1992. Fluidized-bed steam gasification of rice hull. Bioresource Technology 40 (3), 235–239.

Bower, T., 2009. Oil: Money, Politics, and Power in the 21st Century. Grand Central Publishing, New York.

Brammer, J.G., Lauer, M., Bridgwater, A.V., 2006. Opportunities for biomass-derived "bio-oil" in European heat and power markets. Energy Policy 34, 2871–2880.
Brendow, K., 2003. Global oil shale issues and perspectives. Oil Shale 20 (1), 81–92.
Brendow, K., 2009. Oil shale — a local asset under global constraint. Oil Shale 26 (3), 357–372.
Bullin, K., Krouskop, P., October 7, 2008. Composition variety complicates processing plans for US shale gas. In: Proceedings. Annual Forum, Gas Processors Association — Houston Chapter, Houston, Texas.
Bunse, M., Dienst, C., Fischedick, M., Wallbaum, H., 2006. Promoting Sustainable Biofuel Production and Use. WISIONS, German Wuppertal Institute for Climate, Environment and Energy, Wuppertal, Germany.
Brage, C., Yu, Q., Chen, G., Sjöström, K., 2000. Tar evolution profiles obtained from gasification of biomass and coal. Biomass and Bioenergy 18 (1), 87–91.
Chadeesingh, R., 2011. The Fischer–Tropsch process. In: Speight, J.G. (Ed.), The Biofuels Handbook. The Royal Society of Chemistry, London, UK, pp. 476–517. Part 3 (Chapter 5).
Chen, G., Sjöström, K., Bjornbom, E., 1992. Pyrolysis/gasification of wood in a pressurized fluidized bed reactor. Industrial & Engineering Chemistry Research 31 (12), 2764–2768.
Collot, A.G., Zhuo, Y., Dugwell, D.R., Kandiyoti, R., 1999. Co-pyrolysis and co-gasification of coal and biomass in bench-scale fixed-bed and fluidized bed reactors. Fuel 78, 667–679.
Crane, H.D., Kinderman, E.M., Malhotra, R., 2010. A Cubic Mile of Oil: Realities and Options for Averting the Looming Global Energy Crisis. Oxford University Press, Oxford, United Kingdom.
Czernik, S., Bridgwater, A.V., 2004. Overview of the applications of biomass fast pyrolysis oil. Energy Fuels 18, 590–598.
Demirbaş, A., 2001. Yields of hydrogen of gaseous products via pyrolysis from selected biomass samples. Fuel 80, 1885–1891.
Demirbaş, A., 2006. Global renewable energy resources. Energy Sources Part A 28, 779–792.
Demirbaş, A., 2008a. Biohydrogen generation from organic waste. Energy Sources Part A 30, 475–482.
Demirbaş, A., 2008b. Studies on cottonseed oil biodiesel prepared in non-catalytic SCF conditions. Bioresource Technology 99, 1125–1130.
Dupont, C., Boissonnet, G., Seiler, J.M., Gauthier, P., Schweich, D., 2007. Study of the kinetic processes of biomass steam gasification. Fuel 86, 32–40.
Durand, B., 1980. Kerogen: Insoluble Organic Matter from Sedimentary Rocks. Editions Technip, Paris, France.
Dyni, J.R., 2003. Geology and resources of some world oil-shale deposits. Oil Shale 20 (3), 193–252.
Dyni, J.R., 2006. Geology and Resources of Some World Oil Shale Deposits. Report of Investigations 2005–5295. United States Geological Survey, Reston, Virginia.
EREC, 2010. Renewable Energy in Europe: Markets, Trends, and Technologies. European Renewable Energy Council (EREC), Brussels, Belgium. Earthscan, London, UK.
Ergudenler, A., Ghaly, A.E., 1993. Agglomeration of alumina sand in a fluidized bed straw gasifier at elevated temperatures. Bioresource Technology 43 (3), 259–268.
FitzRoy, F.R., Papyrakis, E., 2010. An Introduction to Climate Change Economics and Policy. Earthscan, London, UK.
Gabra, M., Pettersson, E., Backman, R., Kjellström, B., 2001. Evaluation of cyclone gasifier performance for gasification of sugar cane residue — Part 1: gasification of bagasse. Biomass and Bioenergy 21 (5), 351–369.
Gary, J.G., Handwerk, G.E., Kaiser, M.J., 2007. Petroleum Refining: Technology and Economics, fifth ed. CRC Press, Taylor & Francis Group, Boca Raton, FL.

Giampietro, M., Mayumi, K., 2009. The Biofuel Delusion. The Fallacy of Large-Scale Ago-biofuel Production. Earthscan, London, UK.

Gudmestad, O.T., Zolotukhin, A.B., Jarlsby, E.T., 2010. Petroleum Resources: With Emphasis on Offshore Fields. WIT Press, Ashurst, Southampton, UK.

Hamilton, M.S., 2013. Energy Policy Analysis: A Conceptual Framework. M.E. Sharpe Inc., Armonk, NY.

Hanaoka, T., Inoue, S., Uno, S., Ogi, T., Minowa, T., 2005. Effect of woody biomass components on air-steam gasification. Biomass and Bioenergy 28 (1), 69–76.

Hotchkiss, R., 2003. Coal gasification technologies. Proceedings of the Institution of Mechanical Engineers Part A 217 (1), 27–33.

Hoogwijk, M., Faaij, A., Eickhout, B., de Vries, B., Turkenburg, W., 2005. Potential of biomass energy out to 2100, for four IPCC SRES land-use scenarios. Biomass and Bioenergy 29, 225–257.

Hsu, C.S., Robinson, P.R. (Eds.), 2006. Practical Advances in Petroleum Processing Volume 1 and Volume 2. Springer Science, New York.

Hunt, J.M., 1996. Petroleum Geochemistry and Geology, second ed. W.H. Freeman, San Francisco.

Jothimurugesan, K., Goodwin, J.G., Santosh, S.K., Spivey, J.J., 2000. Development of Fe Fischer–Tropsch catalysts for slurry bubble column reactors. Catalysis Today 58, 335–344.

Karekezi, S., Lata, K., Coelho, S.T., 2004. Traditional biomass energy-improving its use and moving to modern energy use. In: International Conference for Renewable Energies, Bonn, June 1–4.

Ko, M.K., Lee, W.Y., Kim, S.B., Lee, K.W., Chun, H.S., 2001. Gasification of food waste with steam in fluidized bed. Korean Journal of Chemical Engineering 18 (6), 961–964.

Khoshnaw, F.M. (Ed.), 2013. Petroleum and Mineral Resources. WIT Press, Ashurst, Southampton, UK.

Kumabe, K., Hanaoka, T., Fujimoto, S., Minowa, T., Sakanishi, K., 2007. Co-gasification of woody biomass and coal with air and steam. Fuel 86, 684–689.

Kundert, D., Mullen, M., April 14–16, 2009. Proper evaluation of shale gas reservoirs leads to a more effective hydraulic-fracture stimulation. In: Paper No. SPE 123586. Proceedings. SPE Rocky Mountain Petroleum Technology Conference, Denver, Colorado.

Langeveld, H., Sanders, J., Meeusen, M. (Eds.), 2010. The Biobased Economy: Biofuels, Materials, and Chemicals in the Post-Oil Era. Earthscan, London, UK.

Lee, S., 1996. Alternative Fuels. CRC Press, Taylor & Francis Group, Boca Raton, FL.

Levant, E., 2010. Ethical Oil: The Case for Canada's Oil Sands. McClelland & Stewart, Toronto, ON, Canada.

Li, Y., Xue, B., He, X., 2009. Catalytic synthesis of ethylbenzene by alkylation of benzene with diethyl carbonate over HZSM-5. Catalysis Communications 10, 702–707.

Luciani, G., 2013. Security of Oil Supplies: Issues and Remedies. Claeys and Casteels Law Publishers, Deventer (Netherlands), Leuven, Belgium.

Lv, P.M., Xiong, Z.H., Chang, J., Wu, C.Z., Chen, Y., Zhu, J.X., 2004. An experimental study on biomass air-steam gasification in a fluidized bed. Bioresource Technology 95 (1), 95–101.

Maher, K.D., Bressler, D.C., 2007. Pyrolysis of triglyceride materials for the production of renewable fuels and chemicals. Bioresource Technology 98, 2351–2368.

Malça, J., Freire, F., 2006. Renewability and life-cycle energy efficiency of bioethanol and bio-ethyl tertiary butyl ether (bioETBE): assessing the implications of allocation. Energy 31, 3362–3380.

Mariolakos, I., Kranioti, A., Markatselis, E., Papageorgiou, M., 2007. Water: mythology and environmental education. Desalination 213, 141–146.

Maschio, G., Lucchesi, A., Stoppato, G., 1994. Production of syngas from biomass. Bioresource Technology. 48, 119–126.

McKendry, P., 2002. Energy production from biomass Part 3: gasification technologies. Bioresource Technology 83 (1), 55–63.

Mohan, S.V., Mohanakrishna, G., Sarma, P.N., 2008. Integration of acidogenic and methanogenic processes for simultaneous production of biohydrogen and methane from wastewater treatment. International Journal of Hydrogen Energy 33, 2156–2166.

Mokhatab, S., Poe, W.A., Speight, J.G., 2006. Handbook of Natural Gas Transmission and Processing. Elsevier, Amsterdam, Netherlands.

Ots, A., 2007. Estonian oil shale properties and utilization in power plants. Energetika 53 (2), 8–18.

Overend, R.P., 2004. Thermochemical conversion of biomass. In: Shpilrain, E.E. (Ed.), Renewable Energy Sources Charged with Energy from the Sun and Originated from Earth-moon Interaction. EOLSS Publishers, Oxford, UK.

Pakdel, H., Roy, C., 1991. Hydrocarbon content of liquid products and tar from pyrolysis and gasification of wood. Energy & Fuels 5, 427–436.

Pan, Y.G., Velo, E., Roca, X., Manyà, J.J., Puigjaner, L., 2000. Fluidized-bed co-gasification of residual biomass/poor coal blends for fuel gas production. Fuel 79, 1317–1326.

Pittock, A.B., 2009. Climate Change: The Science Impacts, and Solutions, second ed. CSIRO Publishing, Collingwood, VIC, Australia.

Pütün, E., Uzun, B.B., Pütün, A.E., 2006. Production of bio-fuels from cottonseed cake by catalytic pyrolysis under a steam atmosphere. Biomass Bioenergy 30, 592–598.

Rao, V.U.S., Stiegel, G.J., Cinquergrane, G.J., Srivastava, R.D., 1992. Iron-based catalysts for slurry-phase Fischer–Tropsch process: technology review. Fuel Processing Technology 30, 83–107.

Rapagnà, S., Latif, A., 1997. Steam gasification of almond shells in a fluidized bed reactor: the influence of temperature and particle size on product yield and distribution. Biomass and Bioenergy 12 (4), 281–288.

Rapagnà, S., Jand, N., Kiennemann, A., Foscolo, P.U., 2000. Steam-gasification of biomass in a fluidized-bed of olivine particles. Biomass and Bioenergy 19 (3), 187–197.

Resini, C., Arrighi, L., Delgado, M.C.H., Vargas, M.A.L., Alemany, L.J., Riani, P., Berardinelli, S., Maraza, R., Busca, G., 2006. Production of hydrogen by steam reforming of C3 organics over a Pd–Cu/γ-Al$_2$O$_3$ catalyst. International Journal of Hydrogen Energy 31, 13–19.

Ryan, L., Turton, H., 2007. Sustainable Automobile Transport. Edward Elgar Publishing Ltd, Cheltenham, UK.

Sastry, G.S.R., Krishna Murthy, A.S.R., Raviprasad, P., Bhuvaneswari, K., Ravi, P.V., 2006. Identification and determination of bio-diesel in diesel. Energy Sources Part A 28, 1337–1342.

Schettler Jr., P.D., Parmely, C.R., Juniata, C., 1989. Gas composition shifts in Devonian shales. SPE Reservoir Engineering 4 (3), 283–287.

Scouten, C., 1990. In: Speight, J.G. (Ed.), Fuel Science and Technology Handbook. Marcel Dekker Inc, New York.

Seifried, D., Witzel, W., 2010. Renewable Energy: The Facts. Earthscan, London, UK.

Sjöström, K., Chen, G., Yu, Q., Brage, C., Rosén, C., 1999. Promoted reactivity of char in co-gasification of biomass and coal: synergies in the thermochemical process. Fuel 78, 1189–1194.

Sorokhtin, O.G., Chilingarian, G.V., Sorokhtin, N.O., 2011. Evolutions of Earth and Its Climate. Elsevier, Amsterdam, Netherlands.
Speight, J.G., Ozum, B., 2002. Petroleum Refining Processes. Marcel Dekker Inc., New York.
Speight, J.G., 2007. Natural Gas: A Basic Handbook. GPC Books, Gulf Publishing Company, Houston, Texas.
Speight, J.G., 2008. Synthetic Fuels Handbook: Properties, Processes, and Performance. McGraw-Hill, New York.
Speight, J.G., 2009. Enhanced Recovery Methods for Heavy Oil and Tar Sands. Gulf Publishing Company, Houston, TX.
Speight, J.G. (Ed.), 2011a. The Biofuels Handbook. Royal Society of Chemistry, London, UK.
Speight, J.G., 2011b. An Introduction to Petroleum Technology, Economics, and Politics. Scrivener Publishing, Salem, MA.
Speight, J.G., 2011c. The Refinery of the Future. Gulf Professional Publishing, Elsevier, Oxford, UK.
Speight, J.G., 2011d. Production, properties, and environmental impact of hydrocarbon fuel conversion. In: Khan, M.R. (Ed.), Advances in Clean Hydrocarbon Fuel Processing. Woodhead Publishing, Cambridge, UK.
Speight, J.G., 2012. Shale Oil Production Processes. Gulf Professional Publishing, Elsevier, Oxford, UK.
Speight, J.G., 2013a. The Chemistry and Technology of Coal, third ed. CRC Press, Taylor & Francis Group, Boca Raton, FL.
Speight, J.G., 2013b. Coal-fired Power Generation Handbook. Scrivener Publishing, Salem, MA.
Speight, J.G., 2013c. Shale Gas Production Processes. Gulf Professional Publishing, Elsevier, Oxford, UK.
Speight, J.G., 2014a. The Chemistry and Technology of Petroleum, fifth ed. CRC Press, Taylor & Francis Group, Boca Raton, FL.
Speight, J.G., 2014b. Gasification of Unconventional Feedstocks. Gulf Professional Publishing, Elsevier, Oxford, UK.
Speight, J.G., Singh, K., 2014. Environmental Management of Energy from Biofuels and Biofeedstocks. Scrivener Publishing, Salem, MA.
Stelmachowski, M., Nowicki, L., 2003. Fuel from synthesis gas — the role of process engineering. Applied Energy 74, 85—93.
Stevens, D.J., Wörgetter, M., Saddler, J., 2004. Biofuels for Transportation: An Examination of Policy and Technical Issues. IEA Bioenergy Task 39, Liquid Biofuels Final Report 2001—2003, Paris.
Suban, M., Tuǧek, J., Uran, M.J., 2001. Use of hydrogen in welding engineering in former times and today. Journal of Materials Processing Technology 119, 193—198.
Vélez, J.F., Chejne, F., Valdés, C.F., Emery, E.J., Londoño, C.A., 2009. Co-gasification of Colombian coal and biomass in a fluidized bed: an experimental study. Fuel 88, 424—430.
Wihbey, P.M., 2009. The Rise of the New Oil Order. Academy & Finance SA, Geneva, Switzerland.
Wiltowski, T., Mondal, K., Campen, A., Dasgupta, D., Konieczny, A., 2008. Reaction swing approach for hydrogen production from carbonaceous fuels. International Journal of Hydrogen Energy 33, 293—302.
Weiland, R.H., Hatcher, N.A., 2012. Overcome challenges in treating shale gas. Hydrocarbon Processing 91 (1), 45—48.
Wu, Y., Carroll, J.J., Du, Z. (Eds.), 2011. Carbon Dioxide Sequestration and Related Technologies. Scrivener Publishing, Salem, MA.

Zatzman, G.M., 2012. Sustainable Resource Development. Scrivener Publishing, Salem, MA.

Zanganeh, K.E., Shafeen, A., 2007. A novel process integration, optimization and design approach for large-scale implementation of oxy-fired coal power plants with CO_2 capture. International Journal of Greenhouse Gas Control 1, 47–54.

Zhang, Q., Chang, J., Wang, T., Xu, Y., 2007. Review of biomass pyrolysis oil properties and upgrading research. Energy Conversion Management 48, 87–92.

Solid fuel types for energy generation: coal and fossil carbon-derivative solid fuels

Panagiotis Grammelis, Nikolaos Margaritis, Emmanouil Karampinis
Centre for Research & Technology Hellas/Chemical Process & Energy Resources Institute (CERTH/CPERI), Ptolemaida, Greece

2.1 Introduction

Fossil fuels, including coal, oil and natural gas, are currently the world's primary energy source. Formed from organic material over millions of years, fossil fuels have fueled global economic development over the past century. Coal plays a vital role in electrical generation worldwide (see Figure 2.1). Coal-fired power plants currently produce 41% of global electricity (http://www.worldcoal.org/coal/uses-of-coal/coal-electricity/).

In some cases, indigenous coal is the only major energy source available for power generation, whereas in other cases it is a means to maintain and reinforce industrial and economic growth. Moreover, coal contributes to the reduction of imported energy dependency by maintaining its relative low cost, whereas improvements in coal technologies mitigate its negative environmental impact. Improvements in conventional pulverized coal combustion power station design and development of new combustion technologies have enhanced the thermal efficiency of the power plants. Efficiency gains in electrical generation from coal-fired power stations will play a crucial part in reducing CO_2 emissions at a global level.

2.2 Fossil fuel feedstocks

2.2.1 World availability of coal

Coal is the second leading source of fuel behind oil and the major one for industrial power generation. Types of coals mostly used are lignite (US—United States and FSU—Former Soviet Union), subbituminous coal (China, FSU, Australia and Germany) and bituminous coals (China, US and FSU).

Contrary to oil and natural gas, coal is not usually subject to 'price shocks' for market and/or geopolitical reasons, due to the quite large amount of producer countries and its relatively low extraction cost. These are the main reasons why coal remains a popular fuel in the power sector. The rapid industrialization and economic development of many Asian countries puts pressure on the local power sector to meet a

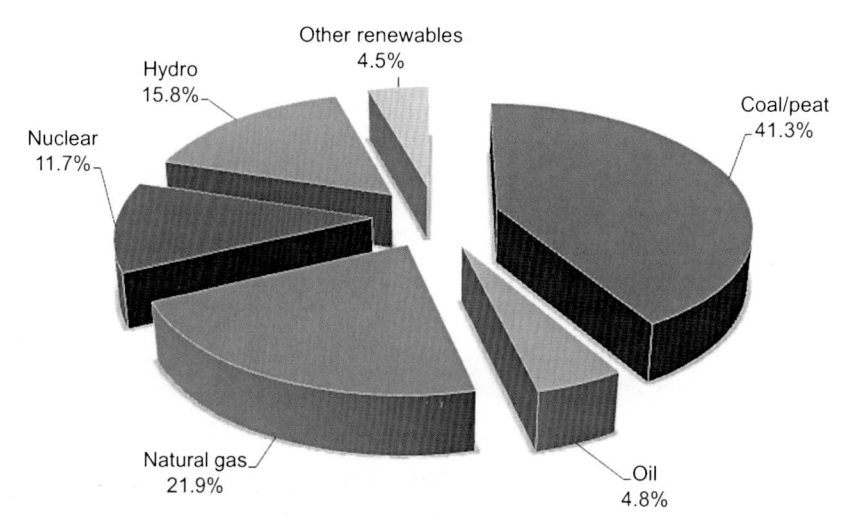

Figure 2.1 Total world electricity generation by fuel, 2013.
Source: Key World Energy Statistics 2011 © Organisation for Economic Co-operation and Development (OECD)/International Energy Agency (IEA), 2013, p. 24.

growing demand for electricity, which currently can only be met economically with coal. The most notable exporting countries for hard coal are Indonesia, Australia, Russia, US, Colombia and South Africa. These represent approximately 87% of total hard coal exports in 2012 (see Figure 2.2).

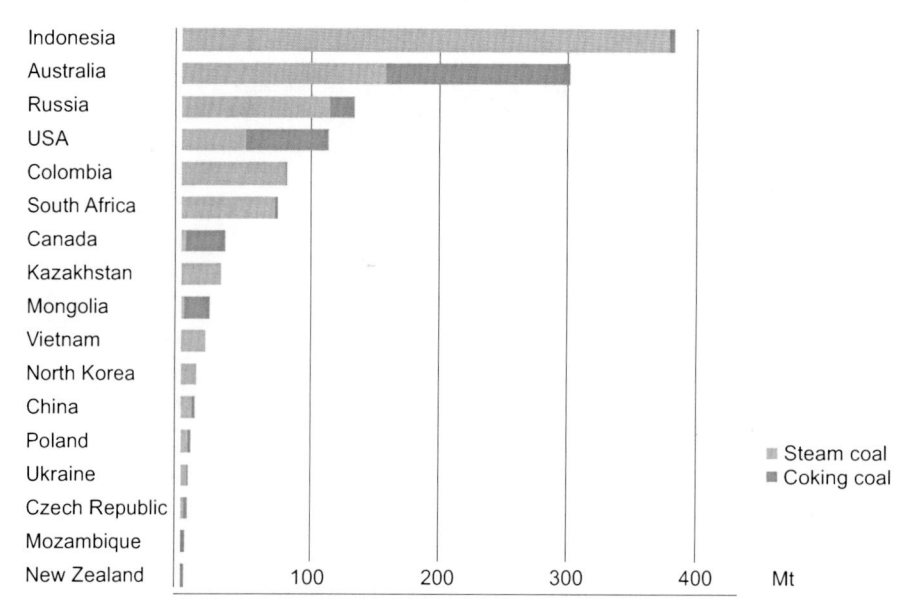

Figure 2.2 Top coal-exporting countries, 2012.
Source: Euracoal (2013).

In 2012, world coal production reached 7.8 billion tn, out of which 0.9 billion tn were lignite. The remaining hard coal production comprised 5.9 billion tn of steam coal and 1.0 billion tn of coking coal. Approximately 100 countries worldwide are part of the coal economy, either as major producers or as importers/end users. World coal consumption is dominated by major actors. European Union (EU) is the fourth largest consumer behind China, US and India. In 2012, China's annual coal production increased by 130 million tn, this being almost identical to the EU's total hard coal production. Nevertheless, at 433 million tn (Euracoal, 2013), the EU remains the world's largest lignite producer (see Figure 2.3).

Every year, the world consumes large amounts of fossil energy raw materials. If raw materials are to be available on demand at all times and in order for the power sector to organize its long-term planning, proven available and sufficient quantities of fuels are needed. For this reason, mining companies secure their annual output for some decades by exploring and developing deposits (Gerling and Wellmer, 2004). Reserves are verified quantities that can be economically extracted using current technology and at current prices. Resources, on the other hand, include (1) known amounts, though at present not economically minable; and (2) assumed amounts, not proven yet by exploration (Mills, 2011). Due to new information on deposits and the development

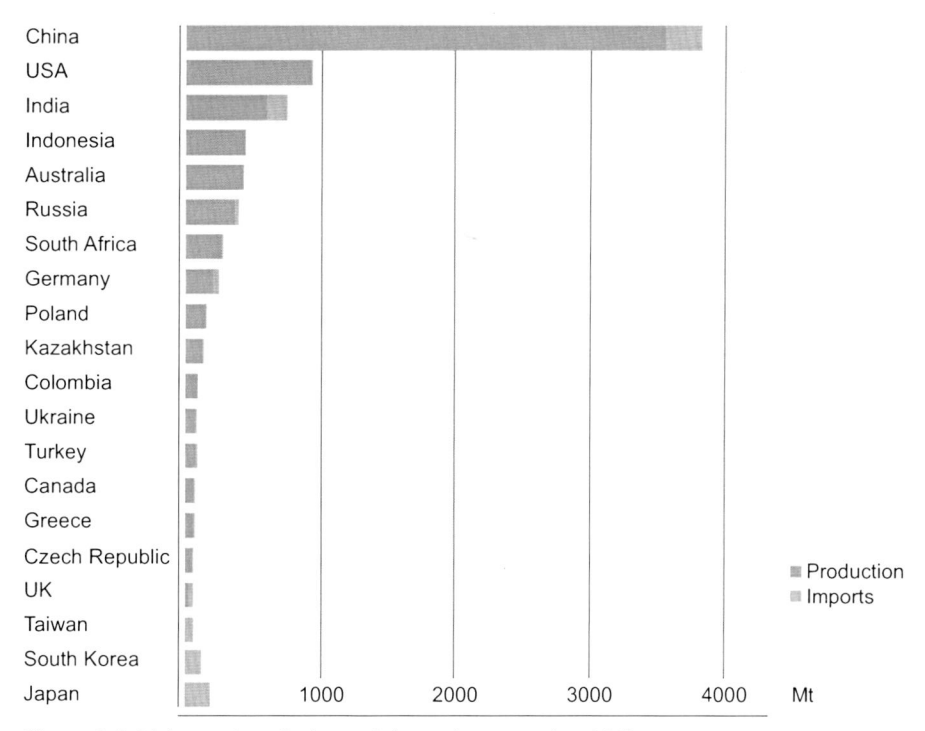

Figure 2.3 Major coal-producing and -importing countries, 2012.
Source: Euracoal (2013).

of new extraction and exploration programmes, resources convert to reserves (Thielmann et al., 2007). Both reserves and resources represent the total coal resources.

Coal and lignite reserves are sufficient for the next 137 years at current production rates. Unlike oil and gas, coal is widely distributed around the world. Particularly large reserves are located in the US, Russia and China (see Figure 2.4). Coal offers a high level of supply security as it is mostly used in the country of extraction. When coal is imported, supply is supported by a competitive market and a well-developed infrastructure.

The EU's share of global energy reserves and resources is rather small, approximately 3%. As at the global level, the reserves and resources of coal and lignite are most significant: together, they account for 94% of the EU's remaining potential. Some coal deposits lie near consumers and can be exploited under very favourable conditions. For example, surface-mined lignite in Bulgaria, the Czech Republic, Germany, Greece, Hungary, Poland and Romania is used mainly for power generation and is often transported to power plants over short distances by conveyor belt to produce some of the lowest-cost electricity in Europe. Hard coal, both indigenously produced and imported, is less expensive than imported oil or gas; thus, the majority of EU member states enjoy the benefits of competitive coal-fired electrical generation. (Euracoal, 2013).

Every year, coal resources convert into reserves, as our knowledge concerning coal deposits grows and new pits or pit sections are developed. This conversion of resources into reserves replaces some of the reserve losses due to annual coal production.

In recent decades, such conversion represented only 5—20% of specific world annual production. In the past, some countries, for example, Germany and Poland, shut down mines, so that access to former reserves was blocked (Thielemann et al., 2007). Measured by the pit openings and extensions (Kopal, 2006) planned for the next few years, it will not be possible to completely replace an annual output by new same-size reserves in the foreseeable future.

2.2.2 Coal classification

Coal is a sedimentary, organic rock, formed in former swamp ecosystems by deposited plant matter, which was covered by acidic water or mud and thus protected from biodegradation and oxidation. It is composed mostly of carbon and lesser amounts of other elements, mostly hydrogen, oxygen, nitrogen and sulfur. Depending on the geological conditions prevailing in the coal formation process, the dead plant material, over geologic time, is progressively transformed into different precursor materials and coal types. A general classification is presented in the list below, starting from the lower to higher ranks, whereas Figure 2.5 provides information on the estimated percentage of the total world coal reserves and typical use for each coal type:

- Peat (precursor to coal)
- Lignite/brown coal
- Subbituminous coal
- Bituminous coal
- Anthracite

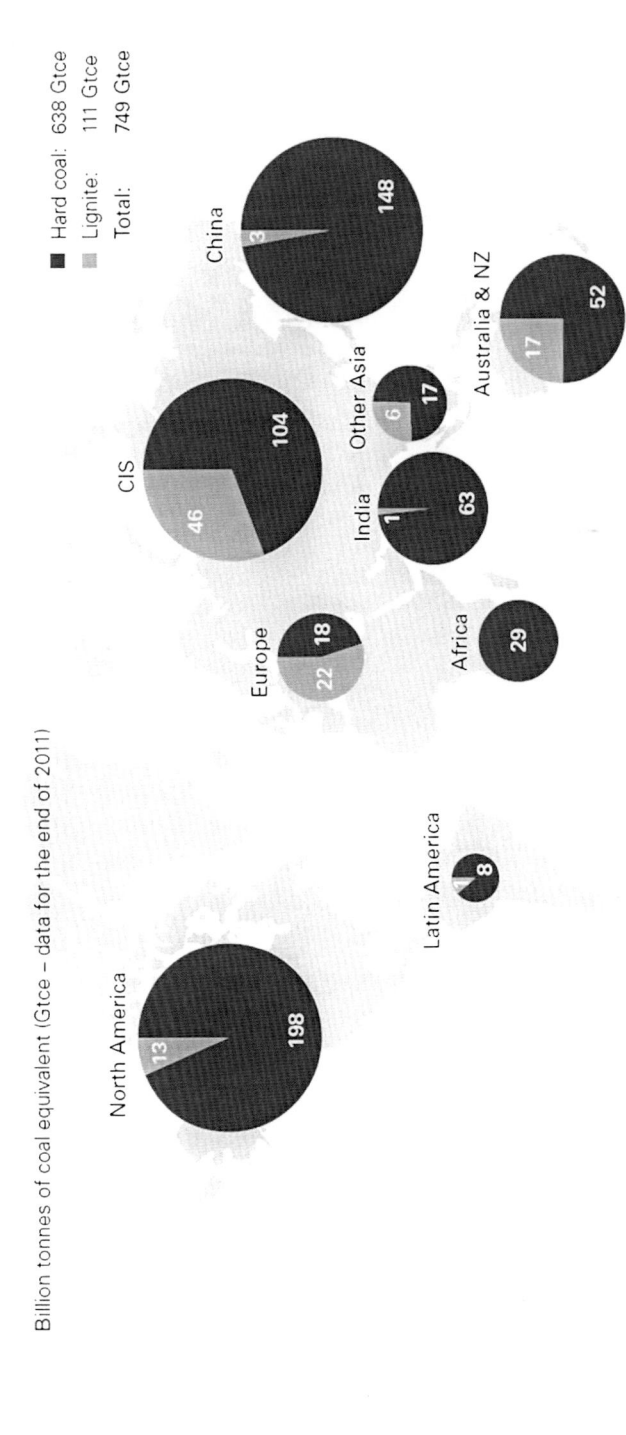

Solid fuel types for energy generation: coal and fossil carbon-derivative solid fuels 33

Billion tonnes of coal equivalent (Gtce – data for the end of 2011)

■ Hard coal: 638 Gtce
■ Lignite: 111 Gtce
  Total: 749 Gtce

China: 148, 3
CIS: 104, 46
Other Asia: 17, 6
Australia & NZ: 52, 17
India: 63, 1
Europe: 18, 22
Africa: 29
North America: 198, 13
Latin America: 8, 1

Figure 2.4 Global hard coal and lignite reserves.
Source: Euracoal (2013).

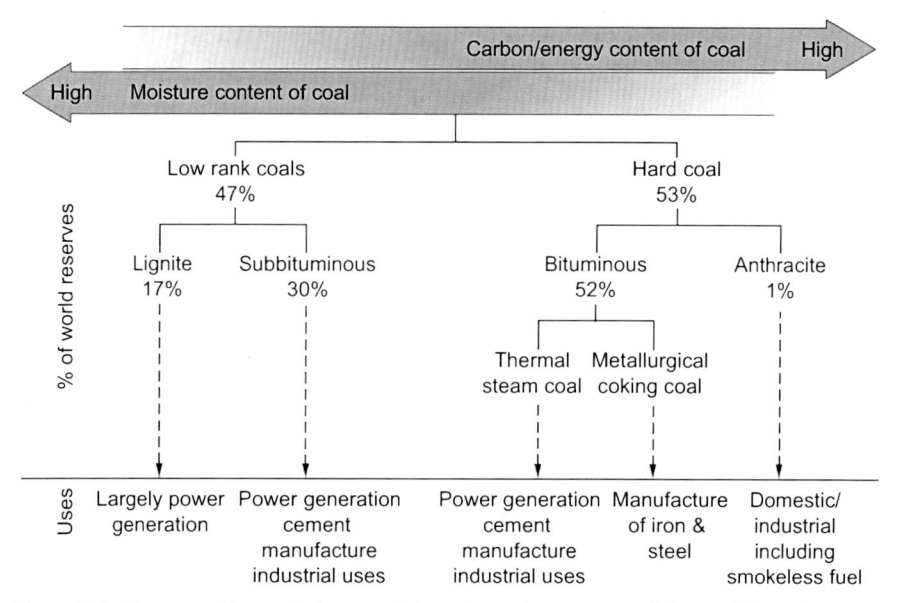

Figure 2.5 Diagram of the typical uses and the estimated percentage of the world's coal reserves for each coal rank.
Source: World Coal Association (2005).

Different systems for the formal classification of coal have been developed at national and international scales depending on scientific (physical, chemical, petrographic), technical (heating value, plasticity, swelling index), commercial, or combined parameters (http://ec.europa.eu/energy/coal/eucores/doc/20120805-eucores-proposalclassification.pdf). For this reason, there are differences in the reported scale of coal deposits and rates of consumption (Mills and International Energy Agency, 2011). The IEA has adopted the United Nations Economic Commitment for Europe (UNECE)/International Organization for Standardization (ISO) 11760 definitions of hard and brown coal in relation with production, trade and consumption. Other systems commonly used are American Society for Testing and Materials (ASTM) D388 and the German Institute for Standardization (DIN) coal classification. However, each country has developed its own criteria for classifying coals. Table 2.1 presents the interrelation between these coal classification systems.

In general, low-rank coals are mostly used in the power sector, whereas higher-rank coals also find other industrial or specialized applications, for example, manufacture of iron and steel (http://www.worldcoal.org/coal/uses-of-coal/coal-electricity/).

2.2.2.1 Peat

Peat is the first step in the geological formation of coal and it is considered as a precursor to it. Peat is a heterogeneous material consisting of decomposed plant and

Table 2.1 Interrelation between coal classification systems

Coal types and peat: UNECE	Coal types and peat: USA (ASTM)	Coal types and peat: Germany (DIN)			Total water content (%)	Energy content a.f.* (KJ/Kg)	Volatiles d.a.f.** (%)	Vitrinite reflection in oil (%)
Peat	Peat	Peat						
					75	6,700		
Ortho-lignite	Lignite	Soft brown coal						
					35	16,500		0.3
Meta-lignite	Lignite / Sub-bituminous coal	Dull brown coal		Hard brown coal				
					25	19,000		0.45
Sub-bituminous coal	Sub-bituminous coal	Bright brown coal		Hard brown coal				
					10	25,000	45	0.65
Bituminous coal	High volatile bituminous coal	Flame coal	Bituminous coal	Hard brown coal				
							40	0.75
Bituminous coal	High volatile bituminous coal	Gas-flame coal	Bituminous coal	Hard brown coal	36,000 hard coking coal			
							35	1.0
Bituminous coal	High volatile bituminous coal / Medium volatile bituminous coal	Gas coal	Bituminous coal	Hard brown coal	36,000 hard coking coal			
							28	1.2
Bituminous coal	Medium volatile bituminous coal / Low volatile bituminous coal	Fast coal	Bituminous coal	Hard brown coal	36,000 hard coking coal		19	
								1.6
Bituminous coal	Low volatile bituminous coal	Low volatile coal	Bituminous coal	Hard brown coal				
							14	1.9
Anthracite	Semi-anthracite	Semi-anthracite		Hard brown coal				
					3	36,000	10	2.2
Anthracite	Anthracite	Anthracite		Hard brown coal				

* *a.f. = ash-free*

** *d.a.f. = dry, ash-free*

UNECE: Ortho-lignite up to 15,000 KJ/Kg
Meta-lignite up to 20,000 KJ/Kg
Sub-bituminous coal up to 24,000 KJ/Kg
Bituminous coal up to 2 % average vitrinite reflection
USA: Lignite up to 19,300 KJ/Kg

Source: http://www.euracoal.org/pages/home.php?idpage=1

Figure 2.6 Peat.

mineral matter (see Figure 2.6). Peat's colour depends on its geologic age, ranging from yellowish to brownish black. According to the ASTM, peat's heating value and moisture content is 6978 KJ/Kg and up to 70%, respectively (The Babcock & Wilcox Company Book, Edition 41, 2005).

2.2.2.2 Lignite

Lignite has the lowest ranking in the coal classification system, and it is mostly used as fuel in the power sector. Lignite's colour ranges from brown to black (see Figure 2.7)

Figure 2.7 Lignite.

and its heating value is less than 19.306 KJ/Kg. Due to the fact that lignite deposits are geologically young they may contain plant debris. Lignite's moisture content and volatile matter are relative high, more than 25 and more than 24% respectively, whereas its ash content ranges from 3 to 15%. The high-moisture content poses several challenges for the utilization of lignite, because it lowers the electrical efficiency of thermal power plants and can pose problems during its handling, such as spontaneous combustion. Moreover, the high-moisture content lowers the energy density of the fuel and makes its transportation uneconomic over long distances. Therefore, lignite power plants are typically constructed next to the mines that supply them.

The largest lignite deposits are located in the US (North to South Dakota, Montana) and in Canada (Manitoba) (The Babcock & Wilcox Company Book, Edition 41, 2005).

2.2.2.3 Subbituminous coal

Subbitunimous coal's colour is black and is associated with brown coals. These coals undergo a small swelling when heating, they have a relative high-moisture content (15−30%) and tend to combust spontaneously upon drying. Compared to lignite, subbituminous coal ignites more easily due to its high volatile matter, approximately 28−45%. Using subbituminous coals is an attractive option for power plants considering their low sulfur content ($\sim 0.3-1.5\%$) and high heating value ranging from 19,306 to 26,749 KJ/Kg (The Babcock & Wilcox Company Book, Edition 41, 2005). Germany's resources consist mainly of subbituminous coal and lignite. With the current utilization rate, these resources are expected to deplete in about 20 years (Spohn and Ellersdorfer, 2005).

2.2.2.4 Bituminous coal

Bituminous coals are the most common fuel in the power sector. Their colour is black or black with layers of glossy and dull black. Their heating values range from 24.423 to 32.564 KJ/Kg. They have a carbon content of 69% to 86% and lower moisture and volatile content than subbituminous and lignite coals, ranging from 2% to 15% and 15% to 45%, respectively. They are easily combusted when pulverized as powder due to their high heating value and high-volatile content, and they burn with a relatively long flame. Yet, in case of improper combustion, bituminous coal is characterized with excess smoke and soot. Spontaneous combustion in storage rarely occurs for bituminous coals. Some types of these coals, when heated without air, release volatiles that form a new hard, black and porous product, called coke (porous, hard and black). Coke is most commonly used for iron making as a fuel (The Babcock & Wilcox Company Book, Edition 41, 2005).

2.2.2.5 Anthracite

Anthracite is a high-rank coal, representing a coal that has been subjected to the highest grade of metamorphism. Anthracite is shiny black, hard and brittle (see Figure 2.8) and has the highest fixed-carbon content (approximately 86−98%). Due to its low volatile

Figure 2.8 Anthracite.

matter (2−12%), anthracite's combustion process is slow. Most anthracites have low-moisture content (about 3−6%) and their heating value is 34.890 KJ/Kg. Anthracite combusts with hot, clean flame, containing low content of sulfur and volatiles. Due to these characteristics, anthracite is sometimes used in domestic applications or other specialized industrial uses that require smokeless fuels (The Babcock & Wilox Company Book, Edition 41, 2005).

2.2.3 Coal characterization

Proximate analysis is the basic process for coal ranking. Except from ranking coal, chemical analysis provides other useful information such as selection of coal for steam generation, evaluation of existing handling, combustion equipment and input for design. Coal analysis is based on ASTM standards (ASTM, 1999) and special tests developed by Babcock & Wilcox Company (B & W).

2.2.3.1 Bases for analyses

There is a variability of ash content and moisture in coals. This leads the results of proximate analysis to be reported on several bases. The most common include (1) as received moisture (ar); (2) moisture-free or dry basis (db); (3) dry, ash-free (daf) basis; and (4) mineral matter/ash-free (maf) basis. As received, analysis reports the percentage by weight of each constituent in the coal as it is received in the laboratory (Figure 2.9).

As received, analysis includes different levels of moisture in the samples. For analysis on a dry basis, the moisture in samples is determined and then used to correct each constituent to a dry level. The ash in coal as determined in proximate analysis is different from the mineral matter in coal. This can sometimes cause problems when ranking coals by ASTM method. The correction for the mineral matter and determination of volatile matter is achieved through appropriate equations (ASTM, 1999).

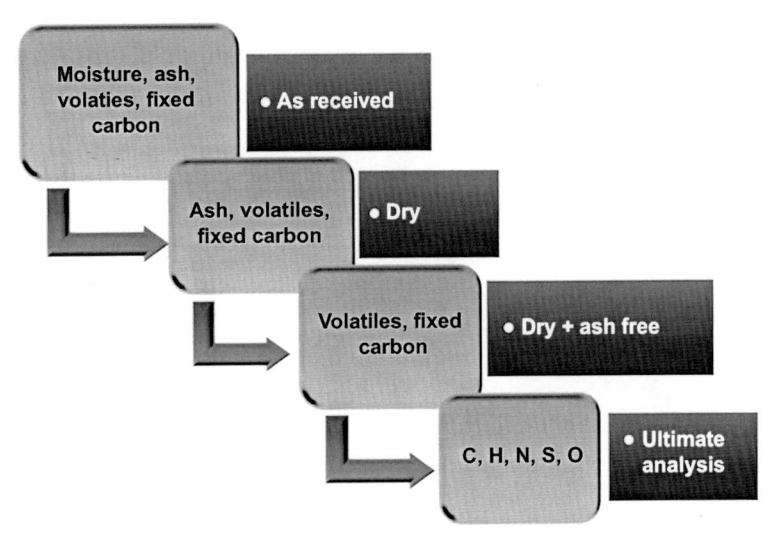

Figure 2.9 Terminology of coal analysis.

2.2.3.2 Moisture determination

Coal contains varying amounts of moisture in several forms. There is inherent and surface moisture in coal. Inherent moisture cannot be removed easily when coal is dried in air and is a natural part of the coal deposit. On the other hand, surface moisture is not a natural part of a coal deposit; therefore, it is easier to remove when exposed to air. There are also other moisture types that characterize coal including equilibrium, free and air-dry moisture. Equilibrium moisture is used as an estimation of bed moisture. ASTM D121 defines the total moisture as the loss in weight of a sample under controlled conditions of temperature, time and air flow.

With ASTM D3302, total moisture is calculated from the moisture lost or gained through air drying and the residual moisture. The residual moisture can be determined by implementing oven drying on the air-dried sample. Because subsequent ASTM analyses (such as proximate and ultimate) are performed on an air-dried sample, the residual moisture value is required to convert these results to a dry basis. The moisture lost due to air drying provides an indication of the drying required in the handling and pulverization portions of the boiler coal feed system (ASTM, 1999).

Coal moisture is a crucial factor for coal boilers, containing both inherent as well as surface moisture, together referred as total moisture. High moisture content in coals causes transportation problems to the power stations as they tend to block the chutes of conveyors. In addition, wet coals cause 'hang-ups' in bunkers, by hindering the free flow of coal. Moreover, wet pulverized coal can also result to the clogging of milling plant and associated pipework.

The temperature of the coal entering a boiler furnace is ± 90 °C. High temperature is used to ensure that combustion is taking place in the shortest possible time and within the confinement of the boiler furnace. When coal is removed from the mills through a blast of hot air, if the coal is too wet, the drying out is ineffective and has a negative effect on the combustion of coal. In other words, if the total moisture content becomes too high, the amount of heat energy required to evaporate the moisture is greater than the boiler's capacity. This limits the amount of coal dried for the milling process and the amount of pulverized coal fired into the boiler and thus limiting the power generated (Generation Communication and Primary Energy, 2013). In conclusion, when coal moisture is reduced, boiler efficiency increases, net unit heat rate decreases and the feed rate of cooling tower makeup water decreases.

2.2.3.3 Proximate analysis

Proximate coal analysis includes determination of moisture, ash, volatile matter and fixed carbon via standard test methods. Proximate analysis is a way to determine the distribution of products when the samples are heated under specified conditions. The basic four group products separated by proximate analysis are: (1) moisture, (2) volatile matter, (3) fixed carbon and (4) ash content.

The basic coal standard test method for proximate analysis is ASTM D3172 and covers the methods connected with the proximate analysis of coal and coke. The determination of moisture, volatile matter and ash is determined by different temperature levels. Moisture and volatile matter can be identified by losses of weight at specific high temperatures. At a final temperature level, the residue remaining after combustion is called ash. The difference between these parameters (moisture, volatiles and ash) is the fixed-carbon parameter. Fixed carbon value in low volatile materials equates approximately to the elemental carbon content of the sample.

The volatile matter content of coals, measured in the absence of moisture and ash, ranges from 2 to about 50%. In domestic stoves and furnaces or in small industrial appliances, coals containing large amounts of volatile matter are easy to ignite, but these coals tend to combust quickly and often with a long, smoky flame. As a rule, coals with higher volatile matter contents have lower heating values.

Concerning ash, the efficiency and the availability of pulverized-coal boilers are strongly affected by typical coal composition, mainly when subjected to high contents of ash. Boiler tube failures caused about 40% of the unplanned outages in coal-fired plants. About 50% of failures in superheater and reheater tubes are due to creep strain at high temperatures, resulting either from poor boiler designs or operation in off-design conditions. Bundle tubes, mainly subjected to overheating, often contain significant deposits of ash layers on them (Lourival et al., 2012). More specifically, the ash generated through the combustion of solid fuels, such as pulverized coal, adheres to tube surfaces during heat exchange and causes problems of heat-transfer inhibition such as slagging and fouling, as well as boiler drive troubles (Naganuma et al., 2009).

The final results of proximate analysis of coal should be reported on a basis of air-dried coal and should be determined to a basis of either dry coal, ash-free coal or as-received coal.

2.2.3.4 Ultimate analysis

Coal ultimate analysis is the determination of the weight percent of carbon as well as hydrogen, sulfur, nitrogen and oxygen. Ultimate analysis is measured using specialized laboratory equipment, such as carbon, hydrogen, nitrogen, sulfur (CHNS) analyzers. Except from these, trace elements such as Cl and F can be included in ultimate analysis. Carbon determination should include organic carbon and any carbon as mineral carbonate. Hydrogen determination should include hydrogen in the organic materials as well as hydrogen in the water associated with the coal.

Sulfur in coal can be found in three different forms:

- as organic sulfur compounds
- as inorganic sulfides that are, for the most part, primarily the iron sulfides pyrite and marcasite (FeS_2) and
- as inorganic sulfates (e.g., Na_2SO_4, $CaSO_4$)

To determine the sulfur value in ultimate analysis, prior methods of coal cleaning, organic sulfur and inorganic sulfur should be included. Moisture and ash should also be determined because the analytical values of the ultimate analysis should be reported on an appropriate basis (ar, db, daf) (Speight, 2005).

The elemental composition of the organic material for coal in ultimate analysis can be represented when corrections are made for any carbon, hydrogen and sulfur derived from inorganic material and conversion of ash to mineral matter can be made. The standard method for the ultimate analysis of coal and coke (ASTM D3176) includes the determination of elemental carbon, hydrogen, sulfur and nitrogen, along with the ash content in the material as a whole. Oxygen is calculated by difference. The test methods for determining carbon and hydrogen are ASTM D3178 and ASTM D3179, for sulfur is ASTM D3177, ISO 334 and ISO 351.

2.2.3.5 Heating value

Heating value of coal can be determined with the use of an adiabatic bomb calorimeter (ASTM D2015) and is expressed in KJ/Kg. This determines the fuel energy available for steam production and represents the quantity of fuel which must be handled, pulverized and fired.

High heating value (HHV) or gross calorific value (GCV) is the quantity measured in the laboratory and is defined as the heat released from combustion of a unit fuel quantity (mass), with the combustion products being at a temperature of 25 °C: ash, gaseous O_2, SO_2, N_2 and water in liquid form. Therefore, the latent heat of vaporization has been reclaimed. On the other hand, the lower heating value (LHV) or net calorific value (NCV) is the heat produced by a unit quantity of a fuel when total water in the products is in vapour form. The method relating to HHV and LHV of coal is ASTM D407. In Europe, LHV is commonly used for heat balance calculations and efficiency determination, whereas in US engineering practice, HHV is generally used (The Babcock & Wilox Company, Edition 41, 2005). One reason for this difference in the method of calculating steam power plant efficiencies is that US electrical utilities purchase coal on an HHV basis and want to calculate their efficiency on the same basis. Conversely, the European practice is based on the realization that the heat of condensation is not a recoverable part of the fuel's energy, because it is not practicable to cool the sulfur-bearing flue gas to below its dew point in the boiler (Beer, 2007).

2.2.3.6 Grindability

Grindability refers to the ease of pulverizing a coal sample compared to certain reference coals. The standard method of coal grindability is called Hardgrove (ASTM D409), though this method does not provide reproducible and repeatable results for hard coal. Another method that can be accepted for petroleum coke is the test method ASTM D5003. According to the Hardgrove method, a prepared and sized sample receives a definite amount of grinding energy in a miniature pulverizer and the size of the pulverized product is determined by sieving. The resulting size distribution is used to produce an index relative to the ease of grinding.

During handling and preparation, physical changes (seam moisture) may occur to high-volatile bituminous and subbituminous coal and lignite. These changes are often sufficient to alter the grindability characteristics of the samples tested in the laboratory, thus producing different indices. The drying conditions and the moisture level can cause inconsistencies in results concerning repeatability and reproducibility.

The grindability test (ASTM D409, ISO 5074) uses a ball-and-ring type of mill in which a sample of closely sized coal is ground for a specified number (usually, 60) of revolutions. Then the ground product is sieved and the grindability index is calculated from the amount of undersize produced using a calibration chart. The results are converted into the equivalent Hardgrove grindability index. High grindability indices refer to easily ground coals. Prior to the experiments, each Hardgrove machine should be calibrated with reference samples of coal. The reference indices used are 40, 60, 80 and 100.

In some cases, data fall outside the experimentally allowable limits. This is due to the following factors: (1) there is no equilibrium with the sample moisture and the laboratory atmosphere; (2) the sample may have been over dried or under-air-dried; (3) excessive dust loss may have occurred during screening due to a loose-fitting pan and cover on the sieve; or (4) the sample may not have had an even distribution of particles (Speight, 2005).

2.2.3.6 Sulfur forms

The ASTM used for measuring the sulfate sulfur, the pyritic sulfur and the organic sulfur in coal is ASTM D2492. The result arises from the difference between the total sulfur, sulfate and pyritic sulfur contents and the organic sulfur. The quantity of pyritic sulfur shows the coal abrasiveness (The Babcock & Wilox Company, Edition 41, 2005).

2.2.3.7 Free-swelling index

The free-swelling index (FSI) is a measure of the increase in volume of coal when heated under specified conditions (ASTM D720; ISO 335). The FSI method is a small-scale test for obtaining information regarding the free-swelling properties of a coal. The results may be used as an indication of the caking characteristic of the coal when burned as a fuel. The volume increase can be associated with the plastic properties of coal; coals that do not exhibit plastic properties when heated do not show free swelling.

The amount of swelling depends on the fluidity of the plastic coal, the thickness of bubble walls formed by the gas, and interfacial tension between the fluid and solid particles in the coal. Greater swelling occurs when the above factors cause more gas to be tapped. FSI increases when the rank for bituminous coals increases. Although, for some individual coals, free swelling indices vary. Results for low-rank coals are lower when compared to bituminous coals (Speight, 2005).

2.2.3.8 Ash fusion temperatures

The ash fusibility test methods (ASTM D1857) are appropriate for simulating the coal ash behaviour when is heated under a reducing or an oxidizing atmosphere. These tests provide information about the ash fusion characteristics, which provide indicators as to the tendency of a coal to form sintered deposits (slags) when combusted.

According to the test method, a laboratory-produced coal ash sample is pressed into a cone-shaped test piece of specific dimensions and is then inserted in a controlled furnace, in which its decomposition is monitored constantly as the temperature increases. The following temperatures are recorded:

1. *Initial deformation temperature* (IT): temperature at which the first rounding of the apex of the cone occurs;
2. *Softening temperature* (ST): temperature at which the cone has fused down to a spherical lump in which the height is equal to the width of the base;
3. *Hemispherical temperature* (HT): temperature at which the cone has fused down to a hemispherical lump, at which point the height is one-half the width of the base;
4. *Fluid temperature* (FT): temperature at which the fused mass has spread out in a nearly flat layer (Speight, 2005).
   A well-known and widely used index based on the temperatures above is the slagging index, which is calculated from the following equation:

$$R_s = \frac{4 \cdot IT + HT}{5}$$

The higher the slagging index, the lower the propensity for formation of strong slagging deposits. Generally, fuels with a slagging index less than 1150 °C have a severe slagging potential, whereas values greater than 1340 °C indicate weak slagging potential. Values in between cover a wide range of severe to moderate slagging potentials.

2.2.3.9 Ash composition

Ash analysis uses a coal ash sample produced according to the ASTM D3174 procedure. The elements in the coal ash are typically reported as oxides. Ash may be defined via ASTM D3682 in which atomic absorption measures silicon dioxide (SiO_2), aluminium oxide (Al_2O_3), titanium dioxide (TiO_2), ferric hydroxide (Fe_2O_3), calcium oxide (CaO), magnesium oxide (MgO), sodium oxide (Na_2O) and potassium oxide (K_2O). The results of the ash analyses permit calculations of fouling and slagging indices and slag viscosity versus temperature relationships (The Babcock & Wilcox Company, Edition 41, 2005).

2.3 Fuels derived from coal

Apart from its direct use in the power or other industrial sector, there has been considerable interest in the production of cleaner and more efficient coal-derived fuels, such as coal-refined liquids and gases, coal slurries and chars, as substitutes for oil and natural gas (The Babcock & Wilcox Company, Edition 41, 2005).

2.3.1 Coke

When coal is heated in the absence of oxygen at temperatures as high as 1832 °F (1000 °C), the lighter constituents are volatized and the heavy carbons crack and liberate gases and tars that leave a residue of carbon. Some of the volatilized portions crack on contact with the hot carbon, leaving an additional quantity of carbon. The carbon residue containing both ash and sulfur is called coke. Both the ash and sulfur content of coke depend on the quality of the coal used for its production. The main byproducts of this conversion are (1) coal-tar, (2) ammonia, (3) light oils and (4) coal gas. The amount of sulfur and ash in the coke depends on the coal sample. Coke from coal is grey, hard and porous and has a heating value of 29,540 KJ/Kg. Coke is mainly used for pig iron in blast furnace production and charging of iron foundry cupolas. Coke in smaller dimensions (under 0.0158 m) is called coke breeze and is used for steam generation, instead for charging blast furnaces. Approximately 4.5% of the coal supplied to slot-type coke ovens, is recovered as coke breeze (Vecci et al., 1978).

2.3.2 Gaseous fuels from coal

2.3.2.1 Coke oven gas

During coke production, a major portion of coal is converted to gas. The products recovered from these gases include ammonium sulfate, oils and tars. The amount of non-condensed coal gas is called *coke oven gas*. The heating value of the coke oven gas depends on the type of coal and the carbonization process used. Thus, high-temperature carbonization produces coke oven gas with high NH_3 and H_2, less tar and lower calorific value when compared to low-temperature process (Gupta, 2006). The typical calorific value of coke oven gas is 19,900 KJ/m^3 (Basu et al., 2000; Ray et al., 2005). The constituents depend on the different types of coal. In some cases, an amount of sulfur from coal may be present in coke oven gas as hydrogen sulfide and carbon disulfide; these gases are most commonly removed via scrubbing. Coke oven gas is mainly used to produce coke in the steel industry and is regarded as a significant feedstock for hydrogen separation, methane enrichment and syngas and methanol production (Razzaq et al., 2013). Coke oven gas is not usually preferred for synthesis gas (syngas) production because costs including its compression and the removal of the various impurities are high (Hiller et al., 2012).

2.3.2.2 Blast furnace gas

Blast furnace gas is produced during the iron oxide reduction in blast furnace iron making in which iron ore, coke and limestone are heated and melted in a blast furnace and is an indigenous process gas of the steelworks industry (Pugh et al., 2013). Blast furnace gas has a high carbon monoxide (CO) content and a low heating value, typical 3900 MJ/m^3 (International Energy Agency, 2007). The five primary components of blast furnace gas are N_2, CO, CO_2, H_2O and H_2. The typical blast furnace gas composition in volume is $N_2 = 55.19\%$, CO $= 20.78\%$, $CO_2 = 21.27\%$ and $H_2 = 2.76\%$ (Hou et al., 2011). The water content is removed by demisters following the cleaning process. This gas is used for the furnace mills, in gas engines and for electricity and steam generation. Often, in the steel industry, blast furnace gas is used as an accessional to natural gas (Bojic and Mourdoukountas, 2000). Blast furnace gas deposits adhere firmly; therefore, boilers using these type of fuel should be frequently cleaned.

2.3.2.3 Water gas

The product derived from driving steam through a bed of hot coke is called water gas—the so-called blue-water gas. Carbon in the coke combined with the steam leads to H_2 and CO formation. The reaction of steam on hot coke to produce water gas is an endothermic reaction, which is described as:

$$C + H_2O = CO + H_2 - q$$

The typical calorific value of water gas is 18,900 KJ/m^3 (Basu et al., 2000; Ray et al., 2005). In many cases, water gas is often enriched with oil, and the oil is cracked to gas by heat. For enrichment, refinery gas is used, mixed with the steam or mixed directly with the water gas. This enriched gas is called carburetted water gas, which recently has been replaced by natural gas.

2.3.2.4 Producer gas

Producer gas is the product obtained when coal or coke is burnt with air deficiency and with a controlled amount of moisture. Producer gas is a gas mixture containing carbon monoxide hydrogen, carbon dioxide and nitrogen. The nitrogen in the air remains unchanged and dilutes the gas, so producer gas has a low heating value 5800 KJ/m^3 (Basu et al., 2000; Ray et al., 2005). After removal of ash and sulfur compounds, it is used near its source. This gas may be used to power gas turbines that are suited to fuels of low calorific value. The typical producer gas composition in volume is

$N_2 = 55\%$, $CO = 29\%$, $CO_2 = 5.5\%$ and $H_2 = 10.5\%$. The composition of producer gas depends on (1) the temperature of the process and (2) the effect of the steam. Producer gas has high nitrogen content, which cannot be eliminated at economically justifiable cost, and therefore it is not suitable as syngas. Both water and producer gas were important from 1920 to 1940, yet these fuels are considered of minor use nowadays because higher performance technologies concerning for syngas have been developed.

2.3.2.5 Byproduct gas from gasification

Coal gasification processes are a source of synthetic natural gas, though they are under development. The gas produced from steam−oxygen coal gasification consists of H_2, CO, CH_4, CO_2 and unreacted steam. In cases in which air is used as the oxygen source, the gas is diluted with N_2. The chemical reactions that coal undergoes are complex, but they usually include the simple reaction of steam and carbon that produces H_2 and CO. CH_4 is produced by the reaction of carbon with H_2 and by thermal cracking. CO_2 and heat needed for the process are produced by the reaction of carbon with O_2.

Coal gasification products are categorized as low-, intermediate- and high-Btu gases. Low-Btu gas has a heating value of 100 to 200 Btu (105.5 to 211 KJ)/SCF (SCF: Standard Cubic Foot) and is produced by gasification with air rather than oxygen. Intermediate-Btu gas has a heating value of 300−450 Btu (316.5 to 474.75 KJ)/SCF and is mainly produced with oxygen. Low and intermediate gases are mainly used as boiler fuel at the gasification plant site or as feed to turbines. High-Btu gas has a value greater than 900 Btu (949.5 KJ)/SCF and may be used instead of natural gas. The production process of high-Btu gas is akin to the intermediate gas process, though in the final stage the gas is upgraded by methanation.

2.3.3 Fischer−Tropsch Process

Due to the gradually increasing oil prices and the decreasing availability of its major deposits, a need for the utilization of alternative technologies arises. Moreover, because the environmental concern focuses on global warming, which is related to carbon dioxide emissions, the interest for cleaner technological developments such as coal liquefaction has changed. The major advantage of coal liquefaction lies on the globally large coal reserves, which present an alternative option, both attractive and practical. The production of synthetic hydrocarbons from coal is based on the established Fischer−Tropsch process. According to this process, a catalyzed collection of chemical reactions takes place, in which carbon monoxide and hydrogen are converted into liquid hydrocarbons of various forms (Schulz, 1999). The aforementioned process was first developed in 1925 by Prof. Franz Fischer, founding director of the Kaiser-Wilhelm Institute of Coal Research in Malheim an der Ruhr, and the head of the department, Dr Hans Tropsch (Anderson et al., 1955). The Fischer−Tropsch process was used effectively in Germany during World War II to produce replacement fuels accounting for an estimated 9% of military purposes and 25% of transportation fuels (Leckel, 2009).

Gasification of coal mixtures of carbon monoxide and hydrogen followed by the implementation of the Fischer—Tropsch process allow the production of liquid fuels from coal. The Fischer—Tropsch chemical reaction is described by the following equation:

$$nCO + (2n + 1)H_2 \rightarrow C_nH_{(2n+2)} + nH_2O \qquad (2.1)$$

in which the n value typically ranges between 10 and 20; CO and H_2 may derive from reactions such as the gasification of coal and partial combustion of hydrocarbon. The reaction is highly exothermic and therefore sufficient cooling to secure stable conditions is necessary. The main catalysts used for the Fischer—Tropsch process are based on cobalt (Co), iron (Fe) and ruthenium (Ru). Nickel (Ni) can also be used but it favors the formation of methane which is unwanted (Fischer and Meyer, 1931). Although cobalt-based catalysts have the advantage of higher activity and longevity, iron-based catalysts are preferred for lower quality feedstocks such as coal because they are less expensive ($\sim$1000 times) and more resistant to poisoning by sulfur-containing compounds (Saiba et al., 2010). During the reaction, iron-based catalysts, unlike Co, Ni and Ru, form oxides and carbides, which have negative impact on the catalysis. Therefore, it is essential to control these transformations to prevent the breakdown of catalyst particles and to maintain the catalytic activity. Despite its high catalytic activity, Ru is not preferred for industrial applications, due to its high price and low availability (Schulz, 1999).

Depending on the material of the catalyst, the Fischer—Tropsch process can be divided into two categories: (1) high-temperature Fischer—Tropsch (HTFT), operated at temperatures ranging between 300 and 350 °C, which uses iron-based catalyst; and (2) low-temperature Fischer—Tropsch (LTFT), operated at temperatures ranging between 200 and 240 °C, which uses cobalt-based catalyst (Reichling and Kulacki, 2011). The most important design feature of Fischer—Tropsch reactors for large-scale applications is the efficient removal of heat and hence the temperature control (Speight, 2014). There are mainly four types of Fischer—Tropsch reactors: (1) the multitubular fixed-bed reactor, (2) the entrained-flow reactor, (3) the slurry reactor and (4) the fluidized-bed reactor (with fixed or circulating bed) (Davis, 2002). Multitubular and slurry reactors are used in LTFT, whereas, on the other hand, entrained-flow and fluidized-bed reactors are utilized for HTFT applications.

Several companies apply and develop Fischer—Tropsch processes to exploit the vast amount of coal deposits worldwide. For many years, Fischer—Tropsch fuels from low-grade coals have been extensively used as transport fuels in South Africa, due to the country's limited oil reserves and the international sanctions imposed on the Apartheid regime. Sasol's South African facility is the world leader in Fischer—Tropsch technologies that uses natural gas and coal feedstocks to produce a variety of synthetic petroleum products, including most of the country's diesel fuel. Over the years, Sasol has developed commercial reactors for both iron- and cobalt-based catalysts including fluidized-bed and slurry-phase distillate reactors. Other companies that use coal for the production of Fischer—Tropsch-derived fuels are Shell's pearl gas-to-liquids (GTL) facility, PetroSA, Waste Management and Processors Inc. and Synstroleum.

2.4 Coal supply chain main characteristics

A typical mining supply chain can be seen in Figure 2.10.

2.4.1 Coal mining

There are two mining methods:

- surface or 'opencast' mining
- underground or 'deep' mining

  Selection of the mining method depends on the geology of the coal deposit.

2.4.1.1 Underground mining

The two main methods of underground mining are: (1) room-and-pillar and (2) longwall mining (http://www.worldcoal.org/coal/uses-of-coal/coal-electricity/).

In room-and-pillar mining, coal deposits are mined by cutting a network of 'rooms' into the coal seam and leaving 'pillars' of coal to support the roof of the mine. These pillars can be up to 40% of the total coal in the seam, although this coal can sometimes be recovered at a later stage. Longwall mining involves the full extraction of coal from a section of the seam, or 'face' using mechanical shearers. A longwall face requires careful planning to ensure that the existing geology is appropriate for mining before development work begins.

The coal 'face' can vary in length from 100 to 350 m. Self-advancing, hydraulically powered supports temporarily hold up (support) the roof while coal is extracted. When coal has been extracted from the area, the roof collapses. Over 75% of the coal in the deposit can be extracted from panels of coal that can extend to 3 km through the coal seam (see Figure 2.11).

As mining techniques continue to improve, coal mining has become a safer and more efficient technology. To keep up with technology and to extract coal as efficiently as possible, modern mining personnel must be highly skilled and well trained in the use of complex, state-of-the-art instruments and equipment. Figure 2.12 shows a general description of underground mining.

2.4.1.2 Surface mining

Surface mining is the most common way of mining. It is known as opencast or open cut mining and is economic only when the coal seam is close to the surface. This method recovers a higher proportion of the coal deposit compared to underground mining, as all coal seams are exploited—approximately 90% or more coal can be recovered (http://www.worldcoal.org/coal/uses-of-coal/coal-electricity/).

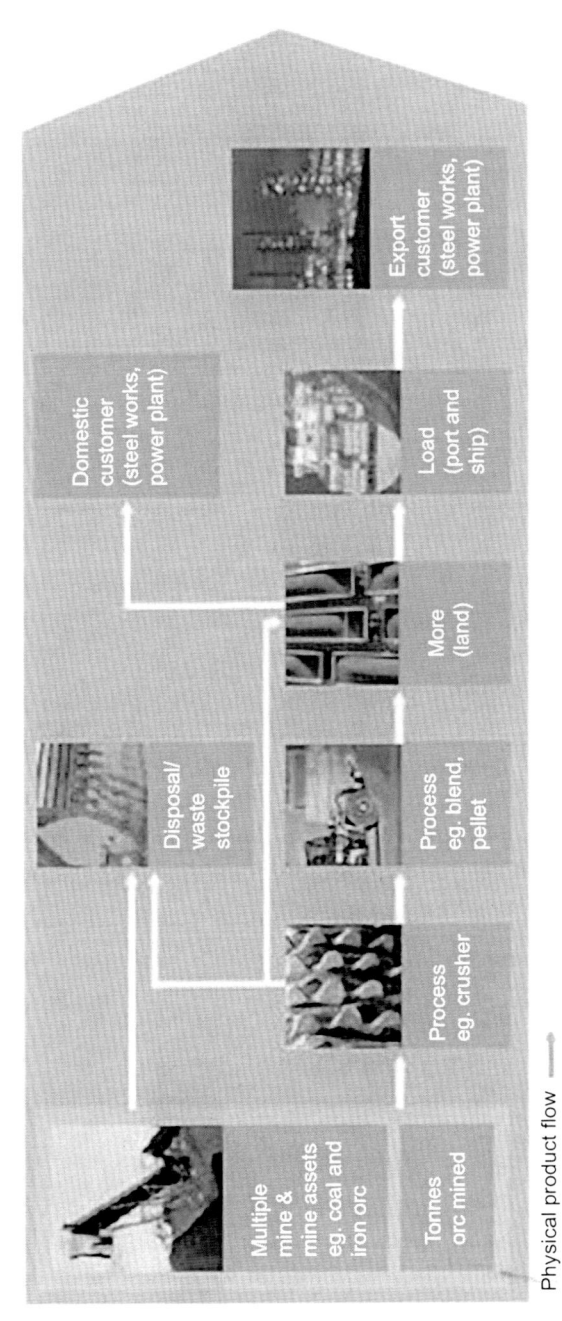

Solid fuel types for energy generation: coal and fossil carbon-derivative solid fuels   49

Figure 2.10 Schematic of a typical mining supply chain.

Figure 2.11 Longwall mining involves the full extraction of coal from a section of the seam using mechanical shearers.
Source: World Coal Association (2005).

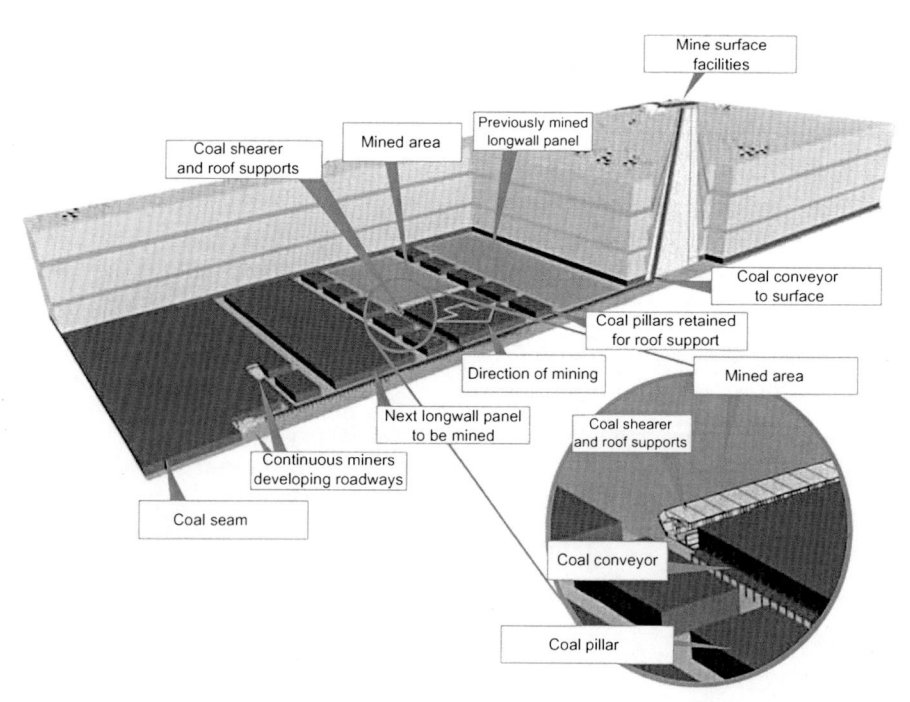

Figure 2.12 Description of underground mining
Source: World Coal Association (2005).

Mining in large opencast mines that may cover an area of many square kilometers demand the use of very large equipment, including:

- draglines, which remove the overburden
- power shovels
- large trucks, which transport overburden and coal
- bucket wheel excavators
- conveyors

Initially, the overburden of soil and rock is erupted by explosives and then removed by draglines or by shovel and truck. Once the coal seam is exposed, it is drilled, fractured and systematically mined in strips. The coal is transported via large trucks or conveyors to plants for treatment (see Figure 2.13).

2.4.2 Coal Preparation

Coal mined straight from the ground, known as run of mine (ROM) coal, often contains unwanted impurities such as rocks and dirt and comes in a mixture of different-sized fragments. Because coal users require coal of a consistent quality, coal preparation is necessary. Coal preparation—also known as coal beneficiation or coal washing—refers to the treatment of ROM coal to ensure consistent quality and to enhance its suitability for particular end uses. Treatment depends on the properties of coal and its intended use. It may require only simple crushing or it may need to go through a complex treatment process to reduce impurities.

To remove impurities, the raw ROM coal is crushed and then separated into various size fractions. Larger material is usually treated using 'dense medium separation' process. According to this process, coal is separated from other impurities by floating in a tank containing a liquid of specific gravity—usually a suspension of finely ground magnetite. Coal, due to its lighter weight, floats and separates from the impurities, whereas heavier rock and other impurities sink and are removed as waste.

Fractions of smaller size are treated in a number of ways, usually based on mass difference, such as in centrifuges. A centrifuge is a machine that rotates a container rapidly, causing solids and liquids to separate. Alternative methods utilize different surface properties of coal and waste. In 'froth flotation', coal particles are removed in a froth produced by blowing air into a water bath containing chemical reagents. The bubbles attract the coal but not the waste and are skimmed off to recover the coal fines. Recent technological developments increased the recovery of ultra-fine coal material.

2.4.3 Coal transportation

Different coal transportation techniques may be utilized depending on distance. Coal is generally transported by conveyor or truck for short distances, whereas, for longer distances, trains and barges are used. Alternatively, coal is mixed with water to form coal slurry and is transported through a pipeline.

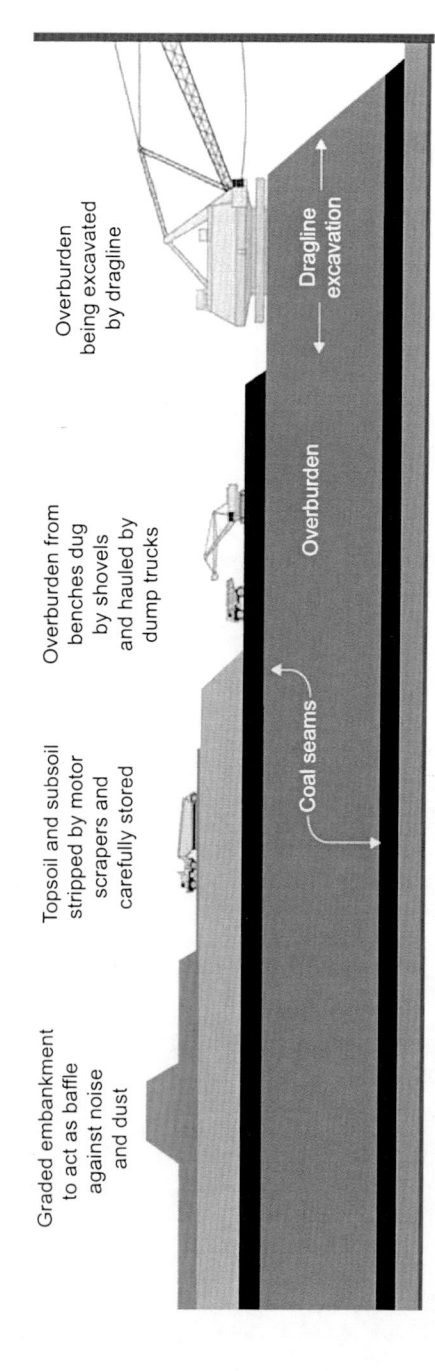

Figure 2.13 Surface coal mining operations and mine rehabilitation.
Source: World Coal Association (2005).

Ships are commonly used for international transportation, in sizes ranging from Handymax (40,000–60,000 DWT[1]), Panamax (about 60,000–80,000 DWT) to large Capesize vessels (about 80,000 + DWT). In 2012, seaborne trade amounted to about 86% of the 1258 million tonnes of the globally traded coal (Euracoal, 2013). Coal transportation can be very expensive and in some cases accounts for up to 70% of the coal-delivered cost (World Coal Institute, 2009). Measures should be taken at every stage of coal transportation and storage to minimize environmental impacts.

2.5 Future trends

The influence of economic growth plays a decisive role in the production and utilization rate of fossil fuels such as coal. Given the current trend of increasing coal consumption worldwide, especially in countries with rapidly developing economies, and stricter environmental legislation and global attempts to stabilize the greenhouse gas (GHG) concentrations in the atmosphere by reducing emissions, companies involved in the coal sector face increased challenges concerning the mitigation of the environmental impacts of the coal value chain. The development of Carbon Capture and Storage (CCS) technologies intends precisely to reduce carbon dioxide emissions from large-scale emitters, such as coal-fired power plants. A key feature of the future energy system is the increased energy production from intermittent renewable energy sources; as a result, coal power plants shift from a base load to a more flexible, load-following operation. Hard coal power plants can already be optimized to operate at a technical minimum load of around 15–20% of the nominal; however, for lignite power plants, the minimum load is usually limited to more than 50% due to flame stabilization issues. Lignite predrying is a technology option that intends to increase the load flexibility of lignite-fired power plants, whereas it can also have a positive impact on plant efficiency and reduction of GHG emissions.

2.5.1 Carbon capture and storage

CCS uses established technologies to capture, transport and store carbon dioxide emissions from large point sources, such as power stations (http://www.ccsassociation.org/). In addition, in the longer term, CCS could be used to reduce emissions from sources, such as intensive industrial processes, natural gas cleanup, hydrogen production, fossil fuel refining, petrochemical industries, and steel and cement manufacturing. The availability of scalable CCS technology by 2020 to 2030 would be most beneficial to lessen the disruption of this transformation by providing low-emission energy services from fossil fuels, whereas alternatives are still developed and scaled-up to meet current and growing energy demands (Benson et al., 2012).

---

[1] Deadweight tonnage (also known as deadweight abbreviated to DWT, D.W.T., d.w.t., or dwt) is a measure of how much weight a ship is carrying or can safely carry.

There are three main technologies currently proposed for CO_2 capture: (1) postcombustion, (2) precombustion and (3) oxyfuel combustion capture.

In postcombustion capture, a new final processing stage, is applied to remove most of the CO_2 from the combustion products, before they are vented to atmosphere. Most commercial advanced methods implement wet scrubbing with aqueous amine solutions. CO_2 is removed from the waste gas by amine solvent at relatively low temperatures (order of 50 °C). The solvent is then regenerated for reuse by heating (to around 120 °C), before being cooled and continuously recycled. The CO_2 removed from the solvent in the regeneration process is dried, compressed and transported to safe geological storage.

Precombustion capture of CO_2 is a method in some ways an oxymoron, as CO_2 is obviously not available for capture prior to combustion. All types of fossil fuels can, however, be gasified (partially combusted or reformed) with substoichiometric amounts of oxygen (and usually some steam) at elevated pressures (typically, 30–70 atm) to give a 'synthesis gas' mixture of predominantly CO and H_2. Additional water (steam) is then added, and the mixture is driven through a series of catalyst beds for the 'water–gas shift' reaction to approach equilibrium:

$$CO + H_2O \leftrightarrow CO_2 + H_2$$

Postcombustion capture of coal is estimated to have higher thermal efficiency for conversion to electricity than the precombustion integrated gasifier combined cycle (IGCC) designs (Gibbins and Chalmers, 2008). Precombustion capture from IGCC plants is estimated to produce low-carbon electricity slightly cheaper, due to the high capital costs of current atmospheric pressure postcombustion absorber designs and the cost of replacing degraded solvent.

Oxyfuel combustion is the third technology that is used for CO_2 capture. This technology focuses on the production of a relatively pure oxygen stream by separating nitrogen from atmospheric air. This is usually achieved through cryogenic air separation processes, which are very energy demanding. The fuel is then combusted in a mixture of oxygen and recycled flue gases (the latter to replace the nitrogen in air and thus moderate peak-flame temperatures to take account of materials and ash-slagging constraints in boiler design, etc.). This gives a flue gas mixture consisting mainly of CO_2 and condensable water vapour, which can be separated and cleaned relatively easily during the compression process. Coal, oxides of nitrogen and sulfur (NO_x, SO_x) and other pollutants must be removed from the product gas before or during the CO_2 compression process. In addition, SO_x may also have to be removed from the recycle stream to prevent high-temperature corrosion in the boiler furnace (Gibbins and Chalmers, 2008).

2.5.2 Lignite pre-drying

As previously mentioned, the high-moisture content of lignite is a major issue in its commercial utilization. High moisture means a decreased heating value and, as a result, a lower energy density. Therefore, the high-moisture content of lignite lowers the plant efficiency, leads to higher CO_2 emissions per unit of energy output and increases the capital costs due to the need for constructing larger boilers. The lignite predrying concept is a step towards optimal lignite utilization and upgrading. Decreasing the amount of moisture in lignite leads to lower energy losses during combustion, low amount of stack gas flow, higher plant efficiency, low transportation cost, but increases the safety measures for their transportation and storage, etc., owed to their higher risk for self-explosion (Jangam et al., 2011). Moreover, the moisture removal from lignite results in a fuel that is more easily combustible; flame stability is thus enhanced and lignite-fired power plants that fire or cofire predried lignite are expected to be able to operate at lower nominal loads.

Technologies that have been developed for lignite predrying fall broadly in two main categories: evaporative drying, in which moisture is removed as steam, and non-evaporative dewatering processes, in which moisture is removed in liquid form (Zhu, 2012). Generally, drying is an energy-intensive process and its application requires careful planning and design to keep high-energy efficiencies. Some examples of its application in the industrial scale include the internal waste heat utilization concept of 'fluidized-bed drying with internal waste heat utilization' of RWE and the DryFining™ technology developed by Great River Energy and used in the Coal Creek Station, North Dakota (Zhu, 2012).

2.6 Summary

Fossil fuels will remain the major source for power generation as energy demands are growing. Coal plays a decisive role in world energy production as coal-fired plants are responsible for 41% of global power generation. Rapid industrialization and worldwide growing energy demands have increased the utilization of coal, as it remains a relative cheap fuel, not associated with market shock prices or geopolitical reasons. Based on the current production rates, coal reserves will be sufficient for 137 years.

Coal is classified in four main categories: (1) anthracite, (2) bituminous coal, (3) subbituminous coal and (4) lignite/brown coal, though each country has developed its own criteria for classifying coals. Therefore, differences appear not only in the scale of coal deposits but in consumption rate. Anthracite coals, due to their high heating value, rank higher, whereas, on the other hand, brown coals (lignite) have the lowest rank. Characterization of coal is important to search means to optimize energy consumption and to provide data influencing its future utilization. Coal characteristics are determined using ASTM procedures or standard analytical methods, including proximate and ultimate analyses, determination of heating value, FSI, ash-fusion temperatures etc.

Regardless of its complexity, a coal supply chain generally includes the following operations: mining, preparation and transportation. Surface and underground are the methods most commonly used for coal extraction, whereas preparation depends on the properties of coal and its intended use. Depending on the distance, numerous ways of coal transportation, such as trains, conveyors, trucks, barges, ships and pipelines, can be selected.

For coal to remain part of the energy portfolio, advanced technology must be developed and commercialized. Recent research has focused more on the technologies concerning mitigation of the environmental issues associated with coal exploitation. Thus, improvements regarding carbon capture storage technologies and lignite predrying techniques are under constant development.

Sources of further information

American Society for Testing and Materials (ASTM Methods)
Euracoal (European Association for Coal and Lignite) http://www.euracoal.org/pages/home.php?idpage=1)
International Energy Agency (http://www.iea.org/)
World Coal Association (http://www.worldcoal.org/coal/coal-mining/)

References

American Society for Testing and Materials, 1999. Gaseous Fuels; Coal and Coke, vol. 05. Annual Book of ASTM Standards, West Conshohocken, Pennsylvania.
Anderson, H.C., Wiley, J.L., Newell, A., 1955. Bibliography of the Fischer-Tropsch Synthesis and Related Processes (In Two Parts). US Bureau of Mines.
Basu, P., Cen, K., Jestin, L., 2000. Boilers and Burners: Design and Theory. Springer.
Benson, S.M., Bennaceur, K., Cook, P., Davison, J., Coninck, H.de, Farhat, K., Ramirez, C.A., Simbeck, D., Surles, T., Verma, P., Wright, I., 2012. Global Energy Assessment — Toward a Sustainable Future. Cambridge University Press, pp. 993—1068.
Beer, J.M., April 2007. High efficiency electric power generation: the environmental role. Progress in Energy and Combustion Science 33 (2), 107—134.
Bojic, C., Mourdoukountas, P., 2000. Energy saving does not yield CO_2 emissions reductions: the case of waste fuel use in a steel mill. Applied Thermal Engineering 20, 963—975.
Davis, H.B., 2002. Overview of reactors for liquid phase Fischer—Tropsch synthesis. Catalysis Today 71, 249—300.
Euracoal — European Association for Coal and Lignite, 2013. Coal Industry across Europe, fifth ed.
Fischer, F., Meyer, K., 1931. Brennstoff-Chemie 12, 225.
Gerling, J.P., Wellmer, F.-W., 2004. Raw material availability — with a focus on fossil energy resources. World of Mining — Surface & Underground 56, 254—262.
Gibbins, J., Chalmers, H., 2008. Carbon capture and storage. Energy Policy 36, 4317—4322.
Gupta, C.K., 2006. Chemical Metallurgy: Principles and Practice. John Wiley & Sons.
Hiller, H., Reimert, R., Stönner, H.M., 2012. Gas Production, 1. Introduction, Ullman's Encyclopedia of Industrial Chemistry. Wiley-VCH Verlag GmbH & Co. KGaA, Weinheim.
Hou, S.S., Chen, C.H., Chang, C.Y., Wu, C.W., Ou, J.J., Lin, T.H., 2011. Firing blast furnace gas without support fuel in steel mill boilers. Energy Conversion and Management 52, 2758—2767.

http://www.ccsassociation.org/ — Carbon Capture and Storage Association (accessed 25.02.14.).

http://www.eskom.co.za/Pages/Landing.aspx — Generation Communication & Primary Energy, 2013 (accessed 04.04.14.).

http://www.worldcoal.org/coal/uses-of-coal/coal-electricity/ — World Coal Association (accessed 15.01.14.).

International Energy Agency, 2007. Tracking Industrial Energy Efficiency and CO_2 Emissions: In Support of the G8 Plan of Action. Energy Indicators, France.

Jangam, S.V., Karthikeyan, M., Mujumdar, A.S., 2011. A critical assessment of industrial coal drying technologies: role of energy, emissions, risk and sustainability. Drying Technology 29 (4), 395–407.

Kopal, 2006. Angebot und Nachfrage am Steinkohlenweltmarkt. Zeitschrift für Energiewirtschaft (ZfE) 30, 67–82.

Leckel, D., 2009. Diesel production from Fischer–Tropsch: the past, the present, and new concepts. Energy Fuels 23, 2342–2358.

Lourival, J., Mendes, N., Bazzo, E., 2012. Characterization and growth modeling of ash deposits in coal fired boilers. Powder Technology 61–68.

Mills, S., 2011. Global Perspective on the Use of Low Quality Coals. International Energy Agency (IEA).

Naganuma, H., Ikeda, N., Kawai, T., Takuwa, T., Ito, T., Igarashi, Y., Yoshiie, R., Naruse, I., 2009. Control of ash deposition in pulverized coal fired boiler. Proceedings of the Combustion Institute 32, 2709–2716.

Pugh, D., Giles, A., Hopkins, A., O'Doherty, T., Griffiths, A., Marsh, R., 2013. Thermal distributive blast furnace gas characterisation, a steelworks case study. Applied Thermal Engineering 53, 358–365.

Ray, H.S., Sing, B.P., Bhattacharjee, S., Misra, V.N., 2005. Energy in Minerals and Metallurgical Industries. Allied Publishers, New Delhi.

Razzaq, R., Li, C., Zhang, S., 2013. Coke oven gas: availability, properties, purification, and utilization. Fuel 113, 287–299.

Reichling, J.P., Kulacki, F.A., 2011. Comparative analysis of Fischer–Tropsch and integrated gasification combined cycle biomass utilization. Energy 36, 6529–6535.

Saiba, A.M., Moodley, D.J., Ciobică, I.M., Hauman, M.M., Sigwebela, B.H., Weststrate, C.J., Niemantsverdriet, J.W., van de Loosdrecht, J., 2010. Fundamental understanding of deactivation and regeneration of cobalt Fischer–Tropsch synthesis catalysts. Catalysis Today 154, 271–282.

Schulz, H., 1999. Short history and present trends of Fischer–Tropsch synthesis. Applied Catalysis A: General 186, 3–12.

Speight, J., 2005. In: Winerforder, J.D. (Ed.), Chemical Analysis: A series of Monographs on Analytical Chemistry and Its Applications, Vol 166.

Speight, J.G., 2014. Gasification of Unconventional Feedstocks. Gulf Professional Publishing, Waltham, USA.

Spohn, O.M., Ellersdorfer, I., 2005. Coal-fired Technologies. EUSUSTEL WP3 Report on Coal-Fired Technologies, November 25.

The Babcock & Wilox Company Book, Edition 41, 2005. Steam — Sources of Chemical Energy (Chapter 9), pp. 9-6 to 9-10.

Thielemann, T., Schmidt, S., Gerling, P., 2007. Lignite and hard coal: energy suppliers for world needs until the year 2100 — an outlook. International Journal of Coal Geology 72, 1–14.

Vecci, S.J., Wagoner, C.L., Olson, G.B., 1978. Fuel and ash characterization and its effect on the design of industrial boilers. Proceedings of the American Power Conference 40, 850—864.
World Coal Association, 2005. The Coal Resource: A Comprehensive Overview of Coal.
World Coal Institute, 2009. The Coal Resource: A Comprehensive Overview of Coal.
Zhu, Q., 2012. Update on Lignite Firing. IEA Clean Coal Centre.

Biomass and agricultural residues for energy generation

Eija Alakangas
VTT Technical Research Centre of Finland Ltd, Jyväskylä, Finland

3.1 Introduction

The term biomass covers a wide range of fuels such as woody biomass, herbaceous, fruit and aquatic biomass and residues from industrial processes. Wood fuels are the most common biomass fuels, and production chains have been developed and well adopted in the market. The wood fuels can be classified according to their traded form (chips or hog fuel, bark, firewood, wood pellets and briquettes and wood charcoal) or their origin (raw material). The International Organization for Standardization (ISO) 17225-1 standard classified woody biomass into three classes by origin and source (Figure 3.1): (1) forest, plantation and other virgin wood; (2) by-products and residues from the wood-processing industry; and (3) used wood (post-consumer wood) (Alakangas and Virkkunen, 2007).

The characteristics of biomass are very different from those of coal (Figure 3.2). The content of volatile matter in woody biomass is generally close to 80%, whereas in coal it is around 30%. Wood char is highly reactive. Nitrogen and sulfur contents of wood are low. This implies that blending wood biomass with coal lowers emissions simply because of dilution. Further, one important difference between coal and biomass is the net calorific value. Biomass fuels often have high moisture content, which results in relatively low net calorific value.

In small-scale heating systems (<1 MW_{th}), the quality of fuel has an important role. The general rule of thumb is that the smaller the system the higher are the quality demands for the used fuel (Table 3.1). The highest-quality chips for small installations can be made from delimbed small wood stems from pre-commercial or commercial thinnings. Pellets are homogeneous biomass fuel that is also used at small scale. When lower-quality chips can be fired, whole-tree chips from undelimbed small-tree stems can also be used.

The properties of biomass fuels set demanding requirements also for heating or power plant operation. These properties include total ash content, ash-melting behaviour and the chemical composition of ash. Alkaline metals that are usually responsible for fouling of heat-transfer surfaces are abundant in wood fuel ashes and will be easily released in the gas phase during combustion.

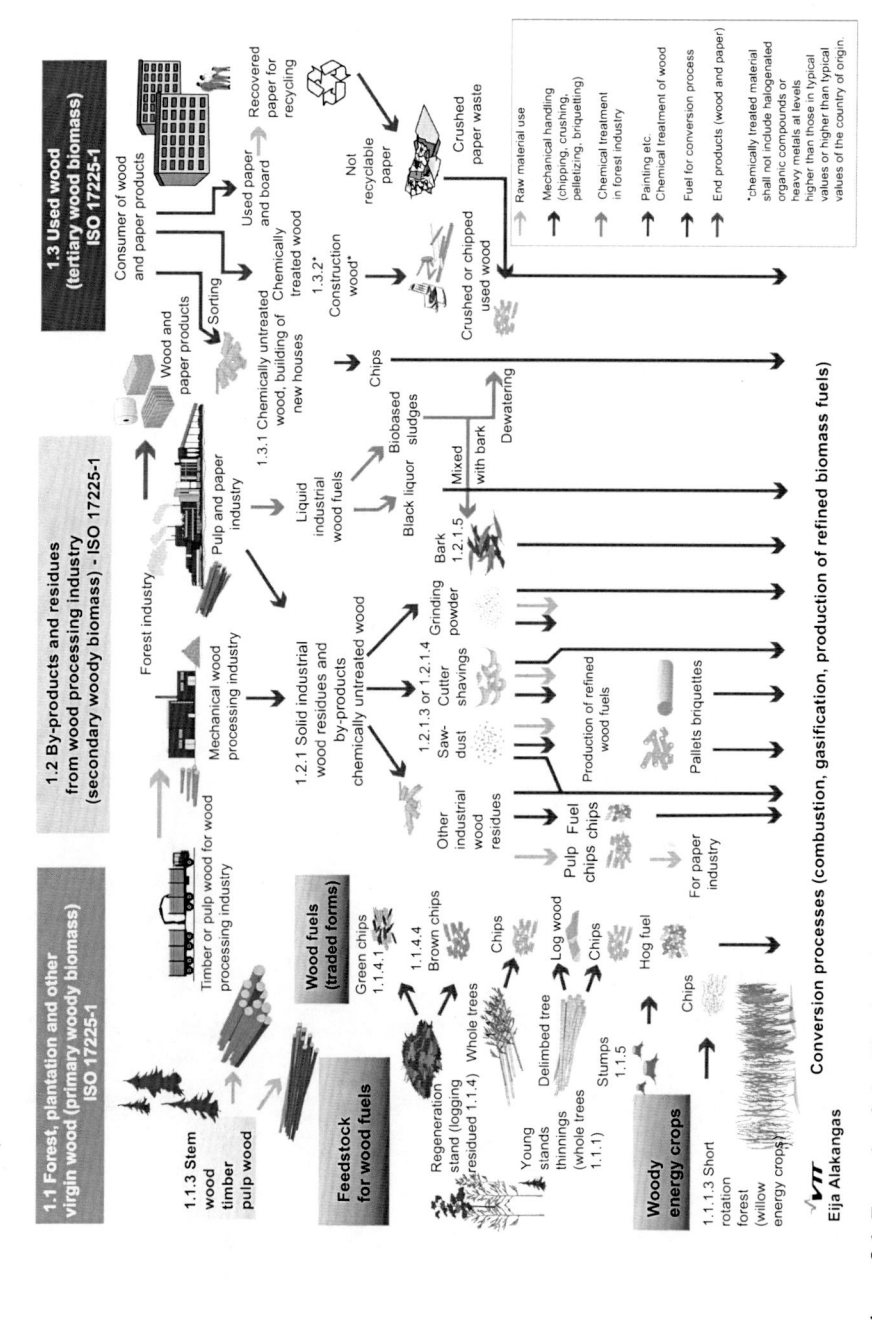

Figure 3.1 Example of classification of Woody biomass based on ISO 17225-1 standard.
Source: VTT.

Biomass and agricultural residues for energy generation   61

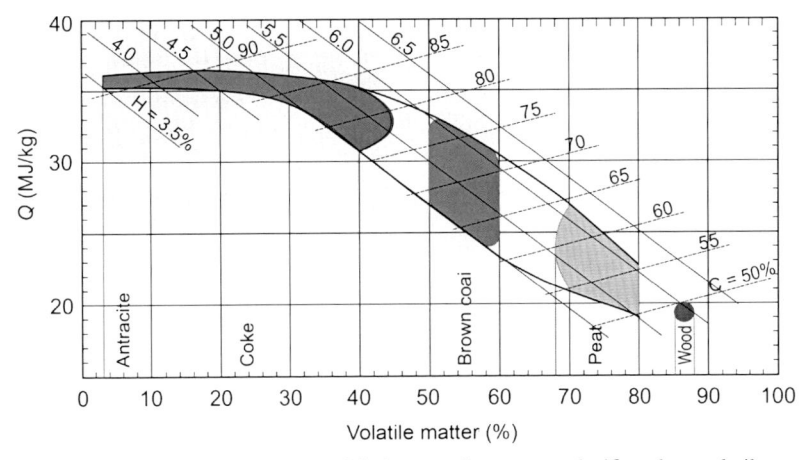

Figure 3.2 Comparison of coal and wood fuels according to net calorific value, volatiles, carbon and hydrogen content.
Source: VTT.

In biomass fuels, these inorganic compounds are in the form of salts or bound in the organic matter, but in peat, for example, inorganic matter is bound mostly in silicates, which are more stable at elevated temperature. The elemental composition of ash (alkali metals, phosphorus, chlorine, silicon and calcium) as well as the chemical concentration of the compounds affect ash-melting behaviour.

During combustion, the behaviour of biomass fuel is influenced by the presence of other fuels. Even a small concentration of chlorine in the fuel can result in the formation of harmful alkaline and chlorine compounds on boiler heat-transfer surfaces. This can be prevented by co-firing fuels such as sulfur and aluminium silicate-containing peat or coal with chlorine-bearing fuels.

Residues from the wood processing industry form one specific group of risky wood fuels. By-products such as plywood and particleboard cuttings are attractive fuels for energy producers: the fuel price may be even negative, as this material should otherwise be taken to a landfill site. Nevertheless, glue, coating and shielding materials may cause bed agglomeration, slagging, fouling and unexpectedly high flue-gas emissions. On the other hand, utilization of solid biofuels and wastes sets new demands for boiler process control and boiler design, as well as for combustion technologies, fuel-blend control and fuel handling systems. Figure 3.3 shows the influence of fuel characterization to boiler design. More challenging fuels are shown on the left side of the drawing.

Characteristics of different kinds of fuels are presented in comparison tables. The material comprises mainly commercial fuels. Definitions for fuels and properties, analysis methods for different properties and formulas for calculation of values are also presented.

Table 3.1 Solid biomass fuel supply chain options according to end-user sector

End-user and average annual fuel consumption	Biomass fuel	Quality requirements	Technology for energy conversion
Households (<50 kW$_{th}$) Annual fuel consumption <30 MWh	Wood pellets	Good mechanical durability	Pellet boilers
		Low ash content	Pellet stoves
	Wood briquettes	Low ash content, packaged	Stoves and fireplaces
	Wood chips	Low moisture content, <35 w%	Stoker boiler
	Log wood	Low moisture content, 15–20 w%	Stoves and fireplaces, boilers
Farms, large buildings (<1 MW$_{th}$) Annual fuel consumption (<3 GWh)	Wood chips from whole trees or delimbed trees	Low moisture content, less than 35 w%	Stoker burners
			Grate firing
	Straw bales	High-quality bales, low moisture content (<18 w%)	Grate combustion, also whole bales
	Wood pellets	Good mechanical durability	Pellet boilers
		Low ash content	Stoker boilers
District heating plants (<5 MW$_{th}$) or power plants (<5 MW$_e$)	Wood chips from forest residues or whole threes	Moisture content usually less than 40 w%	Grate combustion
			Fluidized bed combustion
			Gasification

Annual fuel consumption <35 GWh (DH, CHP) or 85 GWh (power only)	Straw or energy grass bales	Moisture content, less 20 w%	Cigar combustion Grate combustion, also whole bales
CHP and power plants (>5 MW$_e$) Annual fuel consumption from 85 GWh to several TWh	Wood fuels from forest residues, stumps	Boiler and handling equipment based requirements	Usually co-firing with coal Fluidized bed combustion Gasification
	Wood or straw pellets	Boiler and handling equipment based requirements	Co-firing with coal Pulverized combustion
	Herbaceous biomass (straw or energy grasses, like *Miscanthus* and reed canary grass)	Big bales, moisture content less than 20 w%	Cigar combustion Grate combustion Fluidized bed combustion Co-firing with coal
	Olive residues	Boiler and handling equipment-based requirements	Grate firing Co-firing with coal in fluidized bed boiler

Source: Alakangas and Virkkunen (2007).

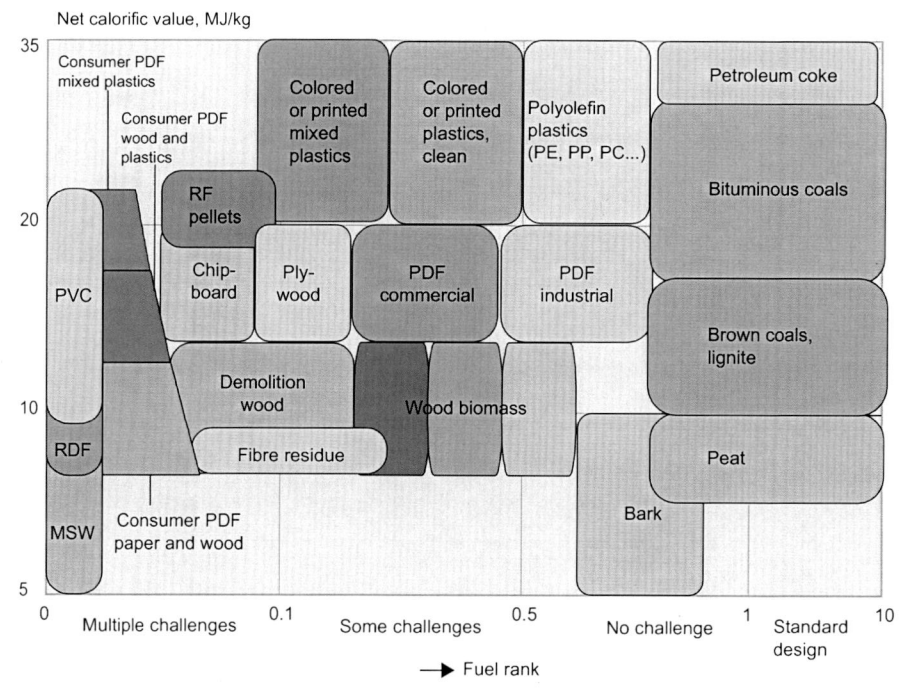

Figure 3.3 Influence of fuel characterization to boiler design.
Source: Veijonen et al. (2003).

3.2 Biomass resources and supply chains

3.2.1 European and global woody biomass resources and supply chains

Forest covers approximately 30% of the Earth's landmass, and the total forestry land area is 3.95 billion hectares. Wood is the most important fuel in developing countries in Africa, Asia and Latin America. The global consumption of wood is likely to increase further in the future because of the continuously growing population as well as the expected increasing demand for biomass for fulfilment of climate policy goals (Alakangas et al., 2014a).

The greatest potential to increase the use of biomass in energy production seems to lie in forest residues and other biomass resources (agrobiomass and fruit biomass) in Europe. The total wood energy potential is about 3700 PJ annually in EU-27. The largest forest energy potential sources are in round wood/stem wood (39%) and forest residues (32%). The biggest forest energy potentials are in the following countries: France, Germany, Finland, Sweden, Italy, Poland and Romania. In 2010, the total wood resource potential was 994 Mm^3 solid and wood demand for material and energy purposes totalled 825 Mm^3 solid, 458 Mm^3 for other uses and 346 Mm^3 solid for energy use. 52 Mm^3 solid of used wood (post-consumer wood) is also available for energy use (Alakangas et al., 2014a; Böttcher et al., 2010).

DBFZ (Thrän et al., 2011) has estimated that the annual global logging residue potential is 10,730 PJ. The utilization of forest residues is often connected with round wood harvesting, so the use of round wood by the forest industry impacts also the exploitation of the forest residue potential. Industrial by-products and residues (bark, sawdust, cutter chips, grinding dust, etc.) are quite well exploited in energy production and pellet or briquette production.

Several different harvesting methods are in use in wood fuel harvesting, depending on different harvesting stands, properties of the harvested wood and different users. The woody biomasses to be harvested are delimbed and whole trees, logging residues and tree stumps Hakkila, 2004.

Harvesting delimbed energy stems often is closely similar to the thinning of pulpwood stands, in which a harvester both fells and delimbs trees and a forwarder takes stems to the roadside. The typical length of the stems is 2.7—5.0 m and the top diameter 4—5.5 cm. This working method suits well for thinning stands in which the average tree size is big enough (e.g. 60 dm^3). The advantages of delimbed stems are good storability and transport efficiency in long-distance transport. Chips made from delimbed stems are suitable for different users from small buildings up to the large power plants. The yield of delimbed stems is about 20—30% lower than for undelimbed stems. Harvesting often requires clearing in advance.

Harvesting of undelimbed stems, whole trees of very small stems, is often done by a harvester or a felling buncher with a special energy wood grab. Harvesting can also be carried out with a normal harvester equipped with a multistem handling grab. The method suits best in young forests, in which the seedling stand management has been neglected. The typical length of harvested stems is 6—7 m. Whole trees are transported with a forwarder to a roadside storage, in which they usually are seasoned and chipped before long-distance transport. In whole-tree harvesting, integrated harvesting machines can be also used, with which trees can be both felled and transported to the roadside.

Energy wood bundling is a new method in energy wood harvesting. In this method, the felling machine is equipped with a bundler which bundles harvested wood directly into bundles. Forest transport takes place with a normal forwarder, which collects the bundles and transports them to a roadside storage. This method enables efficient forest and long-distance transport.

Logging residues are typically collected from spruce-dominated final-felling stands. The harvester delimbs branches and cuts tops to a pile. The forwarder collects residues and transports them to a roadside storage. To improve the transport efficiency, logging residues are often chipped at a roadside storage before long-distance transport.

Harvesting and chipping can be done by the same person or company, but just as well, it can be done as a chain of different operations by different actors (Figures 3.4 and 3.5). The energy wood can be chipped in the terrain, at the roadside landing (roadside storage) or at the plant or a terminal. Chipping at the roadside landing is the most common form of chipping in small-scale applications. When the chipping is carried out by the heat entrepreneur or a member of the cooperative (in some cases also the contractor), the chipping is normally done with a tractor-mounted chipper. This is feasible in especially small sites, where the amount of energy wood is small. Bigger sites, with large amounts of energy wood, use the heavier machinery for comminution. There are also enterprises which are specialized in bioenergy procurement and

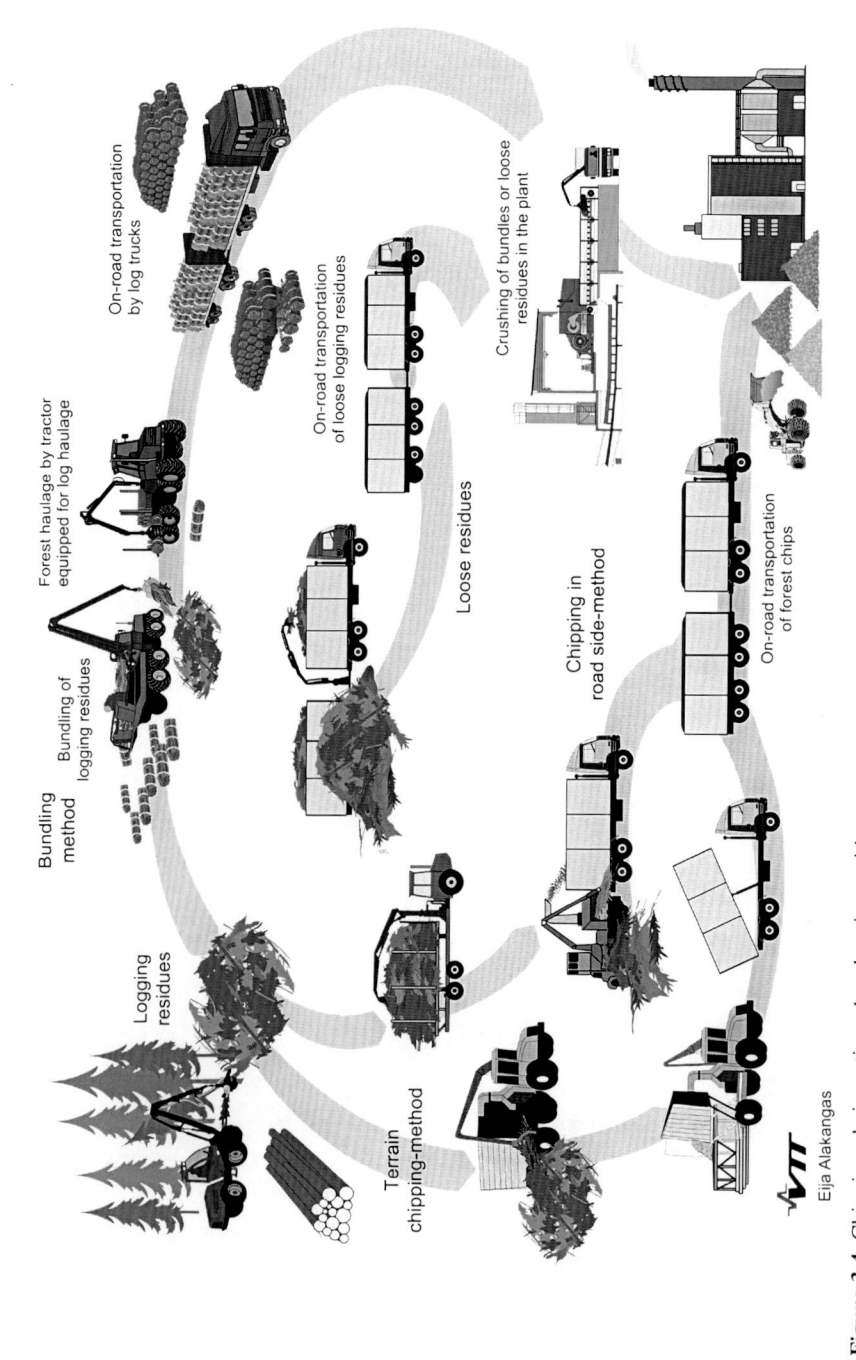

Figure 3.4 Chipping chain options 1 – logging residues.
Source: Alakangas and Virkkunen (2007).

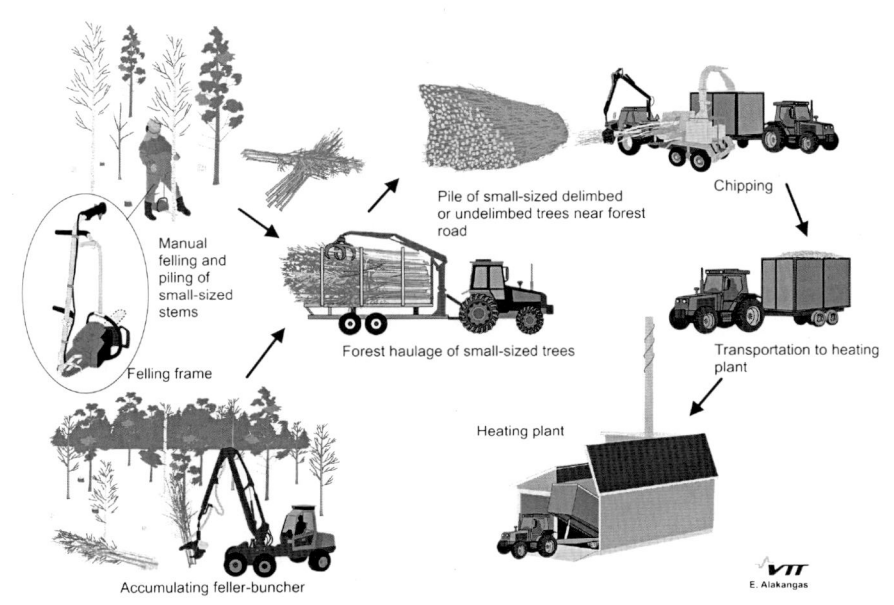

Figure 3.5 Chipping chain options 2 − delimbed stems or whole trees.
Source: Alakangas and Virkkunen (2007).

chipping, and they usually have heavier machinery, which allows larger amounts of chips to be comminuted and transported at a time.

In the Nordic countries, fuel suppliers are paid according to the energy content of the delivered wood chips or hog fuel. For measuring the energy content of the delivered chips, each truckload is weighed at the plant, and samples are taken for defining the moisture content. Based on the weight of the load, the moisture content and the net calorific value of the chips, the energy content of each delivered load can be calculated. Smaller plants do not have any scale so the energy content is calculated based on moisture content, bulk density and net calorific value on a dry basis. In Central Europe, the fuel price for wood chips is determined according to tons delivered. The same is for wood pellets.

Wood pellets and briquettes are usually cylindrical compressed-wood fuel products made from the residues and by-products of the mechanical wood-processing industry. The raw material is dry or moist sawdust, grinding dust and cutter shavings. Pellets and briquettes can also be compressed from fresh biomass, bark and forest chips, but the raw material must be milled and dried before pelletizing.

3.2.2 Herbaceous and fruit biomass resources and supply chains

Agricultural byproducts and residues can be divided into two main categories: herbaceous by-products and residues, and woody by-products and residues. Herbaceous by-products and residues are considered those crop residues that remain in the field after the crop is harvested; their nature is diverse, depending on the crop, method of harvesting, etc. Woody by-products and residues are by definition those produced as consequence of pruning and regenerating orchards, vineyards and olive plantations. Normally, herbaceous crops are cultivated in arable land, whereas woody plantations are considered permanent crops.

The global annual technical potential for straw is around 783 million tons (in dry matter). The largest amounts of straw accrue in the cultivation of maize, sugar cane, rice and wheat. The annual technical fuel potential is 13,317 PJ for 134 countries, with China having by far the largest potential with 2570 PJ/a, followed by India, the USA and Brazil (Alakangas et al., 2014a,b).

Primary agricultural residues remain in fields after harvest. The largest part from the potential comes from common cereal straw, followed by rape straw and corn straw. The largest total potentials are in France, Germany and Spain. Differences in growth conditions, soil quality and soil type and texture complicate estimates of residue potential, but in general 20–30% of the potential straw could be used for bioenergy. Straw potential in EU-27 is reported in different studies and varies from 560 to 982 PJ/a (Alakangas et al., 2014a,b).

Secondary agricultural residues include processing residues generated from the harvested portions of crops during food, feed and fibre production. The largest part of the potential comes from sugar beet bagasse followed by sunflower and rice husks.

In this study, perennial herbaceous potential is estimated for *Miscanthus* and reed canary grass. In total, the cropland area in EU-25 of about 94 million hectares produces on average 12.1 tons biomass per ha from *Miscanthus*, 7.1 from poplar and 5.3 from reed canary grass (Alakangas et al., 2014a,b).

Reed canary grass shows relatively higher productivities in Northern Europe compared to poplar and *Miscanthus*. As a result, experience with the use of reed canary grass as an energy source is focused in Northern Europe (Alakangas et al., 2014a,b).

In the UK, where *Miscanthus* has probably the longest history of agricultural use in Europe, the grass is already grown on several thousand hectares. Other countries in Europe that plan to use *Miscanthus* as an energy crop to a larger degree are Switzerland and Germany. The potentials calculated for Southern and Nordic countries are not considered realistic. In North Europe, *Miscanthus* does not grow because of the arctic climate conditions (Alakangas et al., 2014a,b).

It is estimated that a theoretical potential yield of 7.1 tons biomass per ha from poplar is possible, so the EU-25 could theoretically produce about 12,700 PJ/a bioenergy from this crop. Poplar finds optimal growth conditions in Central and Western Europe (Alakangas et al., 2014a,b).

Residues from palm-oil production arise from the kernels and shells of the palm fruit. Palm fruit theoretical potential is estimated to be 260,000 PJ/a. It is clearly visible that the main potentials are located in Malaysia and Indonesia, with these two countries accounting for 80% of the world palm-oil production (Alakangas et al., 2014a,b).

Large-scale straw handling for energy purposes has developed into an independent discipline in agriculture in which particularly large farms and machine pools make investments. The big bale (width 120 cm, height 130 cm and length 240 cm) is the main type used by district heating plants. The weight is around 500 kg, and new balers can make bales up to 700 kg. In recent years, a midi bale has been introduced with the following dimensions: width 120 cm, height 90 cm and length 240 cm. The resulting weight is 425–500 kg. The advantage is that the bale density (140–185 kg/m^3) is slightly higher, and the tractor/truck can carry three layers of midi bales instead of two layers of big bales. The handling capacity by loading is also increased. The disadvantage is that the straw crane in the plant has to be modified (Alakangas and Virkkunen, 2007) (Figure 3.6).

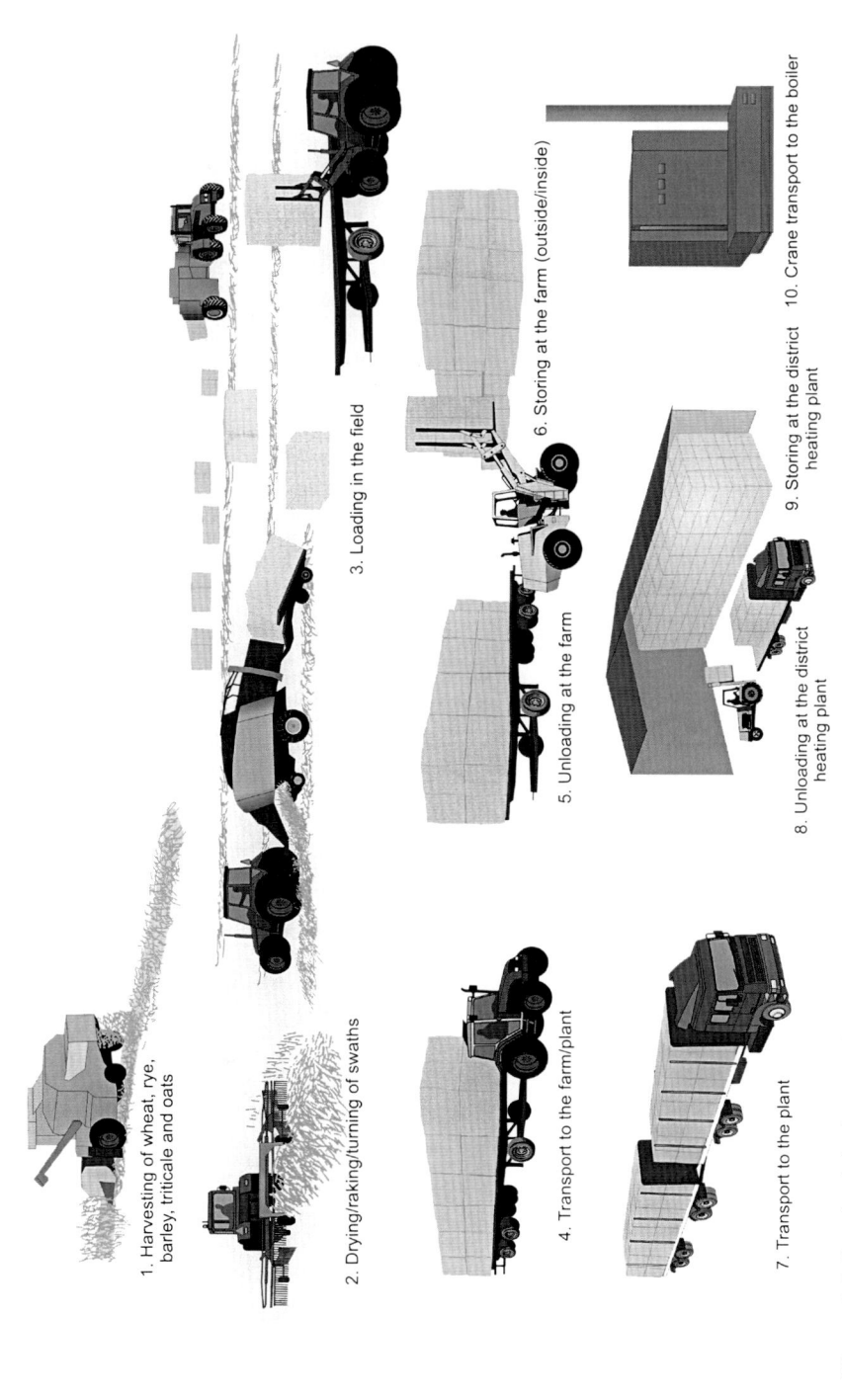

Biomass and agricultural residues for energy generation 69

Figure 3.6 Supply chain for straw bales for district heating plants.
Source: Alakangas and Virkkunen (2007).

Most farmers with straw contracts produce some hundred tons of straw annually. A few large farms and machine pools have developed large-scale handling of 10,000—30,000 tons of straw annually.

Bagasse is the fibrous matter that remains after sugarcane is crushed to extract its juice. Cane trash consists of leafy leftovers of the sugarcane harvest. Pressuring one ton of sugar cane produces 300 kg of bagasse of 50% moisture content.

The rice-husk residue is generated at the rice mills. From one ton of paddy, it is possible to get on average about 0.2 ton of rice husk (Leinonen and Nguyen, 2013).

Maize is harvested by picking and husking the maize cobs. The residues after harvesting include stalks, leaves and corncobs, which are usually collected for fuel and food for livestock.

3.3 Biomass properties and measurement of properties

3.3.1 Properties of biomass

The main significant structural constituents of wood and other biomass are cellulose, hemicellulose and lignin. Wood contains 40—45% cellulose and 25—40% hemicellulose for the weight of dry matter. The hemicellulose content of a coniferous tree is lower (25—28%) than a broadleaf tree (37—40%). The lignin content of a coniferous tree ranges 24—33% and that of a broadleaf tree 16—25%. Lignin binds wood fibres together and gives wood the necessary mechanical strength. Lignin contains an abundance of carbon and hydrogen, that is heat-generating substances. Wood also contains extractives (terpenes, fatty substances and phenols), compounds that can be extracted from wood with neutral organic solvents. For example, wood resin is composed of these substances. The percentage of extractives in wood is usually about 5%, whereas it may be as high as 30—40% in bark (Alakangas, 2005).

The content of volatiles is high in wood, 80—90%. Hence, it is a long-flame fuel and requires a large combustion chamber. The elemental composition of wood comprises mainly carbon, hydrogen and oxygen, the proportion of dry-matter mass being about 99%. The nitrogen content is usually less than 0.2%, alder having the highest nitrogen content. The sulfur content of wood is less than 0.05%. The elementary analysis of different wood species varies only slightly.

The ash content of the fuel is essential for the choice of the appropriate combustion and gas-cleaning technologies. Furthermore, fly ash formation, ash deposit formation, as well as logistics concerning ash storage and ash utilization/disposal depend on the ash content of fuel. Fuels with low ash are wood fuels. The ash content of barkless stem wood is usually less than 0.5%, and higher of broadleaf wood but less than 1%. Significantly higher values are typically found in bark, stumps, straw, grasses, grains and fruit residues. In addition, the harvesting of solid biomass fuels affects ash content; fuel can include soil, sand or other impurities, which increase ash content (so-called extraneous ash) (Alakangas, 2005).

The moisture content of green, fresh wood ranges usually 40—60%. It is dependent on growth site, wood species and age, and also varies in different parts of the tree.

The moisture content of a growing broadleaf tree varies seasonally. In a living tree, the cell wall is impregnated with water, and the cell lumen and spaces are filled with water. When wood is dried, the so-called free water (i.e. that in lumens) escapes. The bound water (i.e. water of cell walls) escapes last. The physical characteristics of wood begin to change as soon as this bound water begins to escape, that is when the saturation point of wood cells is surpassed. The volume of wood is reduced during drying. Usually the conversion method determines how moist wood can be used (see Table 3.1). The larger plants can use moist fuels, whereas, for example household wood should be dry. Households and farms normally dry wood (<25 w%) prior to use. The moisture content of wood fuel burned in fireplaces should be 15–20 w%. In central heating boilers, the storage moisture content of chips should not exceed 25%. If wood chips are used in heating plants of less than 1 MW_{th}, the moisture content should not exceed 40% (Alakangas, 2005).

The net calorific value of wood on dry basis ranges 18.3–20.0 MJ/kg. The calorific value of tops, branches and small trees is slightly higher than that of the whole tree. The greatest differences between different parts of trees have been measured for alder and aspen. The calorific value of wood is low compared to that of other solid fuels. This sets requirements for wood handling and combustion equipment. The wood fuels require more storage space than the other solid fuels.

The basic density of wood (dry-green density) may vary depending on site, genetic genotype and age, and there may also be differences in stem-wood densities of trees grown at the same site. The basic density of birch usually ranges 470–500, pine 380–420, spruce 380–400, grey alder 360–370, aspen 400, rowan 540, oak 600, ash 590, juniper 510 and young willows 380 kg/m^3 (Alakangas, 2005).

Wet wood chips that contain green matter may increase fouling due to higher alkali contents, that is, potassium (K) and sodium (Na) being problematic in combustion. The nitrogen, potassium, phosphorus and calcium contents of needles for dry mass unit are manifold compared to those of stem wood with bark. The sodium content of needles ranges 0.02–0.04% and the chlorine content is <0.4%.

Major (Al, Ca, Fe, K, Mg, Na, P, Si and Ti) and minor (As, Ba, Cd, Co, Cr, Cu, Hg, Mn, Mo, Ni, Pb, Sb, Tl, V and Zn) elements form the ash components together with chlorine (Cl) and sulfur (S). These elements are of relevance for ash melting, deposit formation, fly ash and aerosol emissions as well as corrosion (together with chlorine and sulfur) and the utilization/disposal of the ashes. Calcium (Ca) and magnesium (Mn) usually increase the ash-melting point, whereas potassium decreases it. Chlorides and low ash-melting alkali- and aluminosilicates may also significantly decrease the ash-melting point. This can cause sintering or slag formation in the combustion chamber. Straw, cereal and grain ashes, which contain low concentrations of calcium and high concentrations of silicon and potassium, start to sinter and melt at significantly lower temperatures than wood fuels (Oberberger et al. 2006).

As bark contains a significant amount of lignin, its net calorific value is high. The value is also practically equal at different heights of the stem. On the other hand, the calorific values of different wood species range within wide limits, those of broadleaf being as a rule clearly higher than those of conifers. Aspen is an exception; the net calorific value of its bark being even lower than that of pine and of the same magnitude as that of spruce.

Sawdust used as fuel is obtained as a by-product of timber sawmills and cutter chips from planing machines. Sawdust is usually of wet and light material. However, its moisture content can vary within wide limits (from air-dry to 70%). Cutter chips are usually dry and light, so that they cannot be burnt as such but amongst heavier and moister fuels. Cutter chips are also used by wood-processing industries and heating stations. Compressed products like pellets and briquettes can also be produced from sawdust and cutter chips.

Chopped wood, cut to the length of 20—33 cm and split, is easy to handle and most generally used in fireplaces and ovens. Drying of chopped firewood is dependent on initial moisture content, storage site and weather. The most significant climatic factors are relative atmospheric humidity, precipitation, temperature and wind conditions. In a normal summer, drying of chopped wood made of fresh wood (moisture 40—50%) to combustion moisture content (15—20%) takes at least two months in dry conditions under a shelter outdoors. Fresh chopped wood shrinks by 6—7% during drying. Shrinking starts only at 23—25% moisture content, that is at the so-called saturation point of wood fibres. Storage over winter, for example in a shed, does not essentially reduce the moisture content of wood, as the most favourable drying season of wood is from April to early September.

Wood briquettes and pellets are compressed from dry sawdust, grinding dust and cutter chips. Usually, no binding agents are used, as their own constituents of wood (lignin) stick the briquette or pellet together. Wood pellets and briquettes are homogenous biomass fuels with higher bulk density (600—650 kg/loose m^3), lower moisture content (<10 w%) and higher net calorific value as received (16.7—17.0 MJ/kg) than wood chips.

Thermal treatment includes processes such as torrefaction, steam treatment (explosion pulping), hydrothermal carbonization and charring, all of which represent different exposure to heat, oxygen, steam and water. Drying is not considered thermal treatment in this definition. These technologies are under development and mainly based on woody biomass raw material. Conventional wood pellets tend to absorb moisture from the surrounding humid air. Moistened pellets tend to disintegrate and provide an ideal environment for microbial and biochemical activities. Thermally treated pellets are more hydrophobic (Wilén et al., 2013) and more resistant to water adsorption than conventional wood pellets. Thermally treated biomass pellets usually have high bulk density of 700 kg/m^3. Volatile content is usually more than 75 w% on dry basis and net calorific value on dry basis more than 19 MJ/kg Wilén et al., 2013.

Straw is burnt as big bales, hard bales and chopped, milled and compressed. Some characteristics of straw and wood are very similar. Their elemental analyses and net calorific values are very close to each other. Both fuels contain an abundance of volatiles, burn with a long luminous flame and require a large furnace. However, straw is more problematic compared to wood and other solid fuels due to its low energy density and high ash and alkali content. The calorific value of cereal for kg dry matter is closely to that of wood or straw. In particular, barley cereal burns in the same way as straw, as it contains a lot of grain. Cereal can be burnt as such or milled.

The threshing moisture content of straw is 30–60% and the combustion moisture content usually <20 w%. The straw dries by 2–6% during storage, and hence the moisture content of straw for combustion should not exceed 25 w% at harvest time. The net calorific value at 20 w% moisture content is about 13.5 MJ/kg. The net calorific value on dry basis of straw is 16–18 MJ/kg. The properties of straw ash vary within wide limits depending on cereal species, growth site and fertilization. The calorific value of barley straw is lowest and fusion characteristics poorer (sintering), and the straw is tough. The ash content of cereal straw is 2–10% for dry matter, the highest ash content being measured for wheat straw. The high SiO_2 content of ash increases its fusion temperature, and hence there are no problems in combustion. The content of volatiles ranges 60–70%. The chemical composition is dependent, in addition to plant species, on the age and cultivation conditions of the plant (weather conditions, soil and fertilization). The chemical composition of plant parts may also vary. The harvest time may also affect the composition of biomass. The carbon, hydrogen, and nitrogen contents remain stable. The chlorine and alkali contents of straw are reduced, if left exposed to rain in the field. Straw ash fuses within wide limits of temperature, 350–500 °C. The ashes of different cereals deviate from each other in SiO_2, K_2O and CaO contents. SiO_2 increases fusion temperature, whereas K_2O and CaO decrease it. The fusion temperature of straw harvested late is 150 °C higher than that of straw harvested early. The fusion temperature can be raised by blending oil or coal in combustion. It can also be raised by using an additive like kaolin. When 2% kaolin is added to straw pellets, the deformation temperature rises from 770 to 1100 °C.

The olive cake is a by-product of the olive oil-production process and constitutes a mixture of olive stone, olive pulp and the water added in the olive mills. The moisture content is approximately 55–70 w%. The quantity of stones represents about 1% of the total moist olive cake (Alakangas and Virkkunen, 2007). Table 3.2 lists the main physical and mechanical properties of different biomass fuels. Table 3.3 show chemical composition of biomass fuels.

3.3.2 Sampling and sample reduction

The sampler shall prepare a sampling plan, for example according to the standard EN 14778/ISO 18135 (under preparation). The sampling plan shall include the following: a sample identification number, date and time of sampling, identity of fuel supplier and identification of a lot or sub-lot. A lot may sample as a whole, or be divided into a number of sub-lots. Such division to sub-lots can be necessary to achieve the required precision, maintain the integrity of the sample, for example avoiding bias, keep sample masses manageable, and distinguish different components of moisture of fuels, for example different fuel types within one lot. Figure 3.7 shows an example of sampling and sample treatment for wood fuels.

Table 3.2 Typical properties of different biomass fuels (net calorific value, moisture content, bulk density and ash content)

Biomass	Net calorific value (MJ/kg) $q_{p.net.d}$	Moisture (M_{ar})	Net calorific value as received (MJ/kg) $q_{p.net.ar}$	Bulk density (BD) kg/m^3	Ash content. (A) dry. w%
Sawdust	19.0–19.2	45–60	2.2–10.0	250–350	0.4–0.5
Bark, broadleaf	17.1–23.0	45–55	8.0–11.0	300–400	0.8–3.0
Bark, coniferous	17.5–20.5	50–65	5.0–9.0	250–350	1.5–5.0
Plywood residues	19.0–19.2	5–15	16.0–18.0	200–300	0.4–3.0
Wood pellets	18.9–19.5	6–9	7.0–18.2	600–650	0.1–1.0
Stem wood chips	18.5–20.0	40–55	7.0–11.0	250–350	0.5–2.0
Firewood	18.5–19.0	20–25	13.4–14.5	240–320	0.5–1.2
Logging residue chips	18.5–20.5	50–60	6.0–9.0	250–400	1.0–10.0
Whole tree chips	18.5–20.0	35–55	7.0–12.0	250–350	1.0–2.0
Hog fuel from stumps	17.2–20.9	12–45	6.8–15.5	250–300	0.5–20.0 (average 4.0)
Willow	17.7–19.0	51–53*	8.1–8.5	300–440	0.4–4.0
Poplar	18.1–18.8	35–40	9.9–11.4	250–400	1.5–3.4
Eucalyptus	17.6–18.4	35–40	9.6–11.1	320–400	0.5–4.0
Reed canary grass (spring harvested)	16.5–17.0	10–25	12.6–16.6	60–80	1.0–8.0

Miscanthus	16.0–19.0	10–25	8.6–11.1	60–110	1.0–6.0
Cereal grain	15.0–18.1	11	15.5	600	1.2–4
Straw: wheat, rye, barley	15.8–19.1	17–25	12.4–14.0	80	2–10
Straw, oil seed rape	15.8–19.1	6	14.7–17.8	250 (Bale density)	2–10
Olive cake	13.9–19.2	8–10	12.3–17.9	550–600	3.4–11.3
Fruit stones, apricot, peach, cherry fruit stone	19.5–22.9	11–13	16.7–20.1	n.a.	0.2–1.0
Fruit shells, almond, hazelnut, pine nut	17.5–19.0	8–11	15.3–17.3	n.a.	1.0–3.0
Coconut shells	16.7	10–20	12.9–14.8	n.a.	4–5
Coconut husks	16.7	5–9	15.0–15.7	n.a.	6
Bagasse	15.7	50	7.5	120	1–4
Oil palm shell	18.0–24.8	11–13	15.3–21.8	n.a.	1.4–7.4
Oil palm nut	24.3–24.8	12.5–12.7	20.9–21.4	n.a.	4.0–4.1
Oil palm fibre	18.8–19.6	35–48	8.6–11.9	n.a.	5.0–7.4
Rice husk	14.5–17.5	8–12	12.5–15.9	70–110	13–26
Sunflower husk	17.0–22.0	<10	15.1–20.0	n.a.	1.9–7.6
Coffee husk	16.1–18.2	10–12	15.4–15.8	185–300	1–4

Sources: Alakangas et al. (2012), Garivait et al. (2006), Rice knowledge bank (2013), Youshmione Co Ltd. (2013), Leinonen and Nguyen (2013), DST Technology (2013), Vu et al. (2012), Energy Database of Oil Palm Biomass, Malaysia (2014) ECN, 2013.

Table 3.3 Chemical composition of biomass fuels

Biomass	Carbon, C (w%, dry)	Hydrogen, H (w%, dry)	Sulfur, S (w%, dry)	Nitrogen, N (w%, dry)	Chlorine, Cl (w%, dry)	Sodium, Na (w%, dry)	Potassium, K (w%, dry)
Sawdust	47–54	6.2–6.4	<0.05	0.3–0.4	0.01–0.03	0.001–0.005	0.02–0.15
Bark	48–55	5.3–6.4	<0.02–0.2	0.1–0.8	0.01–0.05	0.007–0.020	0.1–0.5
Plywood residues	48–52	6.2–6.4	<0.05	0.1–0.5	<0.05	0.25–0.50	0.7
Wood pellets	47–54	5.6–7.0	<0.05	<0.2	0.01–0.03	0.001–0.002	0.02–0.15
Stem wood chips	48–52	6.0–6.5	<0.05	0.3–0.5	0.01–0.03	0.001–0.002	0.02–0.15
Firewood	48–52	5.4–6.0	<0.06	0.3–0.5	0.01–0.03	0.001–0.002	0.02–0.15
Logging residue chips	48–52	5.7–6.2	0.02–0.08	0.3–0.8	0.01–0.04	0.002–0.0300	0.1–0.4
Whole tree chips	48–52	5.4–6.0	<0.05	0.3–0.5	0.01–0.03	0.001–0.002	0.02–0.15
Hog fuel from stumps	47–54	5.6–6.5	<0.05	0.1–1.1	0.01–0.04	0.001–0.002	0.02–0.15
Willow	47–48	5.7–6.4	0.02–0.10	0.2	<0.01–0.05	<0.001–0.045	0.17–0.40
Poplar	46–50	5.7–6.5	0.02–0.10	0.2–0.6	<0.01–0.05	0.001–0.006	0.2–0.4
Eucalyptus	46–52.7	4.8–6.2	<0.01–0.11	0.1–1.4	<0.09–0.18	0.002–0.0085	0.15–0.6
Reed canary grass (spring harvested)	45–50	5.7–6.2	0.04–0.17	0.4–2.0	0.01–0.09	<0.002–0.04	<0.08–0.6

Miscanthus	46–52	5.0–6.5	0.02–0.6	0.1–1.5	0.02–0.6	0.0002–0.001	0.1–1.1
Cereal grain	42–50	5.5–6.5	0.05–0.1	2.0	0.05–0.5	0.002–0.005	0.4–0.65
Straw, wheat, rye, barley	41–50	5.4–6.5	0.01–0.2	0.2–1.5	<0.1–1.2	0.01–0.6	0.2–2.6
Straw, oil seed rape	42–52	5.4–6.5	<0.05–0.7	0.3–1.6	<0.1–1.1	<0.3	0.2–2.6
Olive cake	48–52	4.6–6.3	<0.5	1.4–2.7	0.1–0.4	0.02–0.05	1.8
Fruit stones, apricot, peach, cherry fruit stone	51–55	5–7	0.05–0.5	0.2–0.3	0.04	n.a.	n.a.
Fruit shells, almond, hazelnut, pine nut	44–50	5–6	0.04–0.22	0.1–1.2	0.004–0.009	0006–0.007	0.15–0.18
Oil palm shell, nut, fibre	46.3–58.5	5.9–12.6	0.03–0.09	0.04–0.5	0.1–0.25	0.003–0.004	0.3–2.1
Rice husk	38–43	4.3–5.1	0.02–0.1	0.1–0.8	0.03–0.3	n.a.	n.a.
Sunflower husk	51.5–52.9	5.0–6.6	0	0.6–1.4	<0.1	n.a.	n.a.

Source: VTT's laboratory data + Alakangas, E. & Impola, R. Puupolttoaineiden laatuohje - VTT-M-07608-13 (in Finnish), ECN, 2013.

Overall precision (P_L), %

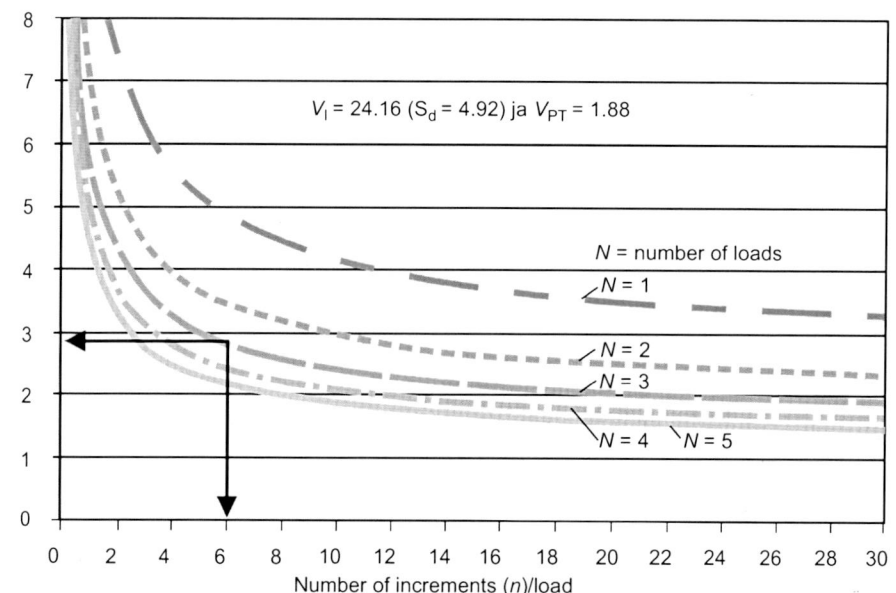

Figure 3.7 Example of sampling and sample treatment for wood fuels. Source: Alakangas, E., Impola, R. Puupolttoaineiden laatuohje – VTT-M-07608-13 (in Finnish).

Calculation of number of increments (n) is determined by Eqn (3.1):

$$n = \frac{4V_I}{NP_L^2 - 4V_{PT}} \tag{3.1}$$

in which

n = minimum number of increments
N = number of sub-lot (for example a truck load)
V_I = the primary increment variance
P_L = the primary increment variance
V_{PT} = a preparation and testing variance

Sampling standard EN 14778/ISO 18135 gives some values for the formula, which is based on the results of empirical values. VTT has also carried out studies for forest chips (Järvinen and Impola, 2012) and has given recommendations for the number of increments for forest chip sampling (Figure 3.8).

Example based on Figure 3.8.

- If six increments for moisture content analysis are taken from a truckload (<120 m³), then overall precision is less than ±3-%-unit if a lot includes three truckloads of forest chips.
- Overall precision will be ±2%-units, if a lot includes five loads.

Biomass and agricultural residues for energy generation

Example of sampling and sample treatment for wood fuels

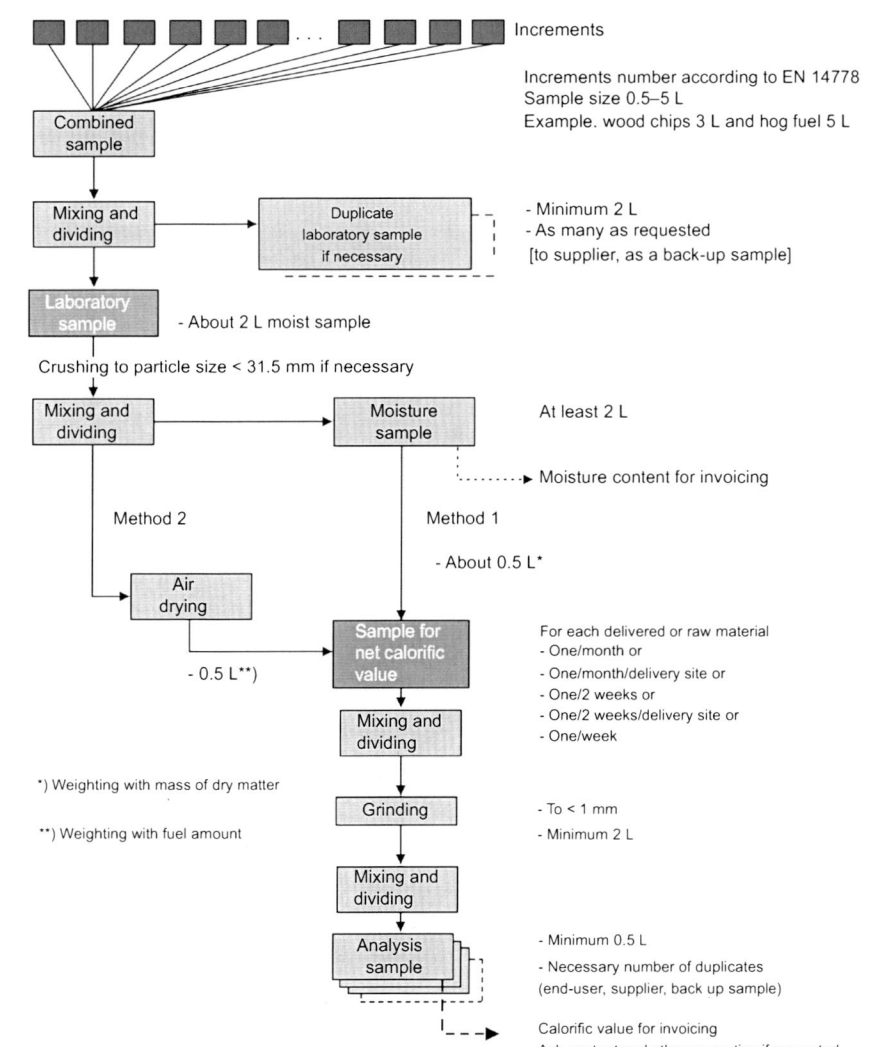

```
Increments ─────────────────────────────
                                         Increments number according to EN 14778
                                         Sample size 0.5–5 L
Combined                                 Example. wood chips 3 L and hog fuel 5 L
sample
   │
   ▼
Mixing and  ──►  Duplicate            - Minimum 2 L
dividing         laboratory sample    - As many as requested
                 if necessary         [to supplier, as a back-up sample]
   │
   ▼
Laboratory
sample          - About 2 L moist sample

Crushing to particle size < 31.5 mm if necessary

Mixing and  ──►  Moisture            At least 2 L
dividing         sample
                    ·········► Moisture content for invoicing

 Method 2           Method 1
                    - About 0.5 L*

 Air
 drying         Sample for            For each delivered or raw material
                net calorific         - One/month or
  - 0.5 L**)    value                 - One/month/delivery site or
                  │                   - One/2 weeks or
                  ▼                   - One/2 weeks/delivery site or
                Mixing and            - One/week
                dividing
*) Weighting with mass of dry matter
                  ▼
**) Weighting with fuel amount
                Grinding              - To < 1 mm
                                      - Minimum 2 L
                  ▼
                Mixing and
                dividing
                  ▼
                Analysis              - Minimum 0.5 L
                sample                - Necessary number of duplicates
                                      (end-user, supplier, back up sample)

                        ──►  Calorific value for invoicing
                             Ash content and other properties if requested
```

Figure 3.8 Example of increments for forest chips for different load numbers.
Source: Järvinen and Impola (2012).

- There are bigger differences for small lots (one to two loads), and the number of increments should be double to keep the overall precision as high as possible.
- Internal deviation does not influence the shape of the curve, for larger internal deviation the curves are lifted upwards and precision is decreased.
- Also delivery time of the year affects the moisture content deviation for some forest fuels.

During winter, the deviation is bigger than in the summer.

Table 3.4 **Typical sample sizes**

Analysis	Size of sample
Basic analysis (net calorific value; Q, ash; A, sulfur; S, carbon; C, hydrogen; H and nitrogen; N)	500 g about 2 L (can be produced from moisture analysis sample)
Moisture; M	At least 300 g (about 2 L)
Bulk density; BD	About 70 L (for wood chips and hog fuel) analysis with 50 L container
	About 7 L (for pellets) analysis with 5 L container
Particle size; P	At least 8 L

Source: Alakangas, E. & Impola, R. Puupolttoaineiden laatuohje - VTT-M-07608-13 (in Finnish).

The size of the increment is calculated according to Eqns (3.2) and (3.3). The minimum volume of the increment shall be:

$$\text{Vol}_{incr} = 0.5 \text{ for } d_{95} < 10 \tag{3.2}$$

$$\text{Vol}_{incr} = 0.05 \times d_{95} \text{ for } d_{95} \geq 10 \tag{3.3}$$

Vol_{incr} = the minimum volume of the increment, litre
d_{95} = the nominal top size, mm

For nominal top size of 45 mm, calculation gives sample size of 2.25 L, but it is recommended to take a 3 L sample. Table 3.4 shows the requested sample size for different analyses.

Examples:

Sawdust	0.5 L
Forest chips	3 L
Hog fuel	5 L
Bark	5 L

Sampling equipment shall enable the sampler to take unbiased increments to provide a representative sample. The opening (W) of the sampling device for manual sampling shall be at least 2.5 times the nominal top size. The volume of the sampling device (box, scoop and shovel) shall comply the minimum increment volume, Vol_{incr}, as described earlier.

Usually manual sampling is only suited for low mass flows. Sampling shall be carried out using a sampling box or other suitable equipment that is passed through the stream of falling material so that it cuts the whole cross-section of the falling stream. Sampling from falling streams can also be done by taking the increments from randomly selected points of the stream. In these cases, careful attention shall be paid to the possible segregation of fuel flow.

Falling-stream samplers (mechanical sampling) are often installed at the end of a conveyor belt or similar. When the material is falling into the cutter, it is important to avoid bouncing and filling the cutter completely. When the cutter has travelled through the stream, it shall be emptied mechanically, and no material should be lost during this operation.

Stockpiles shall preferentially be sampled during build-up or reclaiming as this ensures accessibility to all parts of the lot which in turn minimizes the effect of segregating materials. Only relatively small stockpiles (<40 tonnes) may be sampled while stationary. A scoop, shovel, fork, auger, probe or pipe shall be used to extract increments.

The main purpose of sample preparation is that a sample is reduced to one or more test portions that are in general smaller than the original sample. The main principle for sample reduction is that the composition of the sample as taken on site shall not be changed during each stage of the sample preparation. Each sub-sample shall be representative of the original sample. To reach this goal, every particle in the sample before sample division shall have an equal probability of being included in the sub-sample following sample division. Two basic methods are used during the sample preparation. These methods are sample division and particle-size reduction of the sample. Sample division is the process of reducing the mass of the sample without reducing the size of the particles. Suitable equipment is riffle boxes (Figure 3.9) and rotary sample dividers. A shovel or a scoop is a tool used for manual sample division. Cutting mills are used for reducing the nominal top size of materials. An axe is used for cutting wood logs or coarse material to suitable size to be processed in a cutting mill.

Figure 3.9 Riffle box for sample dividing.
Source: E. Alakangas, VTT.

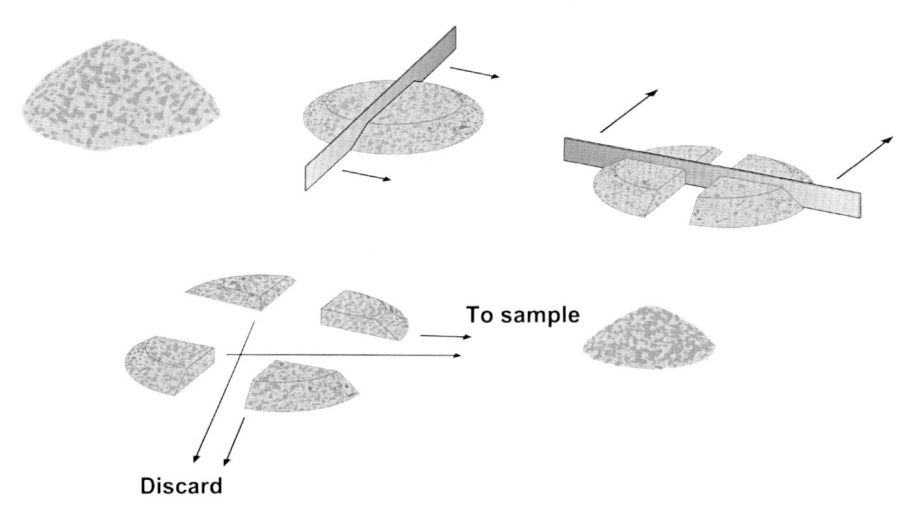

Figure 3.10 Coning and quartering method for manual sample dividing.
Source: Alakangas, E., Impola, R. Puupolttoaineiden laatuohje — VTT-M-07608-13 (in Finnish).

Combined samples may be divided into two or more laboratory samples and laboratory samples are in general further divided into sub-samples (test portions).

Coning and quartering method can be used for materials such as sawdust and wood chips that can be worked with a shovel or, in case of straw, using a fork. It is suitable for producing sub-samples of these materials down to approximately 1 kg. Place the whole combined sample on a clean, hard surface. Shovel the sample into a conical pile, placing each shovelful on top of the preceding one in such a way that the biofuel runs down all sides of the cone and is evenly distributed and different particle sizes become well mixed. Repeat this process three times, forming a new conical pile each time. Flatten the third cone by inserting the shovel repeatedly and vertically into the peak of the cone to form a flat heap that has a uniform thickness and diameter and is no higher than the blade of the shovel.

Quarter the flat heap along two diagonals at right angles by inserting the shovel vertically into the heap. See Figure 3.10. A sheet-metal cross may be used for this operation if available. Discard one pair of opposite quarters. Repeat the coning and quartering process until a sub-sample of the required size is obtained.

3.3.3 Measurement of main properties and applied standards

The most significant properties determined are ultimate analysis, elementary analysis, calorific value, and ash content and ash-melting behaviour. Data are also often required about density, particle-size distribution and other fuel-handling properties. Heavy metal contents and contents of different metals or alkalis either in the fuel or ash are also often determined for environmental or combustion technical reasons. Regarding dust fuels like grinding dust, data are also required on properties related to safety, that is temperature of spontaneous ignition and dust explosion properties.

3.3.3.1 Proximate and ultimate analysis

Proximate and ultimate analyses are main standard laboratory test methods for solid biofuels. Proximate analysis arises from the term "approximate analysis" and is related to the fact that they were not exact. The proximate analysis is used for assessing the quality of fuels and, besides determinations of calorific value, as a basis for fuel trade. Determination is made of moisture, volatile matter, fixed carbon and ash on a mass percent basis. Fixed carbon is the material remaining after the determination of volatile matter, moisture and ash — that is, it is a measure of the solid combustible material in the fuel after expulsion of volatile matter. Figure 3.10 shows the difference between analysis basis and proximate and ultimate analysis. Fixed carbon can be calculated according Eqn (3.4) (Bridgeman et al., 2010).

$$FC = 100 - VM - M - A \qquad (3.4)$$

in which

FC = fixed carbon, w%
VM = volatile matter (w% on dry basis)
M = moisture (w% on dry basis)
A = ash (w% on dry basis)

In ultimate analysis, the amount of C, H, S and N is determined in the combustion products resulting from complete combustion: oxygen can be determined directly, but is more commonly calculated by difference (Eqn (3.5)) (Bridgeman et al., 2010) (Figure 3.11).

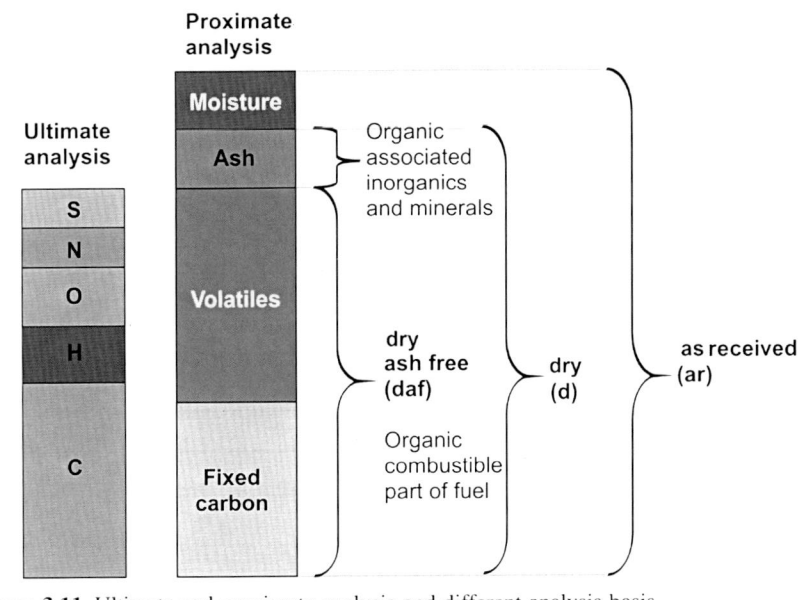

Figure 3.11 Ultimate and proximate analysis and different analysis basis.
Source: Local fuels — properties, classifications and environmental impacts, VTT & Vapo Oy.

$$O = 100 - (C + H + S + N) - M - A \tag{3.5}$$

in which

O = oxygen, w% on dry basis
H = hydrogen, w% on dry basis
S = sulfur, w% on dry basis
M = moisture, w% on wet basis
A = ash, w% on dry basis

3.3.3.2 Calorific value

The net calorific value of solid fuels for dry matter is determined in accordance with standards ISO 1928 and EN 14918/ISO 18125 (for solid biofuels).

Gross calorific value is the absolute value of the specific energy of combustion, in joules, for unit mass of a solid biofuel burned in oxygen in a calorimetric bomb under the conditions specified. The products of combustion are assumed to consist of gaseous oxygen, nitrogen, carbon dioxide and sulfur dioxide, liquid water (in equilibrium with its vapour) saturated with carbon dioxide under the conditions of the bomb reaction and solid ash, all at the reference temperature.

Net calorific value at constant volume is the absolute value of the specific energy of combustion, in joules, for unit mass of the biofuel burned in oxygen under conditions of constant volume and such that all the water of the reaction products remains as water vapour (in a hypothetical state at 0.1 MPa), the other products being as for the gross calorific value, all at the reference temperature.

As the third calorific value, the net calorific value as received is basis of trade. This calorific value is the lowest one, as the energy used for evaporating water contained naturally in the fuel and water formed in combustion is reduced when calculating the calorific value. The calorific value is usually given as megajoules for kg fuel (MJ/kg, 1 MJ = 0.2778 kWh).

In EN 14918/ISO 18125 method, about 1 g $\pm$ 0.1 of air-dry (equilibrium moisture content) analysis sample is burnt in high-pressure oxygen in a bomb calorimeter (Figure 3.12) under specified conditions. The effective heat capacity of the calorimeter is determined in calibration experiments by combustion of certified benzoic acid under similar conditions, accounted for in the certificate. The corrected temperature rise is established from observations of temperature before, during and after the combustion reaction takes place. The duration and frequency of the temperature observations depend on the type of calorimeter used. Water is added to the bomb initially to give a saturated vapour phase prior to combustion, thereby allowing all the water formed, from the hydrogen and moisture in the sample, to be regarded as liquid water.

The gross calorific value is calculated from the corrected temperature rise and the effective heat capacity of the calorimeter, with allowances made for contributions from ignition energy, combustion of the solid biofuels and for thermal effects from side reactions such as the formation of nitric acid. Furthermore, a correction is applied

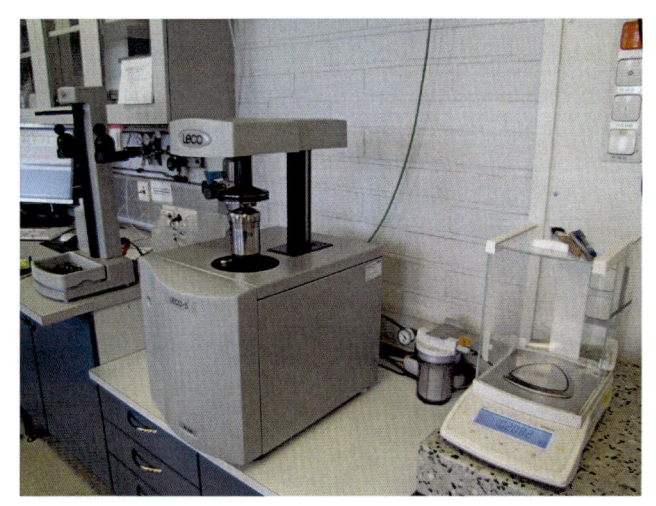

Figure 3.12 Analysis of gross calorific value with a calorimeter bomb.
Source: Labtium Oy. (permission: janne.nalkki@labtium.fi).

to account for the difference in energy between the aqueous sulfuric acid formed in the bomb reaction and gaseous sulfur dioxide, that is the required reaction product of sulfur in the biomass fuel.

The energy of vaporization (constant volume) for water is at 25 °C is 41.53 kJ/mol. This corresponds to 206 J/g for 1% (weight) of hydrogen in the fuel sample or 13.05 J/g for 1 w% of moisture, respectively. The net calorific value at constant volume is derived from the corresponding gross calorific value.

Net calorific value as received is calculated according to Eqn (3.6) or Eqn (3.7). In both of the following cases (a) and (b), the calorific value can be either determined for that particular lot or a typical value can be used. If the ash content of the fuel is low and rather constant, the calculation can be based on the dry basis Eqn (3.6) with a typical value of $q_{p,net}$; however, if the ash content varies quite a lot (or is high) for the specific biofuel then using the equation for dry and ash-free basis (3.7) with a typical value of $q_{p,net,daf}$ is preferable.

1. Dry basis
    The net calorific value (at constant pressure) on as received (the moist biofuel) can be calculated on the net calorific value of the dry basis according to Eqn (3.6):

$$q_{p,net,d} = q_{p,net,d} \times \left(\frac{100 - M_{ar}}{100}\right) - 0.02443 \times M_{ar} \tag{3.6}$$

in which
   $q_{p,net,ar}$ is the net calorific value (at constant pressure) as received (MJ/kg)
   $q_{p,net,d}$ is the net calorific value (at constant pressure) in dry matter (MJ/kg)
   M_{ar} is the moisture content as received (w%)

0.02443 is the correction factor of the enthalpy of vaporization (constant pressure) for water (moisture) at 25 °C (MJ/kg per 1 w% of moisture)

2. Dry and ash-free basis

The net calorific value (at constant pressure) on as received (the moist biofuel) can be calculated from a net calorific value of the dry and ash-free basis according to an Eqn (3.7):

$$q_{p,net,ar} = \left[\left(\frac{q_{p,net,daf} \times (100 - A_d)}{100}\right) \times \left(\frac{100 - M_{ar}}{100}\right)\right] - (0.02443 \times M_{ar}) \quad (3.7)$$

in which

$q_{p,net,ar}$ is the net calorific value (at constant pressure) as received (MJ/kg)
$q_{p,net,daf}$ is the net calorific value (at constant pressure) in dry and ash-free basis (MJ/kg)
M_{ar} is the moisture content as received (w%)
A_d is the ash content in dry basis (w%).
0.02443 is the correction factor of the enthalpy of vaporization (constant pressure) for water (moisture) at 25 °C (MJ/kg per 1 w% of moisture)

The result shall be reported nearest 0.01 MJ/kg.

Energy density (E) can be calculated according to Eqn (3.8). E_{ar} is calculated by net calorific value as received and bulk density.

$$E_{ar} = \frac{1}{3600} \times q_{p,net,ar} \times BD_{ar} \quad (3.8)$$

in which

E_{ar} energy density as received (MWh/loose m^3),
net calorific value as received (MJ/kg),
BD_{ar} bulk density as received (kg/loose m^3) and
$\frac{1}{3600}$ is a factor calculating from MJ to MWh.
The result is reported as 0.01 MWh/loose m^3.

3.3.3.3 Moisture content

The determination methods of moisture content are mainly based on the ISO 589 or EN 14774/18134 standard series. There are three different methods of determining the total moisture content using a drying oven. The sample is dried at a temperature of 105 °C ± 2 °C in air atmosphere until constant mass is achieved and percentage moisture content calculated from the loss in mass of the sample. Reference method — Part 3 includes a procedure for the correction of buoyancy effects. In biomass trade, Part 2 — simplified method is used.

The size of samples used in moisture content determinations is dependent on the weighing accuracy and on the particle size of the fuel concerned. In EN 14774-1/ ISO 18123-1, the sample mass shall be minimum 300 g (about 2 L moist sample). Sample is should have a nominal top size max. 30 mm.

The samples are dried in an air-conditioned heating chamber at 105 ± 2 °C to standard weight. Usually, a drying time of 16 h is sufficient, when the sample layer is not more than 30 cm thick. Drying time shall not exceed 24 h. Dry samples shall be removed from the heating chamber before placing moist samples in it.

After drying, the samples are let to cool to room temperature in a desiccator before weighing. If there is no desiccator available, the samples can be weighed hot immediately after removal from the heating chamber. The accuracy of reporting is 0.1 percentage units.

If moisture determinations are compared with each other, the method applied shall be agreed upon in advance (cooling in a desiccator/weighing hot).

In moisture determinations, it shall be controlled that the vessels used absorb no moisture and that the vessels endure the drying temperature.

The moisture content of the samples is calculated for the mass change during drying in accordance with Eqn (3.9):

$$M_{ar} = \frac{(m_2 - m_3) + m_4}{(m_2 - m_1)} \times 100 \tag{3.9}$$

in which

m_1 is the mass in grams of the empty tray
m_2 is the mass in grams of the tray and sample before drying
m_3 is the mass in grams of the tray and sample after drying
m_4 is the mass in grams of the reference tray before drying (weight at room temperature)

3.3.3.4 Particle-size determination

The particle size of solid biofuels and particle-size distribution are determined for a sample of at least 8 L with sieving methods and sieve series selected according to the standard ISO 3301.

It is recommended to use sieves with hole diameters of 3.15, 16, 31.5, 45, 63 and 100 mm, if the measurement aims at the determination of conformity with ISO 17225 series. Table 3.5 shows an example of particle size of measured forest residue chips. In this example most of requirements of P16 can be fulfilled, only the amount of the coarse fraction (over-sized particle) is higher than requirement of ISO 17225-1 standard, so in this case particle-size class is P16, F25. Fines class (F, particles less than 3.15 mm) is selected separately. In Figure 3.13 the apparatus for sieving is presented and also requirements for P45 and P63 according to ISO 17225-1 for wood chips and hog fuel.

3.3.3.5 Bulk density

Bulk density is an important property for fuel deliveries in volume basis. Energy density can be calculated by bulk density and net calorific value. It also facilitates the space requirements for transport and storage. A test portion is filled into a standard container of a given size and shape in standard EN 15103/ISO 17828 and is weighed afterward. Bulk density is calculated from the net weight per standard volume and reported for the measured moisture content. The container shall be cylindrically shaped and manufactured of a shock-resistant, smooth-surfaced material. For fuels with a nominal top size up to 12 mm, a small container (5 L, 0.005 m^3) shall be used and

Table 3.5 Example of particle size distribution for logging residue chips

	Screen size	Amount, w%	Measured (requirement in standard)	Class[a]	Cumulative amount, %
Fine fraction	< 3.15 mm	24.2	24.2 % (F25)	F25	24.2
Main fraction	3.15 – 8 mm	34.2	64 % (≥ 60 %)	P16	58.4
	8 – 16 mm	29.8			88.2
	16 – 31.5 mm	8.3			96.5
Course fraction	31.5 – 45 mm	0.7	3.5% (≤ 6% more 31.5 mm)	P16	97.2
	45 – 63 mm	2.8			100
	63 – 100 mm	0	All < 150 mm	P16	100
	< 100 mm	0			100

P16 and fine fraction F25 according to EN ISO 17225-1.
[a]Particle-size class, which fulfills the requirement of certain particle-size class which means that smallest class should be selected. In this example P16, F25 is the particle size classification for logging residue chips.
Source: Alakangas, E., Impola, R. Puupolttoaineiden laatuohje – VTT-M-07608-13 (in Finnish).

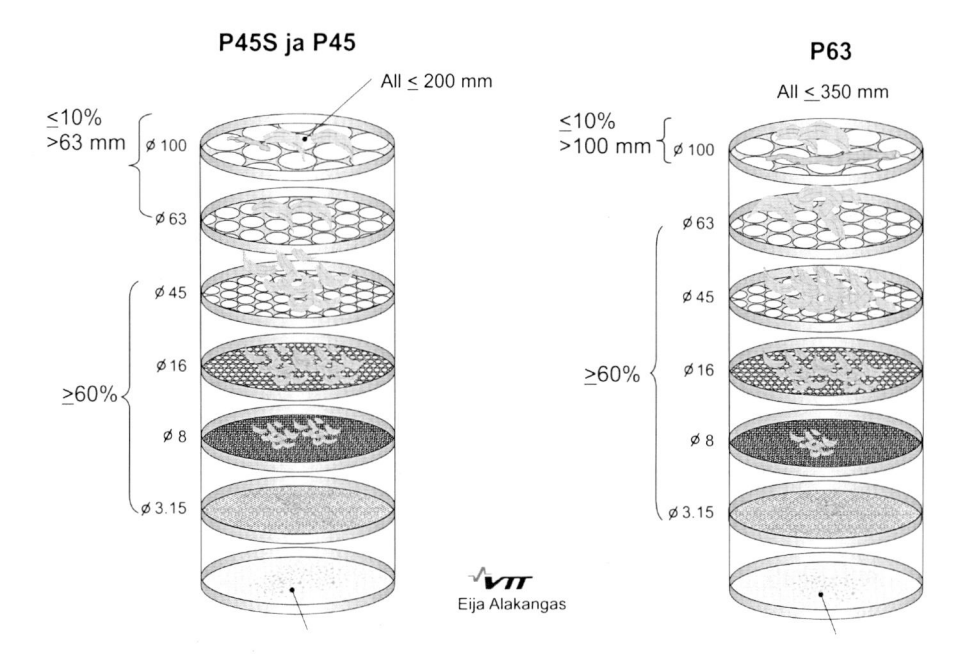

To be stated according to the following classes F05, F10, F15, F20, F25, F30, F30+

For P45S cross cutting area should be < 2 cm²

Figure 3.13 Apparatus and requirements for P45 and P63 according to ISO 17225-1. Note that for logging residues main fraction from 3.15 to 45 mm or 63 mm.
Source: Alakangas, E. Analysis of particle size of wood chips and hog fuel – ISO/TC 238, VTT-R-02834-12.

other fuels a large container (50 L, 0.05 m³, Figure 3.14) shall be used. In standard EN 15103/ISO 17828, the described method includes a defined shock exposure to the bulk material.

Bulk density as received is calculated according to Eqn (3.10):

$$BD = \frac{m_2 - m_1}{V} \qquad (3.10)$$

in which

m_2 = mass of sample and container, kg
m_1 = mass of container, kg
V = volume of a container, m³

The result of each individual determination shall be calculated to 0.1 kg/m³ and mean value of individual calculations shall be rounded to the nearest 10 kg/m³.

Figure 3.15 shows the space requirements of different fuels.

Figure 3.14 Analysis of bulk density for wood chips.
Source: Eija Alakangas, VTT.

3.4 Future trends

Ambitious goals have been set in European energy and climate policy for the year 2020 regarding the promotion of renewable energy sources and the reduction of CO_2 emissions. Large growth scenarios for the use of biomass in second generation transportation fuels as well as production of green electricity have been presented.

Ensuring security of the biomass fuel supply constitutes a key challenge in the provision of reliable and environmentally friendly biomass technologies. The aim is to develop standardized and sector-oriented, sustainable advanced biomass fuels (new bio-commodities, thermally treated biomass fuels, fast pyrolysis bio-oil and upgraded biomethane) including the provision of adequate feedstock at competitive production costs. The development of advanced standardized biomass fuels should focus on ensuring an enlarged raw material portfolio for bioenergy inside Europe, with a particular focus on the use of agricultural and forestry residues as well as biodegradable

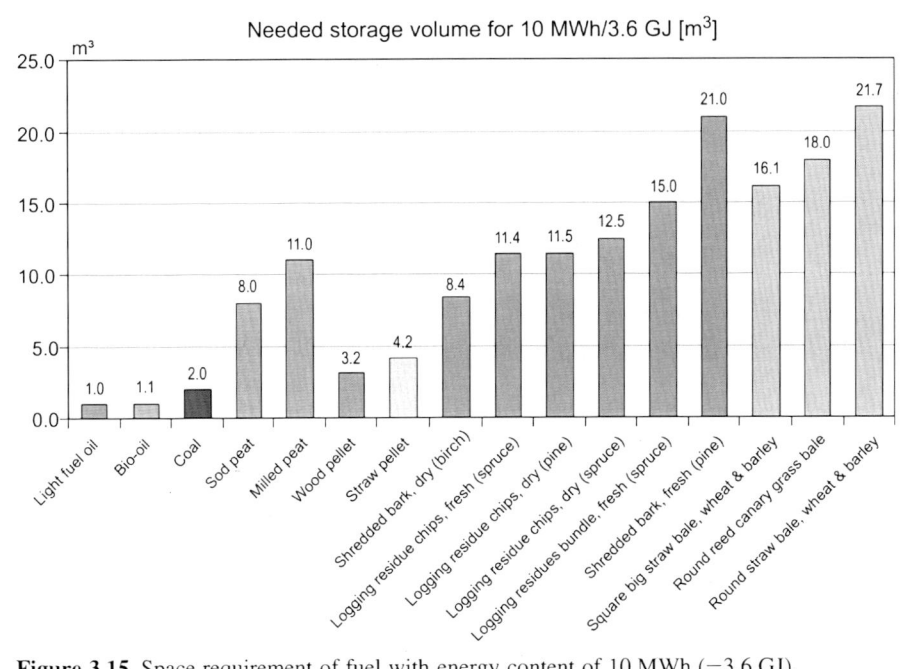

Biomass and agricultural residues for energy generation 91

Needed storage volume for 10 MWh/3.6 GJ [m³]

Fuel	m³
Light fuel oil	1.0
Bio-oil	1.1
Coal	2.0
Sod peat	8.0
Milled peat	11.0
Wood pellet	3.2
Straw pellet	4.2
Shredded bark, dry (birch)	8.4
Logging residue chips, fresh (spruce)	11.4
Logging residue chips, dry (pine)	11.5
Logging residues bundle, fresh (spruce)	12.5
Shredded bark, fresh (pine)	15.0
Square big straw bale, wheat & barley	21.0
Round reed canary grass bale	16.1
Round straw bale, wheat & barley	18.0
	21.7

Figure 3.15 Space requirement of fuel with energy content of 10 MWh (=3.6 GJ).
Source: Alakangas and Virkkunen (2007).

waste and residue streams result from harvesting practices (agriculture, forestry, landscape), feedstock processing, conversion processes (industry) and from end use (Alakangas et al., 2014b).

Biomass is highly fibrous and tenacious in nature, because fibres form links between particles and make the handling of the untreated materials difficult. During thermal treatment, the biomass loses its tenacious nature, which is mainly coupled to the breakdown of the hemicellulose matrix and depolymerization of the cellulose, resulting in the decrease of fibre length. Several thermal treatment methods are under development, for example torrefaction, steam explosion and hydrothermal carbonization.

Physical and chemical properties of biomass can be modified by a torrefaction process closer to the properties of coal to replace large volumes of coal in existing power plants and in coal gasifiers for syngas and transportation fuel production. The main objective is to use torrefied biomass as a fuel, especially as a pellet, with similar grinding properties and storability as coal, for co-firing in power plants. Many pilot- and demonstration-scale plants are in operation in Europe and North America. However, full commercial scale operation is still hampered by numerous technical constraints.

One option of new bioenergy carrier solutions is to integrate a torrefaction plant with forest industry operations in sawmills. Sawmills offer attractive business solutions for solid white or brown pellet production, as well as bio-liquids produced by fast-pyrolysis technology from sawdust and forest residues. There are significant synergies for bioenergy carrier integration due to favourable procurement and logistics, energy and labour benefits (Wilén et al., 2014).

Symbols and abbreviations

The symbols and abbreviations used in this chapter comply with the International System of Units (SI) as much as possible.

d	Dry (dry basis)
daf	Dry, ash-free
ar	As received
A	Designation for ash content (w%, dry basis)*
BD	Designation for bulk density (kg/m^3 loose)*
DE	Designation for particle density as received (kg/dm^3)
D	Designation for diameter (mm)*
DU	Designation for mechanical durability (w%)*
F	Designation for amount of fines (<3.15 mm, w%)*
E	Designation for energy density as received, E_{ar} (kWh/m^3 or kWh/kg, unit is to be stated in brackets)*
L	Designation for length (mm)*
M	Designation for moisture content as received (w%) on wet basis M_{ar}*
P	Designation for particle-size distribution (w%)*
$q_{V,gr}$	Gross calorific value (MJ/kg) at constant volume
$q_{p,net}$	Net calorific value (MJ/kg) at constant pressure
Q	Designation of net calorific value as received (MJ/kg)
VM	Volatile matter, w% dry basis

*Designation symbols are used in combination with a number to specify property levels in the quality tables. For designation of chemical properties chemical symbols like S (sulfur), Cl (chlorine), N (nitrogen) are used and the value is added at the end of the symbol.

Terminology

ash — residue obtained by combustion of a fuel. Depending on the combustion efficiency, the ash may contain combustibles. Adopted to ISO 1213-2:1992.

ash fusibility; ash-melting behaviour — characteristic physical state of the ash obtained by heating under specific conditions. Ash fusibility is determined under either oxidizing or reducing conditions. Adopted to ISO 540:1995.

basic density — ratio of the mass on dry basis and the solid volume on green basis.

bulk density	mass of a portion of a solid fuel divided by the volume of the container which is filled by that portion under specific conditions. Adopted to ISO 1213-2:1992.
bulk volume, loose volume	volume of a material including space between the particles.
calorific value, heating value (q)	energy amount per unit mass or volume released on complete combustion.
combined sample	sample consisting of all the increments taken from a sub-lot. The increments may be reduced by division before being added to the combined sample.
density	ratio of mass to volume. It must always be stated whether the density refers to the density of individual particles or to the bulk density of the material and whether the mass of water in the material is included.
dry, ash-free basis (daf)	condition in which the solid biofuel is free from moisture and inorganic matter.
dry basis (d)	condition in which the solid biofuel is free from moisture. Adopted to ISO 1213-2:1992.
dry matter	material after removal of moisture under specific conditions.
dry matter content	portion of dry matter in the total material on mass basis.
energy density (E)	ratio of net energy content and bulk volume. The energy density is calculated using the net calorific value determined and the bulk density.
fixed carbon (F)	remainder after the percentage of total moisture, total ash and volatile matter are subtracted from 100. Adopted to ISO 1213-2:1992.
foreign material; impurities	material other than claimed, which has contaminated the solid biofuel.
general analysis sample	sub-sample of a laboratory sample having a nominal top size of 1 mm or less and used for a number of chemical and physical analyses. Adopted to ISO 13909.
gross calorific value (q_{gr})	absolute value of the specific energy of combustion, in joules, for unit mass of a solid fuel burned in oxygen in calorimetric bomb under the conditions specified. The results of combustion are assumed to consist of gaseous, oxygen, nitrogen, carbon dioxide and sulfur dioxide, of liquid water (in equilibrium with its vapour) saturated with carbon dioxide under conditions of the bomb reaction and of solid ash, all at the reference temperature and at constant volume. Old term is higher heating value. Adopted to ISO 1928:1995.
increment	portion of fuel extracted in a single operation of the sampling device. Adopted to ISO 13909.
inorganic matter	non-combustible fraction of dry matter.
laboratory sample	combined sample or a sub-sample of a combined sample or an increment or a sub-sample of an increment sent to a laboratory.

lot	defined quantity of fuel for which the quality is to be determined. Adopted to ISO 13909.
mass reduction	reduction of the mass of a sample or sub-sample.
mechanical strength, mechanical durability (DU)	ability of densified biofuel units (e.g. briquettes, pellets) to remain intact during loading, unloading, feeding and transport.
moisture (M)	water in a fuel. See also total moisture and moisture analysis sample.
moisture analysis sample	sample taken specifically for the purpose of determining total moisture. Adopted to ISO 13909.
net calorific value (q_{net})	under such conditions that all the water of the reaction products remains as water vapour (at 0.1 MPa), the other products being as for the gross calorific value, all at the reference temperature. The net calorific value can be determined at constant pressure or at constant volume. Old term is lower heating value. Net calorific value as received ($q_{net,ar}$) is calculated by the net calorific value from dry matter ($q_{net,d}$) and the total moisture as received. Adopted to ISO 1928:1995.
nominal top size	aperture size of the sieve used in determining the particle-size distribution of solid biofuels through which at least 95% by mass of the material passes. Adopted to ISO 13909.
organic matter	combustible fraction of dry matter.
oscillating screen classifier	device containing one or multiple oscillating (flat) screens used to separate material into size classes for calculation of particle-size distribution.
over-sized particles, coarse fraction	portion of particles exceeding a specific limit value.
particle density	density of a single particle.
particle size (P)	size of the fuel particle as determined. Different methods of determination may give different results.
particle-size distribution	proportions of various particle sizes in a solid fuel. Adopted to ISO 1213-2:1992.
proximate analysis	analysis of a solid biofuel reported in terms of total moisture, volatile matter, ash content and fixed carbon measured at specified conditions. Adopted to ISO 1213-2:1992.
sample	quantity of material, representative of a larger quantity for which the quality is to be determined. Adopted to ISO 13909.
sample-size reduction	reduction of the nominal top size of a sample or sub-sample.
solid volume	volume of individual particles. Typically determined by a fluid displaced by a specific amount of material.
stacked volume	volume of stacked wood including the space between the wood pieces.
sub-lot	part of a lot for which a test result is required. Adopted to ISO/final draft International Standard (FDIS) 13909.
sub-sample	portion of a sample.

test portion	sub-sample of a laboratory sample consisting of the quantity of material required for a single execution of a test method.
total ash	mass of inorganic residue remaining after combustion of a fuel under specified conditions, typically expressed as a percentage of the mass of dry matter in fuel. Old term is ash content.
total carbon (C)	sum of carbon in organic and inorganic matter as a portion of the fuel. Adopted to ISO 1213-2:1992.
total hydrogen (H)	sum of hydrogen in organic and inorganic matter. Adopted to ISO 1213-2:1992.
total moisture, M_T	moisture in fuel removable under specific conditions. Indicate reference (dry matter/dry basis, or total mass/wet basis) to avoid confusion. Old term is moisture content. Adopted to ISO1928:1995.
total nitrogen (N)	sum of nitrogen in organic and inorganic matter as a portion of the fuel. Adopted to ISO 1213-2:1992.
total oxygen (O)	sum of oxygen in organic and inorganic matter and in the moisture as a portion of the fuel. For solid biofuels, a calculation method for total oxygen is available.
total sulfur (S)	sum of sulfur in organic and inorganic matter as a portion of the fuel. Adopted to ISO 1213-2:1992.
ultimate analysis, elementary analysis	analysis of a fuel reported in terms of its total carbon, total hydrogen, total nitrogen, total sulfur and total oxygen measured at specified conditions. Adopted to ISO 1213-2:1992.
volatile matter (VM)	mass loss, corrected for moisture, when a fuel is heated out of contact with air under specific conditions. Adopted to ISO 1213-2:1992.
volume	amount of space that is enclosed within an object. It must always be stated whether the volume refers to the solid volume of individual particles, the bulk volume, or the stacked volume of the material and whether the mass of moisture in the material is included.
wet basis	condition in which the solid biofuel contains moisture.

References

Alakangas, E., 2005. Properties of Wood Fuels Used in Finland. Technical Research Centre of Finland, VTT, Project report PRO2/P2030/05, 89 p. + app. 10 p.

Alakangas, E., Asikainen, A., Grammelis, P., Hämäläinen, J., Haslinger, W., Janssen, R., Kallner, P., Lehto, J., Mutka, K., Rutz, D., Tullin, C., Wahlund, B., Weissinger, A., Witt, J., May 22, 2014a. Biomass Technology Roadmap – European Technology Platform on Renewable Heating and Cooling. In: EUREC, Renewable Heating & Cooling – European Technology Platform, 43 p. www.rhc-platform.org.

Alakangas, E., Keränen, J., Flyktman, M., Jetsu, P., Vesterinen, P., Penttinen, L., Tukia, J., Kataja, J., 2012. Biomass resources, production, use, processing and logistics in Central Finland in 2010 and future prospect for year 2020. In: VTT-R-07625–12, 87 p.

Alakangas, E., Virkkunen, M., 2007. Biomass fuel supply chains for solid biofuels. In: EUBIONET II, 32 p.
Alakangas, E., Flyktman, M., Lemus, J., Sanchez Gonzalez, D., Zwart, R., Adler, P., Stelte, W., Pommer, L., Weatherstone, S., Kollberg, V.K., June 3−5, 2014b. Biomass resources and quality requirements for torrefaction − SECTOR project. In: World Bioenergy 2014. Jönköping, Sweden. 6 p.
Bridgeman, T.G., Jones, J.M., Williams, A., 2010. Overview of Solid Fuels, Characteristics and Origin. In: Handbook of Combustion, vol. 4. Solid Fuels, p. 1−30.
Böttcher, H., et al., 2010. Biomass Energy Europe. Illustration Case for Europe.
DST Technology, 2013. Coconut Sell. http://biofuelresource.com/coconut.shell/.
ECN, 2013. Properties of Coffee Husk. http://ecn.nl/phyllis/Biomass/View/2307.
Garivait, S., Chaiyo, U., Patunsawad, S., Deakhuntod, J., 2006. Physical and chemical properties of Thai biomass fuels from agricultural residues. In: The Second Joint International Conference of Sustainable Energy and Environment, 1−23 Bangkok, November 2006, 5 p.
Hakkila, P., 2004. Developing technology for large-scale production of forest chips. In: Wood Energy Technology Programme 1999-2003, Technology Programme Report 6/2004, Final Report, Tekes, 99 p.
ISO 17225−1:2014. Solid Biofuels − Part 1: General Requirements.
Järvinen, T., Impola, R., 2012. Näytteenottostandardin Soveltamisohje. Näytteenotto- Ja Näytekäsittelystandardien (SFS-en 14778 Ja SFS-en 14780) Soveltamisohje Metsäpolttoaineille Suomessa. Implementation of sampling standards (EN 14778 and EN 14780) for forest fuels, Research report VTT-R-03522−12. 21 pp. [in Finnish].
Leinonen, A., Nguyen, D.C., 2013. Development of biomass fuel chains in Vietnam. In: VTT Technology 134, Espoo, 100 p. + app. 15 p.
Obernberger, I., Brunner, T., Bärnthaler, G., 2006. Chemical properties of solid biofuels − significance and impact. Biomass & Bioenergy 30, 973−982.
Rice Knowledge Bank, 2013. Milling Processing of Rice Husk.
Thrän, D., Bunzel, K., Seyfort, U., Zeller, V., Buchhorn, M., Müller, K., Matzdorf, B., Gaasch, N., Klöckner, K., Möller, I., Starick, A., Bandles, J., Günther, K., Thum, M., Zeddles, J., Schönleber, N., Gamer, W., Schweinie, J., Weimar, H., 2011. Global and Regional Spatial Distribution of Biomass Potentials − Status Quo and Options for Specification − DBFZ Report Nr. 7, 122 p.
Veijonen, K., Vainikka, P., Järvinen, T., Alakangas, 2003. Biomass co-firing − an efficient way to reduce greenhouse gas emissions. In: EUBIONET − European Bioenergy Network Project, VTT, 28 p.
Vu, D.H., Wang, K.-S., Chen, J.-H., Nam, B.X., Nam & Bac, B.H., 2012. Glass-ceramic from mixtures of bottom ash and fly ash. Waste Management 32 (2012), 2306−2314.
Wilén, C., Jukola, P., Järvinen, T., Sipilä, K., Verhoeff, F., Kiel, J., 2013. Wood torrefaction − pilot tests and utilisation prospects. In: VTT Technology 122. Espoo, 73 p.
Wilén, C., Sipilä, K., Tuomi, S., Hiltunen, I., Lindfors, C., Sipilä, E., Saarenpää, T.-L., Raiko, M., 2014. Wood torrefaction − market prospects and integration with the forest and energy industry. In: VTT Technology 163. Espoo, 55 p.

Part Two

Fuel preparation, handling and transport

Biomass fuel transport and handling

4

Michael S.A. Bradley
University of Greenwich, Chatham, Kent, UK

4.1 Introduction

4.1.1 The critical importance of fuel handling to cost-effective biomass fuel valorisation

Many biomass energy plants suffer very severe and costly start-up and productivity problems. These difficulties usually revolve around not the combustion or pyrolysis processes, but difficulties with handling and flow of the fuel. To understand why this is, it is necessary to understand the scale and challenges with handling.

Biomass resources are widely disposed and very diverse; nevertheless, four common income streams can be realised from them. Traditionally, biomass has been burned to make steam or electricity for sale, which secondly earns a subsidy through Renewable Obligation Certificates, Contract for Difference, Feed-In Tariff or Renewable Heat Incentive. Thirdly, in some cases it is possible to sell the ash as a fertiliser. Fourthly, for some fuels, a gate fee can be charged instead of paying for the fuel.

Many more process options are now available than simply combustion for steam raising. Domestic waste can be processed into Refuse-Derived Fuel or Solid Recovered Fuel for sale, and the recyclables (steel, glass, aluminium, etc.) recovered for sale. Pyrolysis, to make oil or syngas fuel or even a feedstuff for plastics manufacture, has recently become increasingly commonplace.

Forest residue can be compressed into high-value wood fuel pellets, which are growing in popularity; they are now cheaper than oil for home heating.

All these are commercial reality. Many of the processes are proven and can be purchased easily — but many more are in development and will hit the market in the next couple of years. All are capable of delivering a profit, but one common challenge characterises all these disparate processes, which practical experience has shown is very often when the profits get lost. This is in the handling and flow of the biomass material into and through the process, which always seems to be given far less consideration in research and process design than the actual conversion process, even though in reality it often brings as big, or even bigger, challenges — and certainly always represents a large proportion of the project investment.

A typical biomass process plant is shown in Figure 4.1.

To give an idea of the scale and investment in the handling systems, to convert three of the six units at Drax power station from coal to wood pellets required an investment

[Figure: Flowchart showing Fuel reception (heap, tipping point) → Conveyor → Bulk storage (stockpile, silos) 100s or 1000s of tonnes → Conveyor → Pre-processing (drying, shredding, etc.) → Conveyor → Buffer storage → Feeder → Main process (combustion, reaction, etc.) → Conveyor → Storage of ash or finished product. Caption within figure: "A typical biomass processing plant – note there are more conveyors and storage systems than there are actual processes!"]

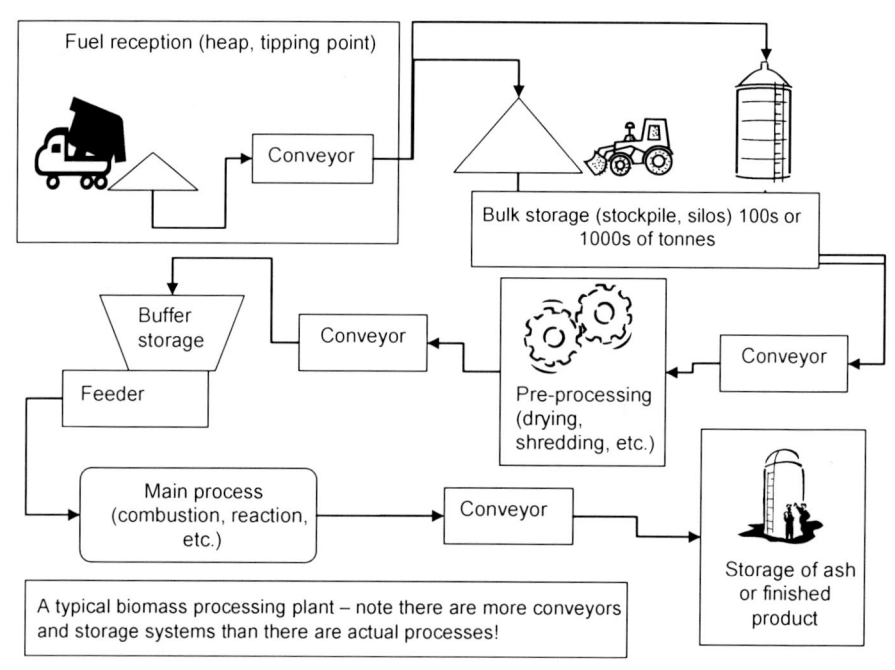

Figure 4.1 Generic illustration of the typical solids-handling processes in a biomass generation plant.

of around £40M in modifications to the boiler and combustion equipment, but £200M in the wood pellet handling system. Whilst Drax is at the extreme upper end of the size scale, the proportion is similar in many projects, in that the investment in fuel handling and storage is often more than that of the conversion process.

Obtaining reliable flow of bulk solids is more difficult than most plant constructors realise. In a study, 60% of novel solids processing plants did not reach full capacity even 2 years after start-up, and such plants cost on average over twice the money budgeted in the business case for construction (Merrick, 1990). However, biomass can be amongst the worst.

4.1.2 The special features of biomass as a fuel

Industry has handled solid fuel for years. The coal-fired power station has been with us well over a century, and designs for the efficient handling of coal are well established. However, the introduction of biomass handled in large quantities has led to major losses and station downtime due to fuel-handling system problems. Precisely, the same difficulties apply to pyrolysis and other plants processing biomass.

The following are a few of the most common problems that The Wolfson Centre has been asked to troubleshoot in biomass handling:

- Poor discharge from silos or hoppers ('arching' or 'rat-holing')
- Irregular or inconsistent feeding

- Dust evolution, and biohazards from this ('Farmer's Lung')
- Breakdown of pellets in handling
- Caking (hardening) of materials in storage, from fermentation or mould formation
- Fires and dust explosions in fuel storage
- Ash-handling problems in conveyors and silos
- Ash self-heating and hardening following conditioning with water

All these and many more common problems lead to unplanned shutdowns and often expensive plant modifications, seriously undermining the marginal profitability of biomass processing and utilisation.

4.1.3 The underlying causes of handling problems with biomass

Recent research has started to throw light on why many biomass materials are so troublesome for flow (Owonikoko et al., 2010). One common cause is the particle shapes; whereas coal and other 'ordinary' bulk solids tend to have particles which are roughly spherical or block-shaped, often with irregularities, many biomass materials are long and thin (chopped straw or *Miscanthus*), or flat and leafy (shredded sheet material like paper, plastic and card). When they are subjected to stress from a weight of material above, they 'knit' or 'mat' together, which makes them hard to move. Many have a low density, so gravity exerts only a small force on them to make them flow. Another very common cause of problems with biomass is its susceptibility to biological attack leading to heating and mould formation. Fire, as a result, is a common occurrence, and dust explosion is a challenge that must be met.

Even after burning, the ash from biomass material behaves quite differently from the ash from coal combustion, so ash-handling systems developed from the coal tradition do not work with biomass ashes.

Another important thing to consider is that many biomass streams are effectively 'waste' materials, even if they are not called by the name. Because they are the rejected leftovers from a primary process, such as, for example, spent grain from brewing, they are often not made to a close specification, and vary from day to day in particle size, dust content, water content, etc. much more than most bulk materials. Many are seasonal, so their properties vary — and so do their prices, which is one of the main reasons why it may be desirable to use different feedstocks at different times of year. Longer-term variation in price and availability is an issue, due to change in demand causing prices to rise when a particular feedstock becomes more widely used.

In summary, the reasons why biomass handling is so problematic, and why these have such a severe effect on project costs and performance, are strongly affected by the following factors:

1. There are usually many more process steps in the handling than in the conversion, all of which have to work smoothly or the handling chain fails.
2. The physical properties of biomass materials that affect the way in which they behave in handling and storage are often far more variable than the properties that affect their conversion (combustion, pyrolysis, etc.).

3. Laboratory studies that show the suitability of the fuel for use in the conversion process do not give any indication of their unsuitability to go through the majority of the handling and storage steps.
4. The ability of any given biomass to flow through a handling plant depends critically on whether the design of equipment is in tune with the flow and storage properties of the biomass.

Added to the above issues, there are a number of factors inherent in the supply and procurement of biomass-handling equipment, which also invariably militate against the success of projects:

5. Most suppliers of storage and handling systems are not specialists in biomass; they have to deal with a wide range of bulk solids, and frequently will not have met the particular bulk solid in question. Even if they have met something with the same name, frequently the behaviour will be significantly different due to changes in moisture, processing, source of supply, etc. so reliance on experience often leads to mis-design.
6. A lack of education is pervasive in bulk solids handling in all aspects of industries that supply and use such equipment. This is because solids handling is not on the educational curriculum of most process, chemical or mechanical engineers. Consequently, many engineers involved in design and specification of solids-handling equipment are ignorant of the differences in behaviour of different bulk solids, the pitfalls of solids handling and the techniques of characterisation and design that can be used to reduce the high technical risks inherent in solids-handling projects.
7. The lack of recognition amongst most purchasers of solids-handling equipment (i.e. system integrators and project promoters) of the difficulties faced by equipment suppliers, struggling with bulk solids which are often ill-defined and variable, without sufficient knowledge of how they may behave, whilst under pressure to provide equipment 'fit for purpose' against the widest material specification the buyer can pressure them to accept.
8. An excessively strong focus on project capital cost and timescale, in the context of a competitive and often adversarial procurement process that puts suppliers in a position in which they cannot afford to spend the time and money to determine the behaviour of the range of materials that may be handled and still provide adequately designed and engineered equipment. Aggressive procurement strategies usually ensure that the equipment supply contract goes to the cheapest bidder, who does not really understand what the problems are; and the procurement team usually does not have the detailed technical abilities to take the offered designs apart and really identify the problems.

These factors mean that achieving success even with a single biomass feedstock can be extremely challenging and in the case of many projects results in large cost and time over-runs before full performance is obtained. It therefore follows that achieving fuel flexibility with dry biomass can be even more challenging. Experience has shown that many biomass-handling systems perform very poorly even when starting the plant up on the fuel they are originally designed for, and changing to a different fuel can bring a whole load of new problems that can set the performance of the plant back months.

4.2 The challenges of biomass handling

To understand the possibilities of implementing fuel flexibility in the use of solid biomass, it is important to understand the key handling challenges with biomass. These will be reviewed in the following.

From the previous discussion, it should be obvious that the understanding of the flow properties of the fuel, and making sure the design of the handling system is 'in tune' to these flow properties, is absolutely the key to any successful biomass conversion project. If the biomass flow properties are not properly understood and accounted for, failure is practically assured.

4.2.1 Effect of low and variable volumetric energy density

The most basic fuel-flexibility challenge with biomass is the difference in the mass needed, and even greater variation in the volume needed, to obtain a given heat output. Biomass materials in an unprocessed form, for example cereal straw, wood chips, etc. have low bulk density compared to fossil fuels (often as low as 200 or 300 kg/m^3), leading to high volumes required for a given mass (see Figure 4.2). Processing of biomass can overcome this to a degree; for example, many pelletised biomass materials range around 550–700 kg/m^3.

However, most biomass materials also have low calorific value (often in the range 10–16 MJ/kg, compared with fossil fuels in the range 30–45 MJ/kg), increasing the mass needed for a given heating value (see Figure 4.3). For a given moisture content, many biomass materials have similar calorific values, and pelletisation does not increase the calorific value even though it enhances the bulk density.

A clearer picture of the above effects when combined can be obtained by looking at the 'volumetric energy density', the heat obtainable from a cubic metre of material. Compared with fossil fuels, this is often very low (for example, see Figure 4.4).

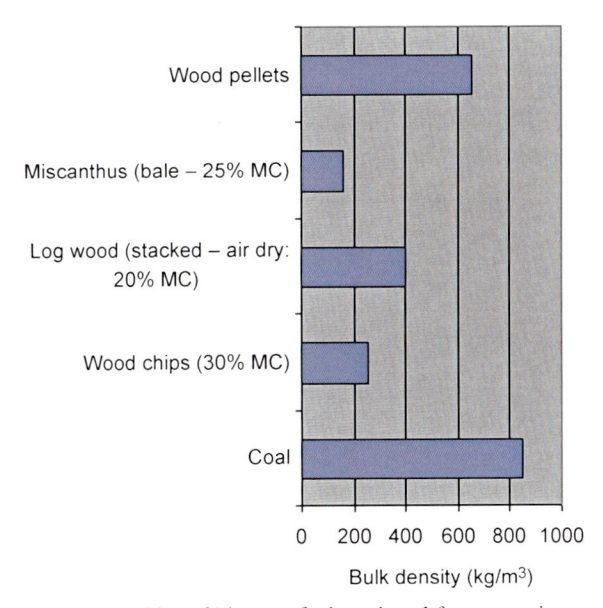

Figure 4.2 Typical bulk densities of biomass fuels and coal for comparison.

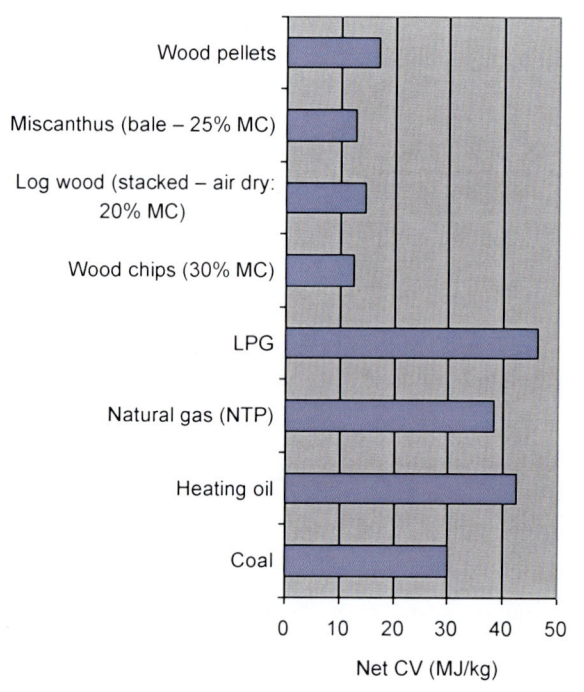

Figure 4.3 Typical calorific values of biomass materials.

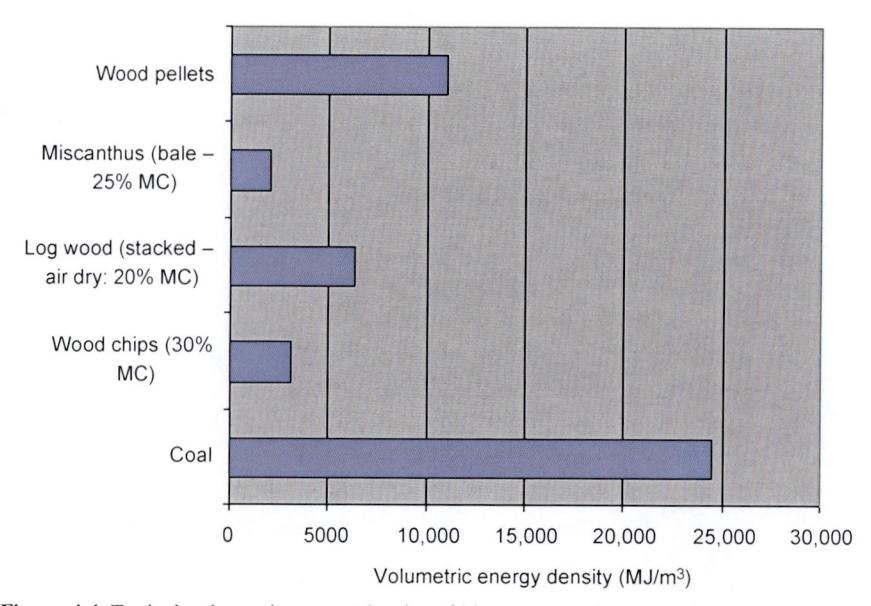

Figure 4.4 Typical volumetric energy density of biomass materials, compared against coal.

Note that even the pelletised materials still have substantially lower volumetric energy density compared with fossil fuels.

Taking an example, half of the Drax Power Station in Yorkshire (the UK's largest power station at 4000 MW) has been converted from coal to wood pellets. It is instructive to consider what this means for the volume of material to be handled, and how different would it be for other biomass feedstocks.

Half of the station (three units) produces 2000 MW of electricity. Running on coal this requires around 570 tonnes per hour of coal, which in round numbers is a little under half of one train load per hour. Converted to wood pellets, producing a similar power output requires around 1080 tonnes per hour, which with a bulk density of around 650 kg/m^3 (compared with coal at about 900) means two trains per hour. Obtaining this increased mass flow or fuel required significant increase in rail unloading capacity, from one track to two. However, what about if other biomasses had been chosen instead of wood pellets? For wood chip (30% MC) and *Miscanthus*, the slightly lower calorific values compared with wood pellets meant slightly increased mass flows, but combining with the lower bulk density would require four trains per hour.

The number of trains per hour required to use raw biomass materials, therefore, would be hugely increased and would have needed unrealistic expansion of the handling facilities. This would also be mirrored in the size of covered fuel storage required; 3 weeks of standby fuel usage requires four huge storage domes (60 m diameter by 50 m high) for wood pellets, but wood chips would need 15 such domes, economically untenable.

At the other end of the size scale, consider a domestic heating installation converted to wood pellets from oil. For a typical house, a 2000 l oil tank will suffice for a year of heating. For oil, density is around 900 kg/m^3 (Lehtikangas, 2000), so 2000 L represents 1800 kg; with a net calorific value (CV) of 42 MJ/kg^3, 2000 L have a heat content of 76,000 MJ. Converting the same house to wood pellet heating and again aiming for a once-a-year fuel delivery, these have a net CV of around 17 MJ/kg, so the same heat content requires 4450 kg, and with the lower density of typically 650 kg/m^3 (Lehtikangas, 2000) this equates to a volume of 6800 L, over three times the fuel volume. However, there is a further complication. Whereas oil will fill level to the horizontal top plate of a tank, and empty to the flat bottom plate, pellets will not flow in this way and require a sloping hopper bottom to promote emptying; they also display an angle-of-repose heap in the top. Each of these creates empty space within the cuboidal envelope around the pellet inventory. Typically, this is around one-third of the total space occupied by the store (greater with wide, shallow vessels and less with tall, narrow ones), so the total volume occupied by the pellet store will be about 10,000 l — in other words, five times as large as the oil store. A superficial comparison of density, pellets being around one-third lighter than oil, gives no clue as to this massive difference in fuel-store volume needed. Consequently, many wood pellet stores have been severely undersized leaving owners having to buy most of their pellets regularly during the peak heating season when prices are at their highest and delivery times most extended.

4.2.2 Effect of variation of volumetric energy density on feeder, conveyor and store design

The low and variable volumetric energy density of biomass also considerably affects the capacity of conveyors, and achieving fuel flexibility can be very challenging. If designed for a denser material, the consequent volumetric throughput limit of feeders and conveyors means that with a less dense material the throughput will be more limited. Conversely, if designed to convey the mass throughput of the less-dense material, when conveying the heavier material the mass load will be higher and the power may not be adequate. Therefore, achieving fuel flexibility with conveyors means that there needs to be sufficiently large turn-down to obtain the same heating value throughput with both the lower and higher energy-density material, but maintaining the drive torque needed for the heavier material at the slow speed. This can be very challenging with variable-speed drives and will often require the inverter, motor and gearbox to all be oversized.

For stores, the store needs to be large enough to give the energy buffer requirement on the lowest energy-density material, yet strong enough to withstand the much greater loads when filled with the heaviest material. However, other considerations regarding self-heating and storage times and quantities, are subsequently discussed in Section 4.4.1.

4.2.3 Moisture effects

Virtually all biomass materials are highly susceptibility to effects of moisture; additionally, many are severely affected in their handling and storage properties by water. For a start, moisture contents are often relatively high. Wood is a useful example to consider. Freshly chipped wood, depending on the species, often runs around 45–60% (wet basis, i.e. water mass as a percent of total wet material mass); leaving it in a pile for a few months will often reduce this to 20–25% depending on the conditions. Keeping wood under cover indefinitely will reduce it to around 10%, so if the wood is reclaimed from demolition or lumber processing this will be its moisture content. This has massive effects on the bulk density — up to half of the initial mass may be lost in drying. However, it also affects the net calorific value (the heat left for energy production after accounting for the moisture being boiled off), as shown in Figure 4.5.

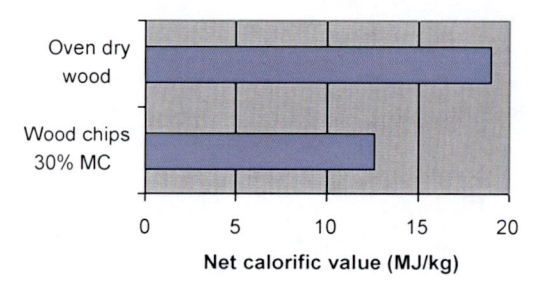

Figure 4.5 Typical net calorific value of wood versus water content (wet basis).

As an aside, it is extremely important to be clear when specifying a moisture content, whether this is quoted as a percentage 'wet basis' (i.e. the moisture as a percentage of the mass of wet material prior to drying) or as a percentage 'dry basis' (the moisture as a percentage of the mass of dry material left after drying). Industry mostly uses the former, whereas the scientific community mostly uses the latter, but not exclusively so. These are significantly different figures, and lack of clarity on this point has often led to confusion and even serious mis-design of equipment.

The variation in bulk density and CV means that for a given heat-production rate, a wide range of different mass-flow rates of wood may be required if the moisture is variable.

However, moisture can have other severe effects on biomass apart from reducing its net CV. Being bio-active, high moisture accelerates decay, leading to self heating and fire as well as loss of heating value. For pelletised biomass, exposure to water in the form of rain or even general dampness must be totally avoided because it will cause the pellets to soften (Lehtikangas, 2000) and break down into wet sawdust; some types of pellets, if thermally processed (steam exploded or torrefied), have some ability to withstand damage from moisture, but nevertheless take up significant quantities of water when exposed to heavy rain, so they still require cover in wet climates such as Northern Europe. Wood chips can be stored outside for a limited period (long-term uncovered storage in Northern Europe will cause decay, reducing heating value). Straw and *Miscanthus*, and most other biomass materials must be covered.

A rather more sinister effect of moisture in biomass is mould formation. Any organic material kept at a moisture above about 14% wet basis (Khan, 2008), will support mould growth. If this mould subsequently dries out in handling or storage, it releases spores which are hazardous to health; inhalation over a period of time can lead to 'farmer's lung' (Anon, 2012) (alveolitis, common in the agricultural industries due to exposure to dust from hay, straw, etc. stored moist). For this reason, biomass dust should be treated as more hazardous than simply a nuisance dust, and spills of biomass and biomass dust should not be allowed to build up outside where it can become mouldy and then release spores in dry weather. Similarly, cleaning of biomass facilities should be undertaken dry and never wet.

Just how severely the handling properties of a biomass change as a result of variation in moisture content is completely dependent on the biomass. A very significant factor is whether the presence of the water causes alteration of the structure within the particles. For example, in wood pellets, any significant exposure to water will cause the pellets to break down into a mass of cohesive fines like damp sawdust, which clearly will utterly change the handling and storage properties; by comparison, many torrefied biomasses will not suffer significant change in structure so exhibit little change in handling properties other than increase in density. Reversibility is also a consideration; in the case of a torrefied biomass, gain then loss of water will often take the material back to where it started, whereas in wood pellets the structural breakdown means that gain then loss of water will leave the material in a completely different physical condition. With materials that do not suffer structural breakdown, changes in handling properties (other than density) usually become significant once water appears on the surface as this tends to encourage adhesion between particles or against constraining surfaces.

Dust emission and particle breakdown tendencies often change with moisture content. Maize has been found to suffer more degradation when handled at lower moisture contents, and the same effect has been observed qualitatively in wood chips. Apart from producing more fine particles as a result of the handling, lower water content also means that dust in the material is much easier to detach from the particle surfaces so tendencies to emit fugitive dust always goes up with lower moisture content in the case of raw biomass materials. Pellets behave differently, however; wood pellets tend to be less durable with increasing moisture content (Lehtikangas, 2000), in contrast to wood chips.

An example was recently seen at a small UK power station (in the 10–100 MW class) handling reclaimed wood, in which wood was being supplied from two different storage sites. One storage site was small with a fast turnaround so the wood was not exposed to much weather; on delivery to the power station, because of its dry condition it gave rise to substantial quantities of fugitive dust which caused mess in the handling chain and built up on exposed sloping surfaces of the bunkers. The other storage site had large quantities of wood stored outside in an exposed position for long periods, so when delivered, it emanated no fugitive dust, and exhibited much poorer flow. This wetter material gave much higher structural loadings on the bunker due to the increased density, and it also showed more self-heating tendency leading to a wet, steamy atmosphere in the bunker so the dust deposited from the dryer feedstock then became wet and sticky from the resulting condensation. Such issues with differential moisture content need to be considered in planning for fuel flexibility.

A final issue with water content can be the effect on processing. Apart from the effect on CV as mentioned previously, and the effect on handling properties, the steam volume emanated in thermal processing can change drastically. This steam volume can make a significant change to the gas volume being processed in pulverised fuel transport, flue gas processing, etc. and the changes to the gas volumes and velocities arising from this should be considered in design.

4.3 Sources and types of biomass, and classifications according to handling properties

Experience of dealing with many different biomass materials in The Wolfson Centre has led to the development of a 'taxonomy' of biomass materials, based upon the way in which they flow. Classifying them in this way is extremely helpful as it helps to identify what sort of handling solutions are likely to be useful for them. The recommended taxonomy is as follows (Owonikoko et al., 2010):

Class 1: Free-flowing particles without extreme shape. Roughly rounded or 'blocky', free-flowing, relatively coarse materials; for example, pellets, chunky wood

Figure 4.6 Wood pellets, a typical 'Class 1' (free-flowing) biomass.

chip, grain, etc. (see Figure 4.6). These materials have favourable flow properties, in that they flow easily under gravity without tendencies to 'hang-up' or arch.

Class 2: Cohesive materials without extreme shape. Roughly rounded particles but with significant fines and/or sufficient oil or water content to make them sticky, so that if picked up and squeezed in the hand they form a 'snowball'. Wet fines, filter cake, milled nut kernels, distiller's spent grain are a few examples (see Figure 4.7). These have more difficult flow properties, with tendencies to arch and hang-up; however, they can be designed for using the characterisation tests and process design models traditionally used for other cohesive bulk solids such as wet coal or iron ore. More care is needed in selection and engineering of handling systems of these materials compared with 'Class 1' biomasses, but measurements of the flow function, bulk density and wall friction of these materials gives data from which can be calculated the necessary geometry of converging hoppers, silos, feeders and flow channels through which they will flow reliably. The calculated necessary geometry will vary with moisture or oil content, particle shape and inherent chemical properties, so such variations, if likely to occur, should be incorporated in samples tested at the project design stage and an envelope around these used for the plant design.

Class 3: Extreme-shape particles. Materials that have extreme particle shapes, such as chopped straw, grasses and other herbaceous materials, forest residues that contain large amounts of thin sticks and pine needles, etc. (see Figure 4.8). Often, wood chips that have been reduced to small size by hammer milling (typically through screens of 2–5 mm) also have long, thin needle-shaped particles. Reclaimed wood, which has

Figure 4.7 Milled palm nut kernels; a typical 'Class 2' (cohesive) biomass.

Figure 4.8 Chopped straw (left) and *Miscanthus* (right); typical 'Class 3' (nesting/entangling) biomass fuels.

usually been crushed from a relatively dry condition (demolition waste, scrap furniture, etc.) also splits very easily along the grain but not so easily across the grain, so breaks into extreme shape particles.

These materials have especially difficult behaviour, because although they are not usually cohesive (i.e. when you pick up a handful, squeeze it and release it, does not remain in a 'snowball'), nevertheless they can have very strong tendencies to arch or nest and resist flow, due to the entangling of the particles as in a bird's nest. A general 'rule of thumb' of whether a material might fall into this category appears, from our limited current knowledge, to be in terms of the ratio of maximum to minimum dimension of the particles; if this ratio is more than about 3 or 4, then nesting or entangling is a danger. Whether nesting actually occurs seems also dependent on the surface texture and flexibility of the particles, but if it is a hazard then a tensile test as described below should be used to assess this.

4.3.1 Identifying the class to which a biomass belongs

The first means of identifying the class to which a material belongs is with a visual and tactile examination, as follows:

First, conduct a simple test for cohesiveness by picking up a handful of the biomass material, and squeezing it with the fingers as if to make a snowball. Open the hand gently and see what happens. If the particles fall apart straight away without any sign of sticking together, then provisionally (subject to the test in the next paragraph) it can be said to be a 'Class 1' material (free flowing). If, on the other hand, it forms a structure, holding together like a 'snowball' on the palm of the hand and needing the touch of a finger to get it to disrupt, it is a 'Class 2' material (cohesive).

Secondly, examine the particles visually. Pick up a number, one by one, and see whether they have extreme shapes. If they are all roughly rounded or 'blocky' with broadly similar dimensions in three directions, then 'Class 1' or 'Class 2' judgement just made, applies. However, if the particles are relatively long and thin, or flat and 'platy', be suspicious. As a rule of thumb, if the ratio of largest to smallest dimension on the particles is more than 3:1, there is a danger it will behave as a 'Class 3' material (Owonikoko et al., 2010). If the material displays this sort of particle aspect ratio or larger, then a tensile strength test should be undertaken; if the material displays a measureable tensile strength, then it is a 'Class 3' material. If it displays no tensile strength, and has no cohesion in the 'snowballing' test, it is a 'Class 1' material.

4.3.2 Typical common biomass materials and their classifications

The following table is derived from biomass materials seen in projects participated in by The Wolfson Centre. It is not definitive, and some materials might sometimes exist in more than one category depending on moisture content and processing. Therefore, do not rely on this table for design purposes; always test as described previously.

Class 1 (free flowing)	Class 2 (cohesive)	Class 3 (nesting/entangling)
Pellets with low fines	Pellets with high fines	Fine milled wood chip
Chunky virgin wood chip	Dust from most biomass materials	Chopped straw, *Miscanthus*, grass
Dried, pelleted sewage sludge	Palm nut kernels	Recycled wood
Cereal grains (wheat, oats, barley, maize)	Sewage sludge in damp form	Municipal shredded waste
	Meat and bone meal	Compost
		Sawdust, wood shavings
		Bran
		Spent distillers grain

Continued

Class 1 (free flowing)	Class 2 (cohesive)	Class 3 (nesting/entangling)
		Corn stover
		Chopped plastic bottles
		Horse bedding
		Waste rags
		Poultry litter
		Bark chips
		Spent grain from distillation
		Brash (twigs, leaves, etc. from arboricultural or horticultural trimmings)

4.3.3 Selection of handling equipment for different biomass materials, and compatibility between different fuels in common systems

The details of what type of feeder, store, conveyor and other equipment to select for specific classes of biomass materials are a complex process which must be done with regard to the quantitative characteristics of the individual fuels in question, and is beyond the scope of this book. For a full treatment, refer to Bradley (2010).

However, in general it can be said that the substitution of fuels within the same class is often possible subject to certain provisions; although substituting fuels of different classes can be difficult, as summed up below:

		The handling system was designed to handle biomass in...		
		Class 1	Class 2	Class 3
...now can it be used to handle another biomass in...	Class 1?	A	A	Unlikely
	Class 2?	Unlikely	B	Unlikely
	Class 3?	Definitely not	Definitely not	C

A − probably, as long as the bulk density, explosion, fire and dusting characteristics are compatible − see below.
B − possibly, as long as the bulk density, explosion, fire and dusting characteristics are compatible AND the cohesive strength (arching dimension) is no higher (this requires a shear test). If the cohesive strength of the new material is significantly higher, then the answer is probably no.
C − possibly, as long as the bulk density, explosion, fire and dusting characteristics are compatible AND the tensile strength is no higher (this requires a tensile strength test). If the tensile strength of the new material is significantly higher, then the answer is probably no.

4.4 Other considerations for compatibility of different fuels with a handling system

4.4.1 Need for stock rotation and limitation of storage time

Almost all biomass materials are to a degree 'time dependent', which is to say that they change with time in residence. Particularly materials with significant water content, and especially with high oil or starch content, have microbes present in the material which can cause degradation — apart from mould formation and rotting, fermentation can cause them to self heat, liberating steam and water and in some cases initiating combustion. Some biomasses even with low bio-activity (such as wood pellets) suffer self-heating (Larsson et al., 2012; Blomqvist and Persson, 2008) due to direct oxidation-reaction with the interstitial air — especially when fresh (or recently handled, exposing fresh surfaces to the air). How fast this happens with a particular material, and how often a store is emptied, will determine whether the store requires to be emptied in a first-in, first-out discharge pattern. Compatibility between different fuels in a system, therefore, requires that none of the potential fuels will self-heat to a situation of thermal runaway within the time they are in storage in the main fuel store or the boiler/process feed bunkers.

For any given biomass material, there is a characteristic period in which self-heating occurs (this is also affected by the size of the vessel — it happens faster in a larger vessel because heat cannot escape so easily). For some materials, it may be several months; often this is the case for wood pellets if kept dry. For wetter wood, such as reclaimed wood stored in an outside stockpile, for example, it may be weeks; although for materials with more food to support microbial life such as sewage sludge and poultry litter it may be days. Some materials are even more sensitive, for example spent grain from distilling can run away in hours. Apart from fire, the effect of mould formation and fermentation can often cause the material to cake hard because the products of the bio-activity sit between the surfaces of the particles and cement them together, sometimes making handling impossible.

In general, it is good practice for all materials to exercise stock rotation — use the older materials first before the newer ones. Management of flat stores (sheds) or stockpiles should be done in this way. However, with hoppers and silos there is commonly a problem — most large silos that have converging bottoms discharge in a flow pattern called 'core flow' (see Figure 4.9).

This flow pattern does not permit stock rotation. The first material put in, at the bottom of the bunker, is the last to come out. Fresh material loaded in on top always comes out before the older material. The oldest material at the bottom ONLY discharges when the bunker is COMPLETELY emptied. However, usually in a process plant we do not want to empty the bunker because it stops the downstream process.

Hence, a requirement for compatibility is that if the fuel has a time dependency causing heating or caking, the quantity that can be stored in such core-flow bunkers must be small enough, and emptied completely often enough, to prevent the problem. If changing, for example, from wood pellets, which may be stored for months before

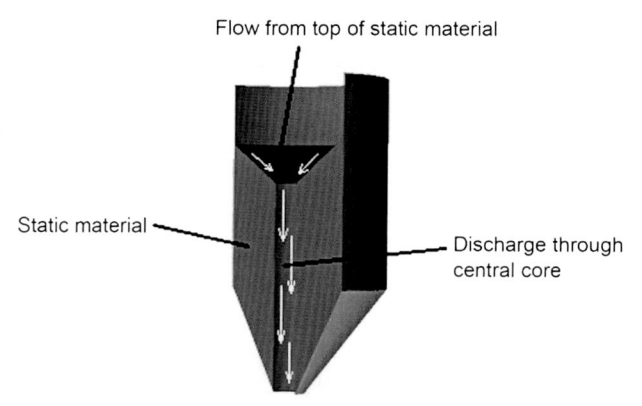

Figure 4.9 Core flow pattern in a bunker, commonly found in many storage silos and mill feed bunkers, etc. which have not been designed to give mass flow.

thermal runaway, to sewage sludge pellets that may be storable only for days, the frequency of emptying will need to be much higher and this will severely reduce the available storage volume.

One way of overcoming self-heating with many materials is the use of nitrogen inerting. Although beyond the scope of this article to discuss in detail, introduction of nitrogen from the bottom of the silo with the outlet valve gas-tight has been found to be effective in preventing self-heating. However, it is expensive to purchase both the equipment and the gas, and introduces another serious hazard on plant − asphyxiation of personnel in areas of nitrogen buildup.

Core flow also has another drawback − if the material is significantly cohesive, it will form a 'rat hole', an empty flow channel above the outlet with the material hung up around the sides.

In summary, if you have a core-flow fuel bunker or feed hopper, you can only put 'Class 1' materials through it, and you have to restrict the quantity stored to that which you can regularly empty COMPLETELY before it self-heats or cakes.

One means of overcoming time dependency of the fuel is to use a 'mass flow' fuel store and feed hoppers (see Figure 4.10).

Mass flow gives first-in, first-out discharge, that is proper stock rotation. It is shown above in a converging bunker; to achieve it requires a wall angle that is steeper and/or a hopper lining of lower friction, than for core flow. Determining the angle for mass flow requires certain special tests, but is very well proven as many thousands of mass-flow bunkers have been built. Mass flow can also be achieved in a parallel bunker or hopper if a mechanical reclaimer is present at the bottom that works over the whole area.

The other advantage of mass flow is that the internal stress pattern has more power to 'break' a hang-up of material than core flow, so it can be used to discharge cohesive ('Class 2') materials. However, two important criteria exist in relation to the fuel:

1. The 'arching dimension' of the fuel. Whereas 'Class 1' (free-flowing) materials will come out of any hole large enough to prevent them mechanically locking together, 'Class 2' materials

Biomass fuel transport and handling 115

All material in motion during discharge

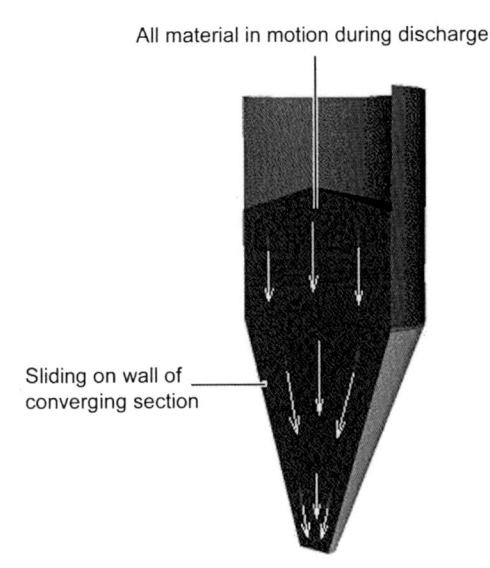

Sliding on wall of converging section

Figure 4.10 Mass-flow pattern in a bunker.

require a certain minimum outlet size to prevent the material arching at the outlet, depending on just how cohesive they are.
2. The 'wall friction' between the fuel and the bunker surface, which is a function not just of the fuel itself but also the materials of which the bunker wall is made, and the way it is finished. For any given geometry of bunker (i.e. convergence angle and shape), there is a critical wall-friction value above which mass flow will not occur, meaning that the stock rotation will be lost and if the material is cohesive, a rat hole and fuel hang-up will occur.

Hence, compatibility between fuels in a converging mass-flow hopper or bunker depends on their individual 'arching dimensions' being smaller than the bunker outlet, and the wall friction being lower than the critical value for the bunker geometry.

Generally, no converging bunker (mass flow or core flow) can be used for 'Class 3' materials because they will arch. Recent research (Owonikoko, 2012) has shown that a limited ratio of convergence from top to bottom of the hopper can in some circumstances be accommodated in mass flow for 'Class 3' materials. However, knowledge is insufficient to exploit this for practical application, so the only safe design for 'Class 3' materials at present is to use a parallel bunker with a discharge device that moves across and reclaims material from the whole of the bottom area. In this case, the main consideration in changing fuels is whether the discharger has the power to reclaim the different fuel, and at the rate required.

The key advantage of mass flow for a time-dependent material, is that all material passes through and out of the store, without having to empty it completely. So a working buffer can be held, determined only by the self-heating time of the material and the rate of use. Again if changing fuel for one with a shorter self-heating time, final-feed hoppers for biomass processes (combustion or pyrolysis) should be designed to promote mass flow (whether converging for Class 1 or 2, or parallel for Class 3, materials) to avoid the need to empty them regularly to prevent self-heating or caking. In addition, they present other benefits in relation to reducing unintentional particle-size segregation and variations in bulk density.

The dangers of self-heating and fire with biomass are invariably high; a fire must always be expected in any storage facility, and appropriate hardware, schemes and training for firefighting must be considered mandatory. For most biomass materials, the requirements will not be greatly different so, whilst the frequency of emptying to prevent fire may be different for different biomasses, the actual fire-management hardware and procedures will not usually need to alter.

4.4.2 Explosion protection

In relation to explosion protection, some biomass materials will not require this because their tendency to emit dust levels is sufficiently low, normally because of high moisture content. If the facility has been, or is to be, built without dust explosion protection, then this will limit the use of other biomass materials to those that also have low dust emission, which will generally preclude the use of dry fuels. However, most biomass materials that are not palpably 'wet' will tend to emit dust, and that will be explosible, so storage vessels and enclosed parts of the handling system will require explosion protection, usually by venting and isolation. The experience of this author in designing many biomass facilities is that many biomass dusts have broadly similar explosion characteristics (explosion severity (K_{st}) values below 200 bar m/s putting them in dust explosion class 'St 1'). Consequently, in general, venting arrangements worked out for one biomass fuel will usually be compatible for others (though this does not have to be checked; no doubt some biomass fuels have dusts in class St 2, even though this author has not seen them).

4.4.3 Special care in relation to large vessels

The business of design and selection of bunkers, hoppers, feeders and ancillary equipment is a specialised area that is beyond the scope of this book; for a detailed treatise on the subject, refer to other references (Various, 2015; Anon, 2003; Rotter, 2001). If contemplating changing fuel feedstock in a large bunker, the potential hazards in relation to flow problems, structural failure, fire, explosion or material spoilage, and the potential consequences in cost, damage to property and loss of life are such that the plan must always be referred to a silo specialist for final evaluation before proceeding.

4.4.4 Dust control, ATEX and DSEAR

The issue of dust explosion protection has previously been discussed; however, dust emission from the handling equipment needs to also be considered. Again, this is linked to the tendency of the biomass to emit dust, and different materials vary greatly in this regard. In general, materials with high moisture content, such as virgin wood chip or wet sewage sludge, will have little tendency to emit dust because the moisture will bind it to the larger particles, and the moisture in the particles will tend to make them relatively elastic so they do not generate small particles when handled. For these materials, there will be little emission of dust at transfer points, heaping in stores, etc. so it may be possible to avoid the need for containment and extraction of dust.

Materials that are dry to the touch (the actual moisture content at which this is the case will vary with the biomass) usually suffer from significant dust emission. For example, pellets, reclaimed wood, straw, etc. will require dust control methods to avoid excessive emission to the workspace and the wider environment.

The other aspect of dust is the potential for explosion. The need for explosion protection for equipment has previously been discussed in Section 4.4.2, but in addition, if there is potential for an explosible concentration of dust in areas of the plant, then this will require an assessment under the Dangerous Substances and Explosible Atmospheres Regulations (DSEAR) in the UK, or the Atmospheres Explosibles (ATEX) 137 Workplace Directive 1999/92/EC in mainland Europe (note that many projects in countries outside Europe now specify application of ATEX principles, even though it is a European standard). If zones are identified in which a recognisable risk exists for the formation of explosible concentrations of dust, then ATEX-rated equipment will be needed in these areas.

Again, a detailed review of DSEAR and ATEX principles and responsibilities is beyond the scope of this book, but very good practical guidance for engineers, plant developers and owners can be found in Anon (undated).

Overall, therefore, compatibility in regard to dust will usually mean whether the potential fuel has the ability to emit a significant quantity of dust, and whether the handling has appropriate measures for dust control, ATEX-rated equipment when necessary and explosion protection. Usually, as discussed above, this will be determined by whether the fuel is wet or dry to the touch. Note that if a material is dry to the touch but without any dust present — like clean wood pellets for example — it can still be relied upon to suffer breakage leading to dust emission in handling. If any possibility exists that a system may be required to handle dry biomass, all these features should be built in from the start.

4.4.5 Other tests always required to check compatibility

Bulk density; in all cases, change in bulk density from old material to new will need to be checked. If the new material has a lower bulk density, then the structure will probably be alright but the mass it holds when full, and the mass throughput, will be reduced. If the bulk density is higher, then structural loadings on silos and power availability in conveyors will need to be checked, and the allowable inventory and throughput may have to be restricted.

If large or very large silos are used for storage (say more than a few tens of cubic metres capacity), then the wall friction and internal friction of the new fuel will also need to be measured (using a shear cell). The combined effects of these with the changed bulk density will need to be assessed against the structure by an engineer who specialises in silo structures using EN1991:2006 part 4. Note that most good general structural engineers do NOT have the specialised knowledge to know how to predict actions on silo structures; several cases have been seen in which non-specialist structural engineers of long experience in other structures have used inappropriate principles for predicting loads on silo structures, resulting in structural distress or even failure.

4.5 Conclusions

4.5.1 Choosing the right solutions

Plenty of solutions are available for moving, conveying and feeding biomass materials at least one machine option exists that works for every material. However, many of these machines only work with a narrow range of materials, and it is hard to be sure that you choose the right option. Those with the ability to handle the widest range of materials are much more expensive, and may make the project uneconomic. From this, it will become obvious that to make a system reliable yet affordable requires a careful choice of the right equipment balanced against the range of potential fuels that together make sense economically.

Furthermore, experience shows that the equipment manufacturers are not always as well informed as the buyer expects them to be, when it comes to advising on the right 'tools for the job'. They may be experts in equipment design and manufacture, but they cannot really be expected to know about the way in which every possible material they might meet will behave — *it is up to the buyer* to make sure he selects suitable solutions, whatever the contract may say about contractual responsibilities and process guarantees. When a project is a 'design-build' or 'turnkey' project, special care needs to be taken to ensure that the engineering, procurement and construction (EPC) contractor building the plant takes the necessary steps to accurately bottom the range of fuels to be used and spends enough time and money on their characterisation based on realistic samples. Furthermore, this information should be agreed upon and used intelligently by all parties to the design.

Finally, because of the unpredictability of the demands for fuel change and the natural variations even in a fuel that is nominally 'the same' over time, the chances are that the plant will be called upon to handle fuels for which it was never designed. In some cases, this will require minor modifications to the handling equipment, but in many cases the modifications will be more severe. If the plant has been designed with space around the equipment, rather than packing everything in tightly, it will be much easier and more cost-effective to make the necessary changes.

4.5.2 The need to 'know your enemy'

The key messages are:

- All biomass materials handle differently; many are inherently variable
- Most conveying, feeding and handling systems can cope only with a restricted range of materials; those with wider capability are more expensive

- Often, facilities designed around one feedstock will have to change to another to maintain profitability
- It is up to the buyer, not the equipment supplier, to make sure he chooses the right equipment for the materials he is to handle

Many, even most, systems that handle biomass do not start up and run straight away — many need an extended period of development during which retrofit and lost opportunity costs are incurred, often for a year or two before they get to full operation.

To give the best chances of success:

- Assess, before embarking on a development, what the range of feedstock is likely to be — not just now, but in the future. Consider the influence of other developments on availability and price. Draw up a possible portfolio of material types and sources classified as highly, moderately and less likely to be used.
- Recognise the importance of ensuring the feedstock will flow reliably between reception and process. Do not make the mistake of spending all the time and effort on the conversion (combustion/pyrolysis, etc.) and leave the material handling to the engineering contractor.
- Above all — GET THE FEEDSTOCK CHARACTERISED FOR FLOW, not just the favoured material but the range of other options as well. This will identify the range of flow properties that the handling equipment will have to service from the outset. If some of the potential feedstocks require significantly different size or type of handling equipment, it may well be more economical to exclude these from the acceptable list.
- Ensure the contractor takes account of flow property characterisation in the equipment they buy, because experience shows they often buy more on price than on technical suitability.
- Before changing the feedstock, get the proposed new material characterised, to see if will go through the handling system you have bought — if not, it is probably best to look elsewhere for suitable feedstock instead of persevering trying to put a 'round peg in a square hole'.
- In particular, look carefully at:
  - Bulk density
  - Volumetric energy density
  - Flow classification (Class 1, 2 or 3)
  - More detailed flow properties (arching tendency and wall friction in particular)
  - Variation in moisture content
  - Self-heating and fire characteristics
  - Dust emission and explosion characteristics

4.5.3 Future trends

Future trends in biomass are unpredictable. One of the most important things to understand is that most biomass processes are only economical because of the subsidies on heat and electricity production. These are politically driven, and we have in recent times seen ever-increasing pressure and funding for 'green' issues. However, justification for this cannot be relied upon going forwards especially in a world in which, globally, carbon emissions from developing countries are rocketing vastly faster than the ability of developed countries to control emissions. Given also the ever-increasing pressure to re-use or re-purpose waste materials, and to redesign manufacturing processes to promote a 'circular economy' in which wastes are minimised, availability of specific waste streams cannot easily be relied upon. All of these factors will affect feedstock costs or gate fees and process economics. For these reasons, flexibility in feedstock utilisation does help to ensure a good future for the project, but only if this can be achieved economically.

References

Anon, 2003. Actions on Silos and Tanks. EN 1991−4.
Anon, 2012. Health Aspects of Burning Biomass in Power Generation Electricity Industry. Occupational Health Advisory Group Guidance Note 4.3. Available at: http://www.energynetworks.org/modx/assets/files/electricity/she/occ_health/OHAG_guidance_notes/OHAG%20-%20Biomass%20-%20Cover%20-%20web%20site%202013.pdf.
Anon, undated. Practical Guidance for Suppliers and Operators of Solids Handling Equipment for Potentially Explosive Dusts: Compliance with Legislation Implementing the ATEX Directives. Solids Handling and Processing Association. Available at: www.shapa.co.uk.
Blomqvist, P., Persson, H., 2008. Self-heating in storages of wood pellets. In: Proceedings of the World Bioenergy Conference and Exhibition on Biomass for Energy, Jönköping, Sweden, 27−29 May 2008, ISBN 978-91-977624-0-3, pp. 172−176.
Bradley, Berry, 2010. Handbook of Biomass Handling. The Wolfson Centre, University of Greenwich.
Khan, N.S., 2008. Handling Characteristics of Coal/Biomass Mixes: Measurements and Establishing Benchmarks (Ph.D. thesis). University of Greenwich.
Larsson, S.H., Lestander, T.A., Crompton, D., Melin, S., Sokhansanj, S., April 2012. Temperature patterns in large scale wood pellet silo storage. Applied Energy 92, 322−327.
Lehtikangas, P., November 2000. Storage effects on pelletised sawdust, logging residues and bark. Biomass and Bioenergy 19 (5), 287−293.
Merrick, E., 1990. Understanding Cost Growth and Performance Shortfall in Pioneer Process Plants. The Rand Corporation.
Owonikoko, A., Berry, R.J., Bradley, M.S.A., 2010. Characterisation of extreme shape materials: biomass and waste materials. In: Bulk Solids Europe 2010 Conference and IMechE Symposium in New Frontiers in Bulk Materials Handling, 9−10 September 2010, Glasgow.
Owonikoko, A., 2012. Predicting Storage Vessel Geometry Requirements for Discharge of Extreme Shape Materials (M.Phil. thesis). University of Greenwich.
Rotter, 2001. Guide for the Economic Design of Circular Metal Silos. CRC Press.
Various, 2015. Storage and Discharge of Bulk Solids. The Wolfson Centre, University of Greenwich.

Fuel pre-processing, pre-treatment and storage for co-firing of biomass and coal

5

Michiel C. Carbo, Pedro M.R. Abelha, Mariusz K. Cieplik, Carlos Mourão, Jaap H.A. Kiel
Energy Research Centre of the Netherlands (ECN), Petten, The Netherlands

5.1 Handling and storage of biomass at coal-fired power plants

Dedicated storage and handling equipment for bulk biomass encompasses a broad array of technologies (Solids Online, 2015; SolidsWiki, 2015). This subject is quite comprehensive; therefore, this chapter only describes the most important rules of thumb for design and use. In principle, most biomass handling and storage equipment is designed to cope with the two most crucial aspects of biomass: its susceptibility to water-induced degradation as well as mechanical degradation. Both aspects lead to physical loss of material and pose safety risks. Each stage of the on-site delivery, transport and storage with its corresponding equipment will be discussed, with emphasis on the aforementioned two aspects.

5.1.1 General considerations

Untreated biomass is prone to decay relatively quickly in case it is exposed to water in the form of precipitation or condensation. Hence, the first priority when receiving biomass on site is shielding it from any exposure to water. Keeping biomass dry at moisture content typically below 12 wt% diminishes the risk of moisture-induced biological decay and the corresponding formation of methane and/or carbon monoxide emissions, self-heating and spontaneous ignition of the material. However, it should be noted that bone-dry biomass, with a moisture content below 2 wt%, increases the risk of electrostatic charge build-up, which could lead to ignition and therefore seriously increases the risk of fire and/or explosion.

Limiting the exposure to water is often done by the combination of a proper, sheltered or fully contained handling and storage environment, as well as simple operating measures like avoiding discharging during precipitation periods. Once unloaded and dry, the material should be stored in a fashion that minimises contact with moisture, particularly in the form of condensate, and eliminates the risk of aggregation of large volumes of combustible gaseous decay products. Depending on the exact climate settings and storage times, this is facilitated by proper ventilation and temperature control inside the handling and storage facilities. In case storage for an extended period is desired, the use of more advanced features like blanketing gases and active moisture control should be in place (Dorp, 2015). Just-in-time delivery, combined with limited storage duration and volumes, is common practice during continuous co-firing of solid biomass fuels that are not thermally pretreated.

The second priority during handling and storage of biomass is mitigation and control of dust formation, which originates from mechanical degradation of the pellets. This is partly related to the control of the moisture content, because both insufficient and excessive moisture levels weaken biomass, which leads to increased dust formation. The most crucial factor in the mechanical degradation of the pellets is the mechanical forces that are imposed on the biomass during handling and storage. Therefore, dedicated handling equipment is designed, which allows smooth processing and impact limitation.

5.1.2 Unloading/discharge

Biomass is most of the time delivered by bulk transport through rail or barges, although trucking, even in containers, is used occasionally (Carbo, 2014). Rail car or ship cargo haul discharging occurs in various ways. The first simply involves the classical scooping out of the cargo, using the same infrastructure that is often in place for coal. This method of discharging does not affect the biomass pellets very adversely, provided the free drop distance in on-site handling equipment is limited. The latter can be resolved by using dedicated hoppers that limit the distance that pellets drop (Berry, 2015). The use of scoop unloading often causes local dust formation, which is difficult to mitigate by measures that are typical during coal handling, such as the use of water spraying. The use of local suction hoods could prevent further dust dispersion, as well as dust precipitation and aggregation on nearby surfaces.

For rail delivery, the discharging preferably takes place using gravitational transport. Side-discharge hopper wagons are most commonly used in sheltered rail bulk solid transport, and often in transport of pellets. More tailor-made wagons could be desired to obtain optimal throughput while minimizing mechanical impact. An example of such a dedicated infrastructure is used at the Drax power plant in the United Kingdom (UK) (Griffin, 2011). These top-loading, bottom-discharge, fully sheltered wagons are designed bearing in mind maximising the volume combined with smooth discharge.

Large-haul discharging, both for ocean- and river-going ships, is often tackled by using pneumatic vacuum conveying (Dorp, 2015). This solution provides very good dust control options, but, due to the relatively low-solids loading of the flow, it tends to be rather energy intensive. Moreover, this mode of discharge leads to substantial attrition of the pellets and therefore requires proper explosion-mitigating measures.

5.1.3 Conveying

Upon discharge, the biomass needs to be conveyed to the storage environment. Typically, short-distance conveying is performed by using fully enclosed chain conveyors, particularly if steep height gradients need to be covered. Alternatively, screw conveying can be used, although this tends to lead to additional mechanical degradation. The conveying with high-solids loading in a steel infrastructure with good heat-conducting properties limits the possibility of ignition of the conveyed material.

Long-distance conveying of biomass typically takes place in belt conveyors, but unlike coal these must be completely sheltered, either by placing hoods over the existing belts or by using a dedicated fully enclosed infrastructure. To limit the possibility of ignition due to bearing failures and associated heat accumulation, so-called air-ride conveyors that are suspended on air cushions are used. Dedicated hoppers are used when discharging the material from one conveyor onto the next or at the storage silo, to limit drop impact and dust formation during conveying. Because the conveyor infrastructure is one of the most critical parts of the plant with respect to fire and explosion risks, it is often equipped with temperature, gas composition, infrared monitors and camera monitoring, while the perimeter is off limits for personnel during operation (Marshall, 2015). In addition, strict housekeeping rules regarding prevention of dust accumulation particularly apply in this part of the handling infrastructure.

5.1.4 Silos/storage

Biomass storage mostly takes place in fully enclosed silos, using a wide range of designs. Conventional solutions include classical frame sheet metal-covered steel structures (Dorp, 2015), as well as self-supporting concrete domes (Griffin, 2011). The vast majority of the designs involve top loading, and, like other handling operations, special attention must be paid to limit the drop distance and associated impact during loading. This can be done, for instance, by using spiraling — instead of straight-discharge columns, which limit the drop velocity. Silo discharge is mostly done via an active discharge system in the floor, which can vary from screw and chain conveying to pneumatic systems. To facilitate discharge, vibrating floors are used, which are a complex yet effective measure to prevent bridging in large systems while limiting the required number of intake points.

Silos are particularly vulnerable to explosions and fires. The vast volumes of these systems require the use of advanced ventilation, inert-gas blanketing, temperature and moisture control, as well as gas monitoring. Large pressure-release valves are indispensable to prevent major structural damage to the silo in case of an explosion. In such an unfortunate event, the relatively light sheet-metal constructions have some advantages; however, the light construction may not be strong enough to contain fire and the water used to extinguish it (European Pellet Council, 2014). When using fuels with alternate chemical and physical characteristics, special attention should be attributed to long-term storage, proper housekeeping and any limitations that regulations may pose for storage and handling.

5.1.5 Hardware modifications to convert existing dedicated coal-handling infrastructure

When implementing biomass at an existing coal-fired power plant, part of the coal infrastructure may be adopted for biomass, sometimes at a marginal cost, which increases the flexibility of the facilities. In this section, several examples of such adaptations are given.

As discussed earlier, conveying infrastructure may be adopted by installing hoods, sheltering the material from exposure and by installing additional infrastructure for dust and temperature control.

Coal mills are generally ill-suited to pulverise biomass that has not been thermally pre-treated. Hence, most biomass-upgrading techniques focus on improving the grinding characteristics of the biomass. Once this objective is achieved, often only operational measures are necessary to replace coal with biomass. For the raw, woody biomass, coal pulverisers can be used to comminute the fuel but only up to a certain extent. It should be noted that this does not occur by actual grinding, but instead the pellets are simply disintegrated into their primary pre-ground particles. Certain modifications to the mills are needed to facilitate grinding and transport of these particles. The internal construction is made more gastight to accommodate higher gas velocities, and higher pressures are needed to carry larger biomass particles out. This is achieved by placing additional, fixed cylinder-shaped steel structures around the grinding assembly and in the classifier section. The latter part can be adjusted during operation, which maintains fuel flexibility. Threaded wheels and/or grinding dishes are used to improve traction and provide some cutting action, often in combination with scrapers that eliminate the risk of build-up of a tenacious layer on the grinding bed. Additional active-roller pressure control is installed to provide better control during the pulverization, whereas the classifier rotational speed is often reduced to prevent accumulation in the milling chamber.

Biomass typically requires higher velocities during pneumatic transport because of the larger particle sizes. This leads to increased impaction of particles onto the conveying tubes surface, particularly in the bends, and erodes the pipe walls. Reinforced-tube sections are placed at critical points in the system, such as bends and splitters, to counteract this erosion. These sections can be simply thick-walled or plate-reinforced steel or a ceramics-lined infrastructure. The latter comes at a much higher cost, but is also more durable. Quick response fire-extinguishing systems based on CO_2 injection are mounted at critical points to prevent the spreading of fire from the mill to the pneumatic conveying system. The online monitoring of the mass transport can be used to minimise the required amount of the transport air, which could reduce the energy consumption and optimise the firing conditions.

The co-firing of biomass through a dedicated or coal-adapted infrastructure may require certain additional operating measures to mitigate fire and explosion risks, as well as to achieve optimal firing conditions. This particularly applies to the mills/pulverisers; regardless of the technology, the temperatures of the pre-heated transport and/or drying air as well as those at the mill exit are very critical while processing biomass. Too high mill-exit temperatures will inevitably result in a fire, but an increased mill temperature may also result in plastic softening of biomass. This self-amplifying process may lead to stickiness and increase the accumulation of material in the mill, which leads to increased friction and heat generation. Allowing higher moisture content of the biomass does not lead to reduced mill-exit temperatures, but instead results in an increase of the transport air temperature to evaporate the additional moisture, which increases the risk of fire. The typical advisable mill temperature should be lower than 80 °C and the moisture content of biomass lower than 12%, to allow transport air pre-heating to temperatures below 110 °C (Marshall, 2015).

5.2 Biomass pre-treatment technologies

5.2.1 Torrefaction

Torrefaction is a mild thermochemical treatment used for the upgrading of biomass into a high-quality solid fuel. It is performed at temperature ranges between 250 and 320 °C and in the absence of oxygen. Upon torrefaction, the hygroscopic and tenacious nature of the biomass is largely destroyed, with the extent of destruction depending on operating temperature and residence time. Torrefied biomass tends to become more resistant with respect to biological degradation and more straightforward to pulverise. The destruction of the tenacious behaviour of the biomass is a very welcome improvement when considering size reduction. Loss of the tenacious nature of the biomass is mainly coupled to the breakdown of the hemicellulose matrix, which bonds the cellulose fibres in biomass. Depolymerization of cellulose decreases the length of the fibres. These properties make torrefied biomass an attractive feedstock to produce biomass fuel pellets. Hemicellulose is mainly decomposed during torrefaction, leaving cellulose and lignin virtually intact. This potentially enables the production of high-quality fuel pellets from raw materials other than feedstocks that are currently economical, such as sawdust, without the use of an additional binder (Verhoeff et al., 2011).

5.2.2 Steam explosion

Steam explosion is a process that has been available for a long time, and was developed as an alternative process for wood pulping. During steam explosion biomass is treated with hot steam with a temperature of 180–240 °C and pressures of 10–35 bars followed by an instant flash to ambient pressure (Stelte, 2013). This explosive decompression of the biomass results in the rupture of the fibrous structure, and in analogy with torrefaction, the combination of the temperature and residence time determines the extent of destruction. During the explosion, the lignin content is softened and distributed across the surface of remaining material, making it available as a binder during pelleting. Heat reclamation is essential to minimise the production costs of biomass fuel pellets.

5.3 Industrial-scale experience with pre-treated biomass

A significant number of industrial-scale trials with pre-treated biomass have taken place in coal-fired power stations; a few of these are described in the following including the most important findings.

Vattenfall conducted a logistics and co-firing test with 2400 tons of steam-exploded pellets at the Reuter West power station in Berlin in 2011 (Khodayari, 2012). This test indicated that dust formation during handling could be effectively counteracted by dust-suppression systems, and that co-firing rates of 20—50 wt% are feasible without making any modifications to the existing power plant. In 2012, 1200 tons of torrefied wood pellets were co-fired in the Buggenum integrated coal gasification combined-cycle power plant in the Netherlands (Padban, 2014). During this test, a co-firing rate of 70% on energy basis could be reached at approximately 90% of the nominal load, again without any major hardware modifications. Dust formation was again effectively counteracted by dust-suppression systems.

Late in 2013, 2300 tons of Topell torrefied forest residue pellets were co-fired in RWE/Essent's AMER-9 pulverised coal-fired power plant (Dorp, 2015). The test was conducted by a consortium of Topell, RWE/Essent, NUON/Vattenfall, GdF Suez/Electrabel and Energy research Centre of the Netherlands (ECN). The torrefied pellets were deposited on top of one of the coal conveyors leading to the mills. During the two-month trial a maximum 25 wt% co-milling was established, which corresponds to 5 wt% co-firing, without any major hardware modifications or significant issues.

In March 2014, 200 tons of Andritz/ECN torrefied spruce pellets were co-fired in DONG Energy's Studstrup-3 pulverised coal-fired power plant (Carbo, 2014). During the test, the torrefied pellets were fed to a dedicated roller mill that is normally used to pulverise either coal or white wood pellets. A co-firing share of 33 wt% was established during the 8 h trial without any major issues.

This shortlist of industrial experience with pre-treated pellets highlights the main advantage of biomass pre-treatment, being the ability to achieve increased co-firing rates without any major hardware modifications, and as such at lower investment and operating costs compared with white wood pellets. In the following paragraphs relevant lab-scale assessments of transport-handling aspects are presented for torrefied pellets that were produced in the ECN pilot torrefaction plant.

5.4 Biological degradation

The biological degradation behaviour of torrefied biomass pellets, white wood pellets, coal and a mixture of coal and torrefied biomass pellets was investigated during climate chamber experiments. Some of the torrefied wood pellet samples were exposed during a small-scale outdoor storage test for a period of 84 days. Prior to the biological degradation tests all the samples were dried overnight at 105 °C, and subsequently

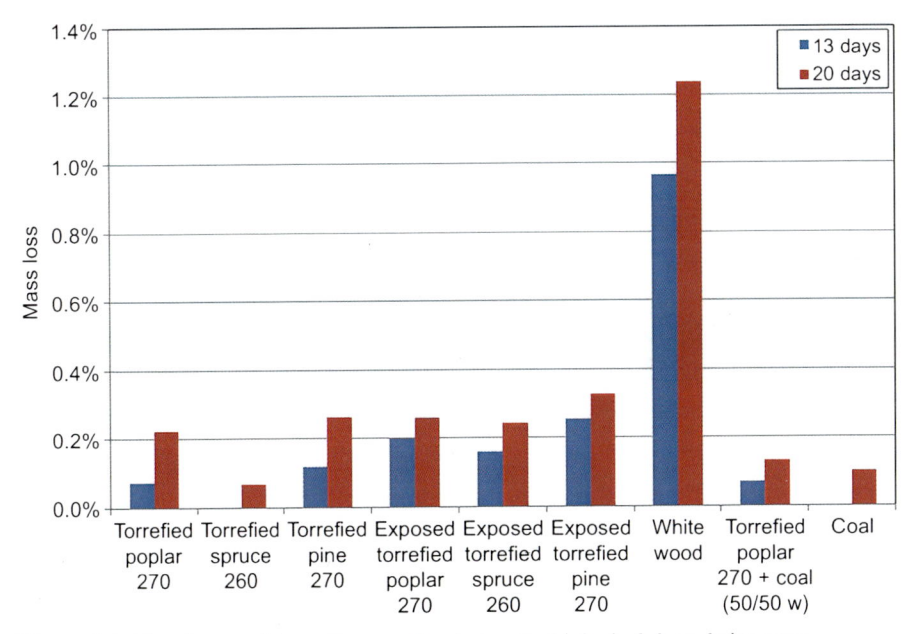

Figure 5.1 Mass losses of the pellet samples during the biological degradation tests.

these were stored for 20 days at a temperature of 22 °C and a relative humidity of 95%. The samples were removed after 13 and 20 days, dried at 105 °C overnight and weighed. The dry-matter loss was used as a measure for the biological degradation. The resulting mass losses are displayed in Figure 5.1, and indicate that white wood pellets are most prone to biological activity with a mass loss in excess of 1.0 wt%. All other tested pellet samples display mass losses below 0.3 wt%. This comparison indicates that torrefied biomass pellets are much more resistant to biological degradation than white wood pellets.

5.5 Pneumatic conveying

In a coal-fired power plant, pulverised coal is pneumatically transported from the mills to the burners using air. Typically, the air should have a velocity of approximately 20 m/s, whereas the mass loading of coal should be approximately 0.5 kg coal/kg air (Storm and Reilly, 1987). These requirements ensure that the pneumatic transport of coal takes place in dilute phase, and that the required thermal input can be reached given the available infrastructure.

During the design of a dilute-phase system it is necessary to determine the saltation velocity. This is the minimum velocity at which the material starts to settle in a horizontal pipe, and depends on the mass loading, tube diameter and particle diameter. For a dilute-phase system, the velocity should be higher than the saltation velocity; as a design rule the actual velocities should be 1.5–1.6 times the saltation velocity to ensure dilute-phase transport. At lower velocities, particles may settle and could cause blockage of the transport line. The saltation velocity can be calculated through the Rizk equation (Holdich, 2002):

$$U_{\text{salt}} = \left[\frac{4 \cdot M_p \cdot 10^\alpha \cdot g^{\beta/2} \cdot D^{(\beta/2)-2}}{\pi \cdot \rho_f}\right]^{1/(\beta+1)}$$

in which

$\alpha = 1440 d_p + 1{,}96$
$\beta = 1100 d_p + 2{,}5$
and M_p is the mass-flow rate (kg/s), D is the pipe diameter (m), ρ_f the transport-gas density (kg/m³) and d_p the particle diameter (m).

The phase diagram for pneumatic conveying was determined for the transportation of pulverised coal with a particle diameter of 100 μm through a line with a diameter of 500 mm, and is displayed in Figure 5.2. The lower limit is based on the saltation velocity, whereas the upper limit corresponds to velocities 1.5 times higher than the saltation velocity.

This phase diagram shows the velocities that should be used to ensure dilute phase according to the respective mass loading. For example, to obtain a mass loading of 0.5, the air velocity should be at least 19 m/s to mitigate settling of the material and consequently the blockage of the line. This corresponds to the requirements of the boiler and illustrates how the design of the system is used to optimise pneumatic

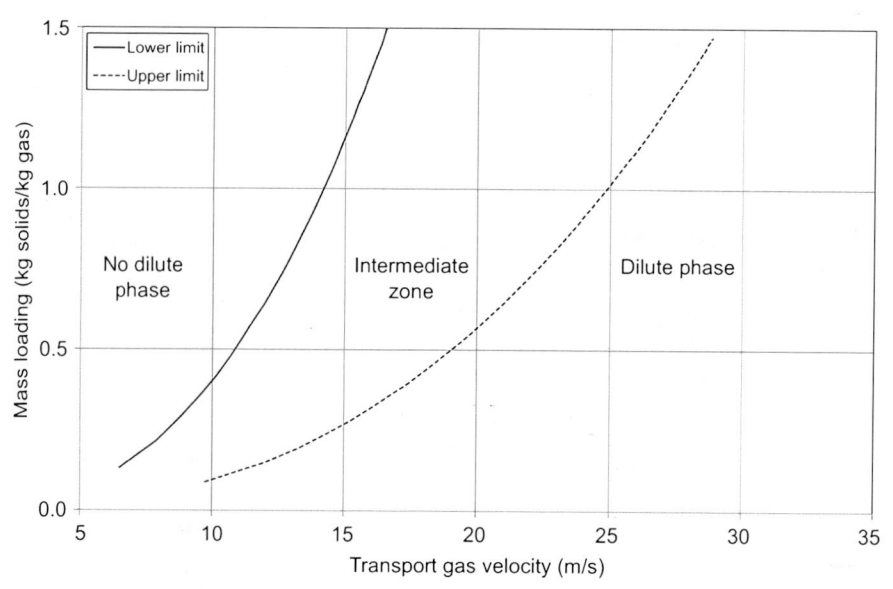

Figure 5.2 Pneumatic conveying phase diagram for pulverised coal with particle diameter of 100 μm transported in a 500 mm pipeline.

Fuel pre-processing, pre-treatment and storage for co-firing of biomass and coal   129

transport. Higher gas velocities or lower mass loadings could be used without compromising the pneumatic transport, for instance for alternative fuels. However, this will likely lead to limitations in existing infrastructures, such as a lower thermal input, increased amount of unburnt fuel and/or a lower efficiency.

The same theoretical approach was used to determine the phase diagram for pneumatic conveying of untreated biomass with a particle diameter of 1 mm, to be transported in the same line with a diameter of 500 mm. The results are presented in Figure 5.3 and can be directly compared with those in Figure 5.2 for pulverised coal.

The pulverised untreated biomass requires a much higher gas velocity to achieve the same mass loading as coal; a gas velocity of 28 m/s should be used to obtain a mass loading of 0.5 and ensure proper dilute-phase transport. Moreover, the heating value of biomass is generally lower than that of coal; therefore, larger volumes of biomass have to be used to reach the same thermal input. As such, air velocities higher than 30 m/s have to be used for pulverised untreated biomass to meet the same thermal input. These results also indicate that if pulverised untreated biomass is used in existing coal transport lines at similar conditions as coal, particle settling and blockage of the line are highly probable.

The theoretical analyses presented in Figures 5.2 and 5.3 show that the particle size plays an important role in the design of a dilute-phase system. Larger particle sizes require higher air velocities to avoid blockage of the transport line from the mills to the burner. However, the shape of the pulverised materials may have an influence as well during pneumatic transport. This influence has been assessed in a lab-scale setup with a smaller line diameter of 28 mm. Because the decrease of the diameter of the

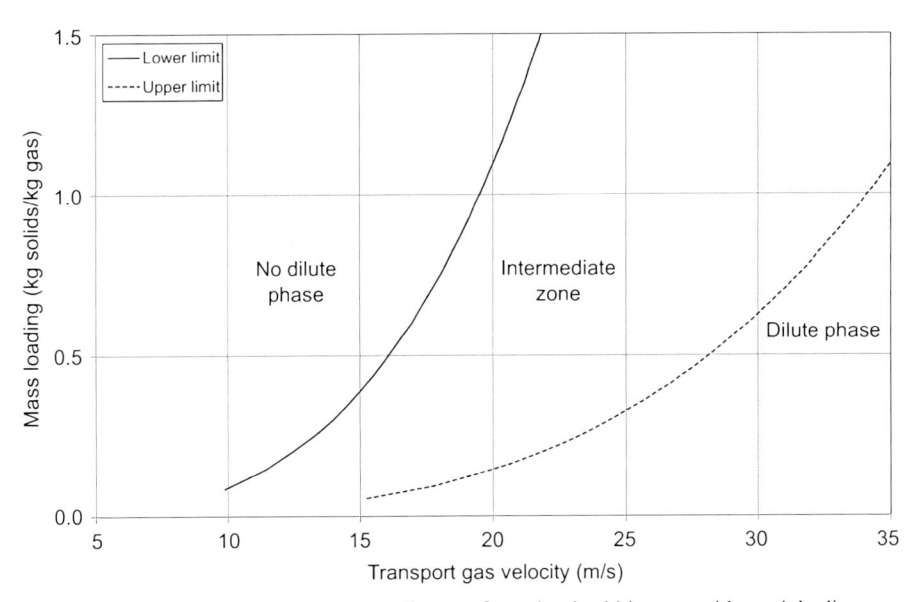

Figure 5.3 Pneumatic conveying phase diagram for pulverised biomass with particle diameter of 1 mm transported in a 500 mm pipeline.

Table 5.1 **Tested samples in the lab-scale pneumatic transport setup**

Material	Size (μm)	ρ_{bulk} (kg/m^3)
Pulverised coal (Coal)	<100	640
South African coal (Coal SA)	<350	750
Torrefied *Eucalyptus* pellets (270 °C)	<500	690
White wood pellets	<500	480

transport line leads to a reduction of the saltation velocity, lower transport-gas velocities were used to obtain dilute-phase transport. Four different types of materials were tested: a pulverised coal sample obtained from a coal-fired power plant with a particle size smaller than 100 μm; a coarser ground South African coal sample with a particle size smaller than 350 μm; a torrefied (270 °C) and pelletised *Eucalyptus* sample with a particle size smaller than 500 μm; and pulverised white wood pellets with a particle size smaller than 500 μm. Table 5.1 summarises the tested materials including the bulk powder densities.

The results of the pneumatic transport experiments are shown in Figure 5.4. The lines for the saltation velocity and design velocity in this figure were calculated for pulverised coal with a diameter below 100 μm which is transported in a tube of

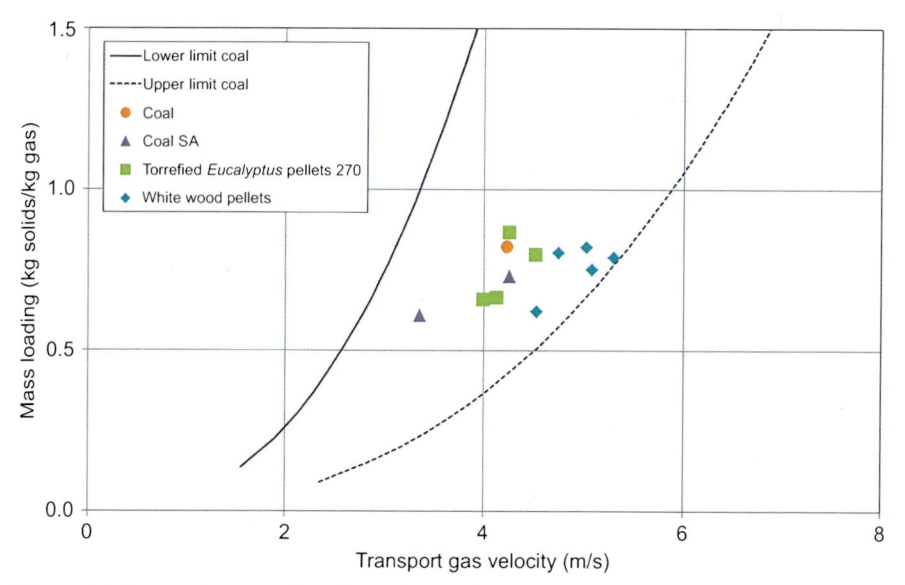

Figure 5.4 Pneumatic conveying phase diagram for pulverised coal 100 μm transported in a 28 mm pipeline including maximum transport-gas velocities at which saltation still occurred for the four materials presented in Table 5.1.

28 mm diameter. A number of tests at different conditions were performed with the four materials to assess when saltation occurred, the maximum transport gas velocities at which saltation still occurred are presented by the markers in Figure 5.4.

The results imply that the morphology of the pulverised samples also plays a role during pneumatic transport. The torrefied *Eucalyptus* pellets and white wood pellets were pulverised in the same mill, to a similar maximum particle size and with a similar particle-size distribution, however both pulverised samples display different behaviour in terms of saltation. The pulverised white wood pellets already start to settle at much higher transport gas velocities compared with pulverised torrefied *Eucalyptus* pellets. The results demonstrate that in case pulverised white wood pellets are used at the same conditions as pulverised coal (the upper limit on the right side of the phase diagram), the likelihood that settling and blockage of the transport line may occur is large. The pulverised torrefied *Eucalyptus* pellets could be transported at the same conditions as coal without posing any risks of settling or blockage. The theoretical limits presented in Figure 5.4, which were calculated for a transport line with 28 mm of diameter, are representative for the dilute-phase limits presented in Figure 5.2 for a transport line with a diameter of 500 mm.

Optical microscopic images of the tested samples were used to confirm the influence of the particle morphology. Figure 5.5 illustrates that the particle morphology

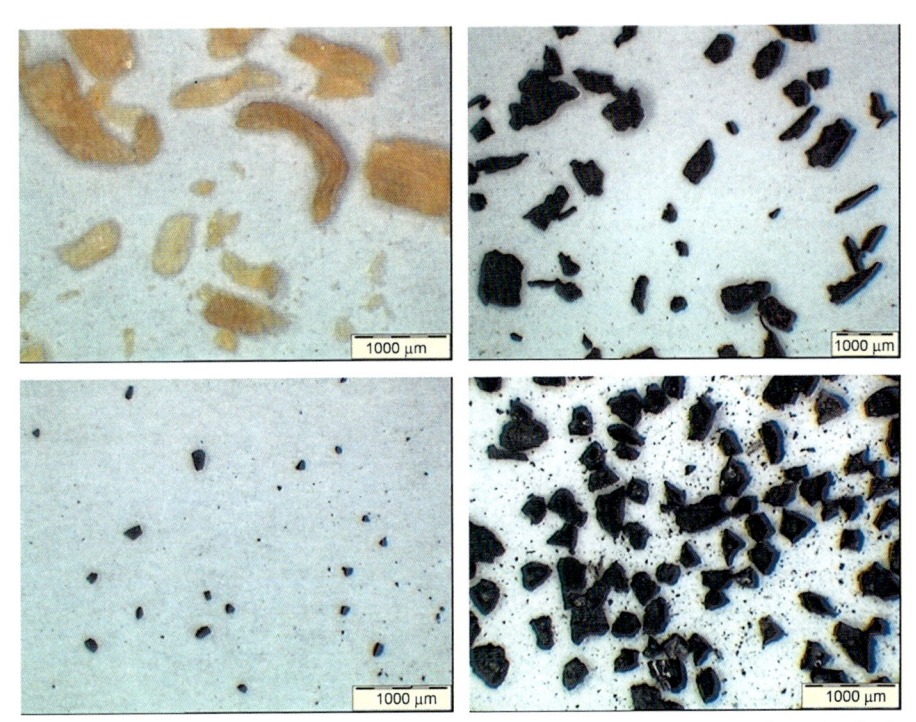

Figure 5.5 Optical microscope images of pulverised white wood pellets (top left), pulverised torrefied *Eucalyptus* pellets (top right), pulverised coal from an industrial mill (bottom left), and pulverised South African coal (bottom right).

of pulverised white wood pellets is very heterogeneous and contains the needle-shaped particles that stem from the original material of which the pellets were composed. The pulverised torrefied *Eucalyptus* particles are quite homogenous in shape and more "spherical" than the pulverised white wood pellets. These images seem to prove indeed that shape plays a role in the pneumatic transport besides the particle-size distribution and higher bulk powder density for pulverised torrefied pellets as displayed in Table 5.1. It also shows that the pulverised sample obtained from torrefied biomass pellets resembles coal much closer in terms of morphology than the pulverised white wood pellet sample. Both images at the bottom of the figure show that the coal particles are angular and homogenous, needle-shaped particles with lengths that largely exceed the diameter are practically absent.

5.6 Mechanical durability and storage

Industrial experience demonstrated that the handling of solid fuel biomass materials can result in the release of small particles as airborne dust, which have proved to be an issue of major concern during processing, transport, handling as well as end use (Hedlund and Astad, 2013; Hedlund et al., 2015; Huéscar Medina, et al., 2013). The hazard level presented by a particular fuel dust depends on the total amount of dust that is released during fuel handling, the sensitivity of the dust to cause an explosion and the severity that the explosion can reach (Eckhoff, 2003).

The production of pellets is one of the most common ways to process a solid biomass, because both the transport and handling becomes more efficient. When these pellets display insufficient mechanical strength, partial degradation during transport and handling can lead to dust formation. The extent of the mechanical strength is quantified by a standardised durability test (EN15210-1, 2009). This standard test evaluates the tumbling of 500 g pellets from which the native dust (particles with diameter below 3.15 mm) is removed prior to the experiment. Tumbling occurs during 10 min in a standard revolving machine at 50 rotations per minute. After tumbling, the pellets are collected and the produced dust fraction below 3.15 mm is separated by sieving and quantified. The fraction of the remaining pellets is known as the pellet durability index (PDI) and is usually presented as a weight fraction of the initial mass of the pellets tested. Figure 5.6 displays a number of values for the PDI of several torrefied pellets produced by ECN, and commercial white wood pellet as a reference. The observed durability for torrefied biomass pellets is typically in excess of 96%, which does not tend to lead to dust formation problems upon transport and handling. The torrefied pine pellet sample with relatively low durability was obtained prior to optimizing the pelleting conditions, and was included in this study to illustrate the effect of low mechanical durability on safety aspects. In general, the produced dust fraction decreases upon increasing pellet durability.

Small-scale outdoor storage tests with torrefied samples of approximately 2 kg revealed that the extent of weather resistance of the pellets is directly proportional to the mechanical durability, i.e. the degradation or weathering of the pellets proceeds

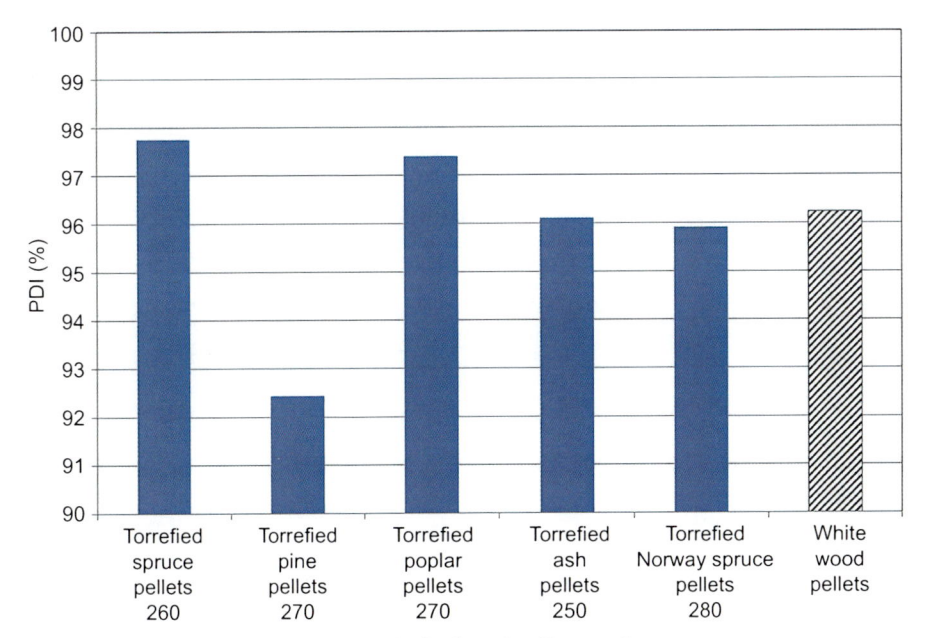

Fuel pre-processing, pre-treatment and storage for co-firing of biomass and coal

Figure 5.6 Pellet durability index (PDI) of selected pellet samples.

relatively slowly if the initial durability is high. It should be noted that tests with these relatively small sample sizes merely represent the outer surface of the stored piles. The first results with larger quantities indicate that the initial 10 cm below the surface are affected in time, although the majority of the pellets inside the pile demonstrate no mechanical degradation. Rain exposure during unloading as well as on-site uncovered storage for periods of two to three weeks seems perfectly feasible, if the mechanical durability of the pellets is sufficiently high, at approximately 96% or higher.

5.7 Explosivity

The sensitivity of a dust to explode can be assessed by determining the minimum ignition energy (MIE) following a standard method like EN13821 (2002). The dust sample is dispersed in air under controlled conditions in a Hartmann tube apparatus, and the dust cloud is subjected to a discontinuous spark discharge from a capacitor at energy levels of 1–3–10–30–100–300–1000 mJ. The MIE is a function of the dust–air mixture and of the dynamics or turbulence. The MIE should be measured at the optimum dust concentration and at the lowest turbulence level experimentally possible by extending the ignition delay time. The MIE lies between the highest energy level at which ignition fails to occur in 10 successive attempts, and the lowest energy level at which ignition occurs within up to 10 successive attempts.

5.7.1 Native dust and dust formation upon handling

Experiments on MIE determination using the pellet dust fraction below 500 μm demonstrated that the native dusts are hardly ignitable because MIE values were obtained in excess of 1000 mJ. However, the tumbling dusts are more prone to ignite. Figure 5.7 displays that there is a significant difference between the sensitivity of the tumbling dust materials with respect to explosivity; low pellet durability appears to lead to increased dust formation and easier ignition of the obtained dust.

The dust obtained from tumbling white wood pellets and the torrefied spruce pellets appears less problematic, because relatively high-energy levels are required to cause an ignition. The particle-size distribution of the tumbling dusts presented in Figure 5.7 indicated that for the torrefied spruce pellets only 10 vol% of the particles had a diameter below 63 μm, whereas torrefied poplar and pine pellets presented 20 and 30 vol%, respectively. Therefore, dust formation with increased levels of particle-size diameters below 63 μm clearly leads to a reduction in the MIE, and increased explosivity risks.

5.7.2 Explosivity of raw biomass chips versus torrefied biomass pellets

The torrefied biomass pellets and the corresponding raw biomass chips were also pulverised using a cutter mill, and the obtained dust samples were used to determine the MIE. These tests demonstrated that pulverised torrefied spruce pellets were the most sensitive to ignite for the dust fraction below 63 μm, as displayed in Figure 5.8.

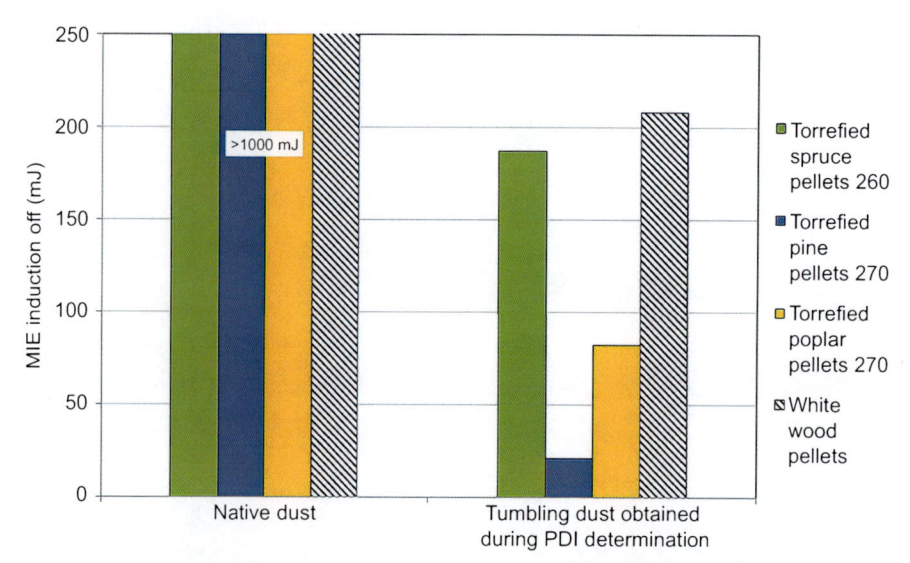

Figure 5.7 Minimum ignition energy (MIE) of native dust and tumbling dust of pellet samples (fraction below 500 μm and dried at 75 °C).

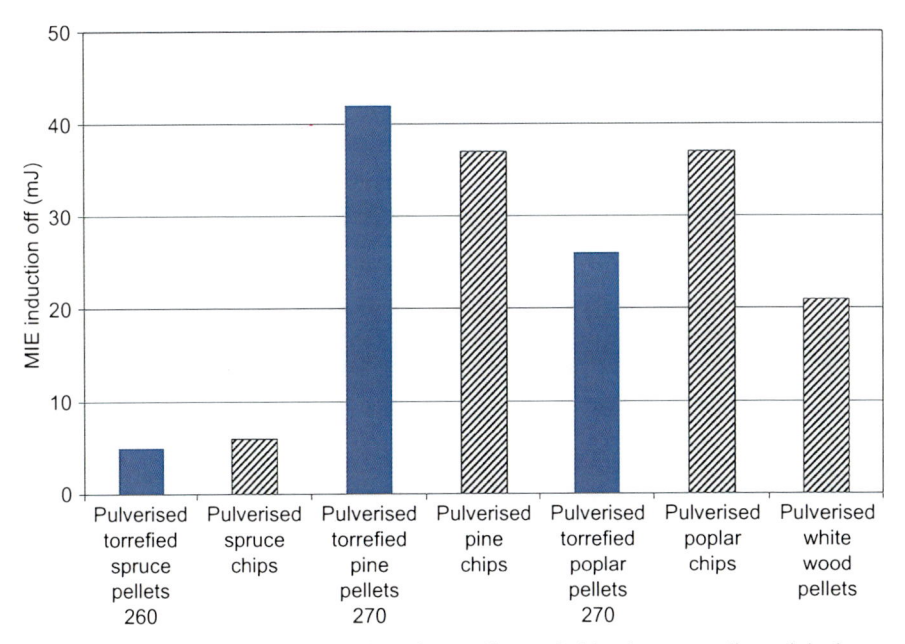

Figure 5.8 MIE of dust samples obtained from pellets and chips (corresponding original material) through a cutter mill (fraction below 63 μm and dried at 75 °C).

This figure also shows that pulverised torrefied biomass pellets are not more sensitive to explode in comparison with the pulverised biomass chips that were used as original raw material during torrefaction, but that this sensitivity appears directly related to the material that is used as feedstock. Additional experiments that were conducted with the dust fractions between 63 and 125 μm presented MIE values in excess of 1000 mJ, clearly indicating that, indeed, the fraction below 63 μm dictates the sensitivity of these materials with respect to explosivity. It should be noted that fibrous materials like raw biomass tend to produce needle-shaped particles that are more difficult to disperse completely and uniformly in a dust cloud, which could pose issues during the experimental determination of explosivity data (Huéscar Medina et al., 2013). Although it should be noted that the latter particularly holds for the larger particle-size ranges.

Another observation was that the method through which the dust sample is obtained appears to influence the obtained MIE. Although the MIE determination for the tumbling dust of the torrefied pine pellets led to a value of 25 mJ, the dust obtained by cutter milling the same material showed an MIE value of 42 mJ. Besides any potential uncertainties in the MIE determination, this could indicate that the method to produce dust samples can have a significant impact on the result, for instance because of obtaining alternate particle morphologies. A relatively high mechanical durability is required to reduce explosivity during transport and handling, whereas the explosivity of the torrefaction feedstock appears to be more important to reduce explosivity during milling and pneumatic conveying.

5.7.3 Moisture content

In case different materials are compared great care should be given to ensure that the results were obtained under exactly the same conditions. Besides the particle size that was discussed earlier, also the moisture content has a significant impact on the MIE. At moisture contents between 6 and 10%, which is a fairly normal range for the equilibrium moisture content attained by torrefied biomass pellets, the typical values for MIE were found to be between the energy levels of 100 and 300 mJ, as displayed in Figure 5.9. For the pulverised white wood pellets it turned out to be impossible to obtain reliable values for the MIE at moisture content levels in excess of 6%, due to increased stickiness of the particles on both the wall of the Hartmann tube and the surfaces of the electrodes, as such preventing any spark to occur.

5.7.4 Minimum explosible concentration (MEC)

The earlier observation that pulverised torrefied biomass pellets are equally ignitable as the raw chips used as torrefaction feedstock was further substantiated by determining the minimum explosible concentrations (MEC). It is often assumed that a material presents a greater hazard at a lower MEC, because the required concentration of powder that is required for the mixture to ignite is lower. During this study, it was found that for dust samples with particle diameters below 63 µm the dust could be ignited prior to reaching the top of the Hartmann tube, and that depending on the material between 50 and 100 m were needed for complete dispersion. The consequence of these findings is that the actual concentration cannot simply be calculated on the basis of sample weight and the Hartmann tube volume. Instead, it has to be

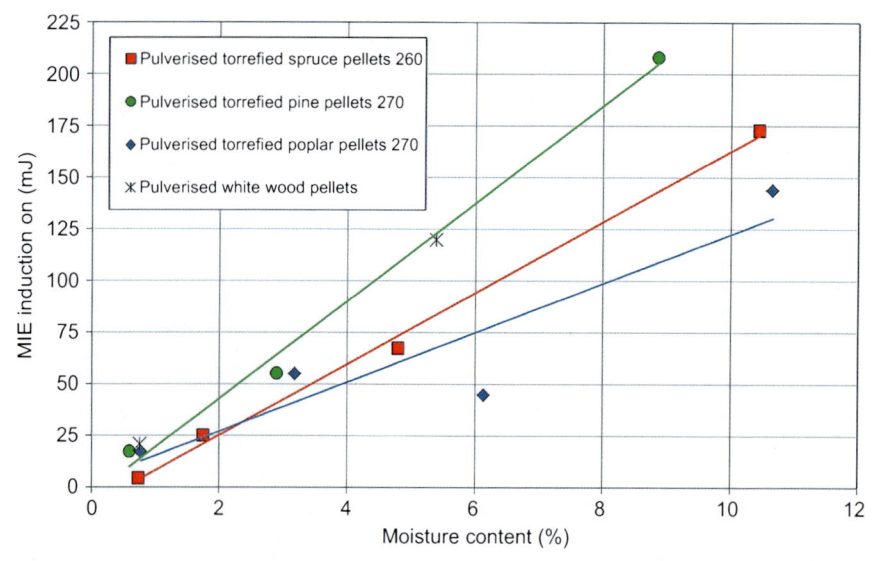

Figure 5.9 Influence of the moisture content on the MIE of dust samples obtained from pellets through a disc impaction mill (fraction below 63 µm and dried at 75 °C).

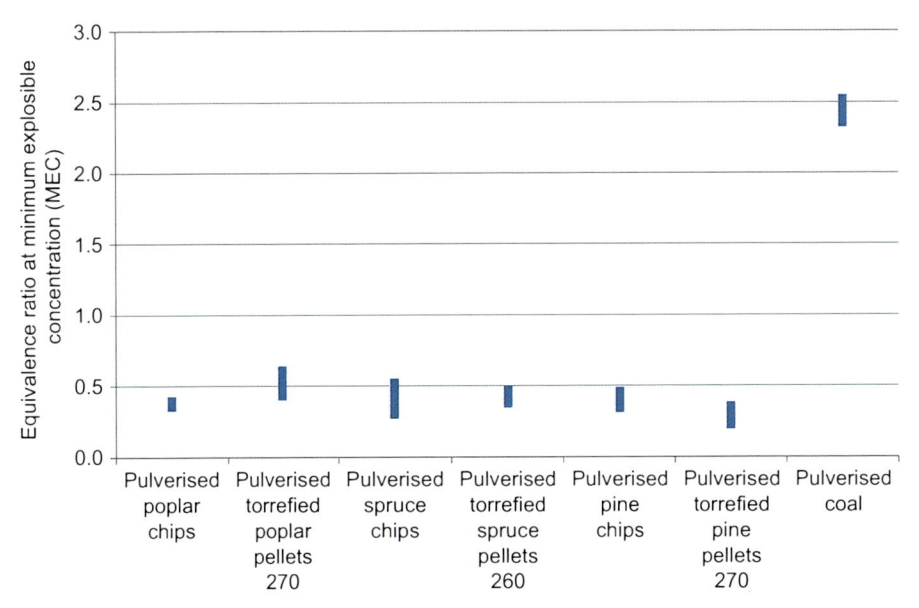

Fuel pre-processing, pre-treatment and storage for co-firing of biomass and coal

Figure 5.10 Minimum explosible concentration (MEC) of dust samples obtained from pellets and chips (corresponding original material), and coal, obtained through a disc impaction mill (fraction below 63 μm and dried at 75 °C).

corrected based on the actual volume occupied by the particles at the moment that the ignition occurs, which on most occasions is significantly smaller than the Hartmann tube volume and thus leads to higher actual concentrations. The actual concentration is therefore determined using a high-speed camera. Figure 5.10 illustrates that there is no significant difference between the MEC, expressed here as the equivalence ratio (the fuel–air ratio between the tested and the stoichiometric conditions) for the pulverised raw biomass and torrefied biomass pellets. The chart also highlights the difference between the raw and torrefied biomass materials compared to coal.

5.7.5 Flame-front velocity

A dedicated method has been developed to assess the severity of a dust explosion through determination of the flame-front velocity. More-reactive materials combust faster and develop higher flame-front velocities. These can be correlated to the maximum rate of rise of the explosion pressure, which is an important parameter in explosion hazard evaluations. For this purpose, a continuous spark system with a 15–20 W power source is used in the Hartmann tube apparatus, and coupled with a high-speed camera that allows the monitoring of the explosion flame development. The average velocity between the ignition point and the end of the Hartmann tube is calculated using the video footage and plotted as a function of the dust concentration, whereas duplicates of each concentration are performed. A maximum value of the flame velocity is thus obtained for the optimum concentration for which the dust sample is more reactive, as shown in Figure 5.11. In this case, pulverised raw pinewood was compared with the pulverised torrefied pine pellets. The pulverised raw pine produced a maximum flame velocity that was significantly higher than the pulverised torrefied pine pellets. The maximum velocity was obtained at an equivalent ratio between 1 and 2. In addition, materials like poplar and spruce wood were tested and it was concluded that the torrefied materials explode with maximum flame velocities that are at best equal but mostly lower than their raw peers.

The flame-front velocity of pulverised torrefied forest residue pellets and pulverised coal was also investigated, together with a mixture of both with 25 wt% pulverised torrefied forest residue pellets. The mixture only slightly increased the flame-front velocity when compared to the pulverised coal, as displayed in Figure 5.12. The maximum flame velocity of the mixture decreased almost twofold when compared to the pulverised torrefied forest residue pellets. Furthermore, the maximum velocities for pulverised coal appear to be reached at higher equivalence ratios between 2 and 3.

One other important finding is that the higher explosible concentration (or the higher flammable limit) of these types of materials could be far beyond an equivalence ratio of 10, although higher values could not be assessed due to safety reasons and the limited volume of the sample holder.

A parametric study was conducted to determine the influence of the moisture content, the particle diameter and sample temperature on the MEC and flame-front velocities, comparing both pulverised white wood pellets and pulverised torrefied wood pellets.

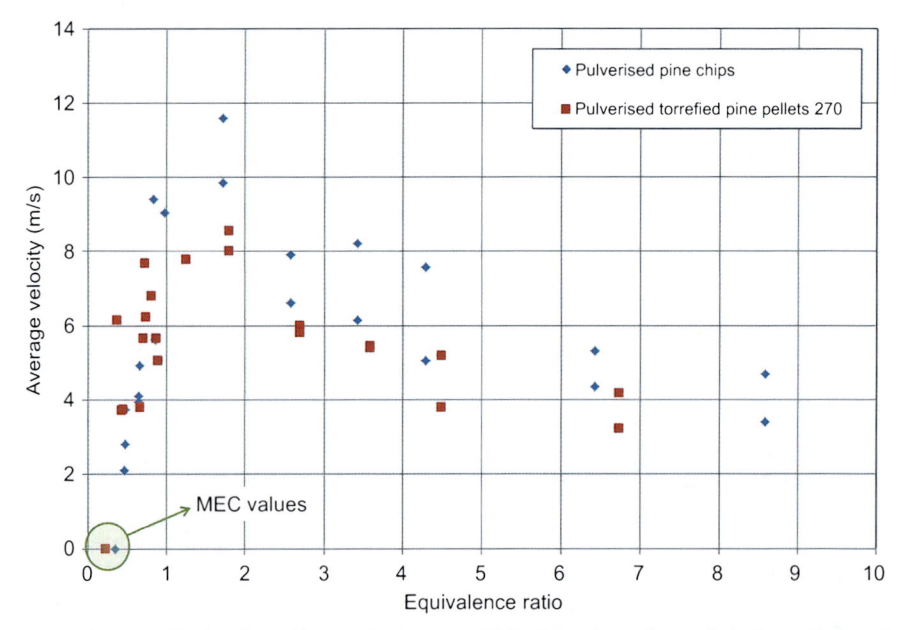

Figure 5.11 Explosion flame-front velocity and MEC of the dust of torrefied pine pellets and pine chips (corresponding original material) obtained through a disc impaction mill (fraction below 63 µm and dried at 75 °C).

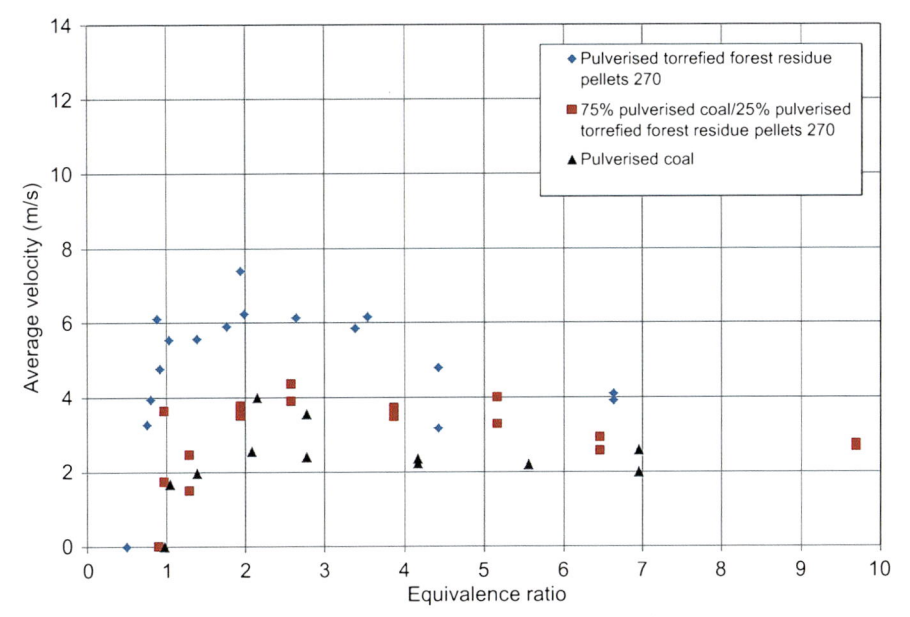

Figure 5.12 Explosion flame-front velocity and MEC of pulverised torrefied forest residue pellets, and pulverised coal sample obtained through a disc impaction mill (fraction below 63 μm and dried at 75 °C).

An increase of the moisture content for pulverised white wood pellets from bone-dry to 10% decreased the maximum velocity by 35%, shifted the maximum to slightly higher concentrations and increased the MEC twofold. For pulverised torrefied spruce pellets, an increase of the moisture content to 10% decreased the maximum velocity 45% whereas the MEC increased 25%.

The effect of the particle diameter was assessed using three different particle-size ranges: below 63 μm; between 63 and 125 μm, and between 125 and 250 μm. An example of the results is presented in Figure 5.13. For pulverised torrefied ash pellets an increase of the particle diameter range by a factor of 2 resulted in a decrease of the maximum flame velocity of 70% and an increase of the MEC of roughly 60%. Furthermore, the pulverised torrefied ash pellet sample was not ignitable for particle diameters between 125 and 250 μm. The same effects were observed for pulverised white wood pellets in Figure 5.14; when incrementing the particle diameter range by a factor of 2, the maximum flame velocity of the pulverised white wood pellets decreased 30% and the MEC increased 40%. When the maximum particle diameter was increased by a factor of 4 the decrease of the maximum velocity was about 80% and the MEC increased by a factor of 15. The increase of the maximum particle diameter clearly leads to the fact that the maximum velocity is lower and reached at higher concentrations.

The increment in the temperature from ambient to 80 °C affected the maximum flame velocity to a similar extent. Pulverised white wood pellets and pulverised torrefied wood pellets showed increased maximum flame velocities of 33 and 30%, respectively. Furthermore, a 40% decrease of the MEC value was observed.

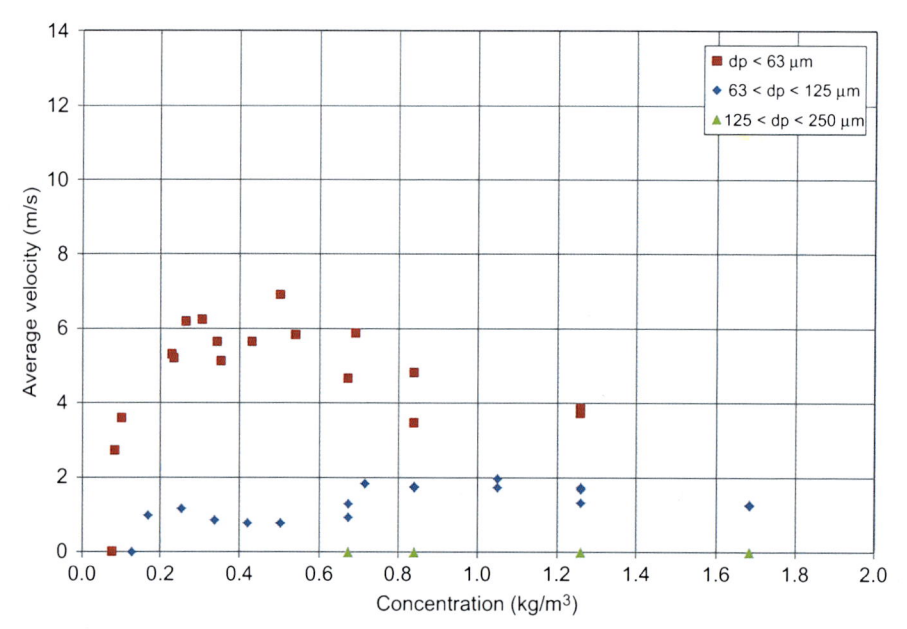

Figure 5.13 Explosion flame-front velocity and MEC of pulverised torrefied ash wood pellets obtained through a disc impaction mill (dried at 75 °C).

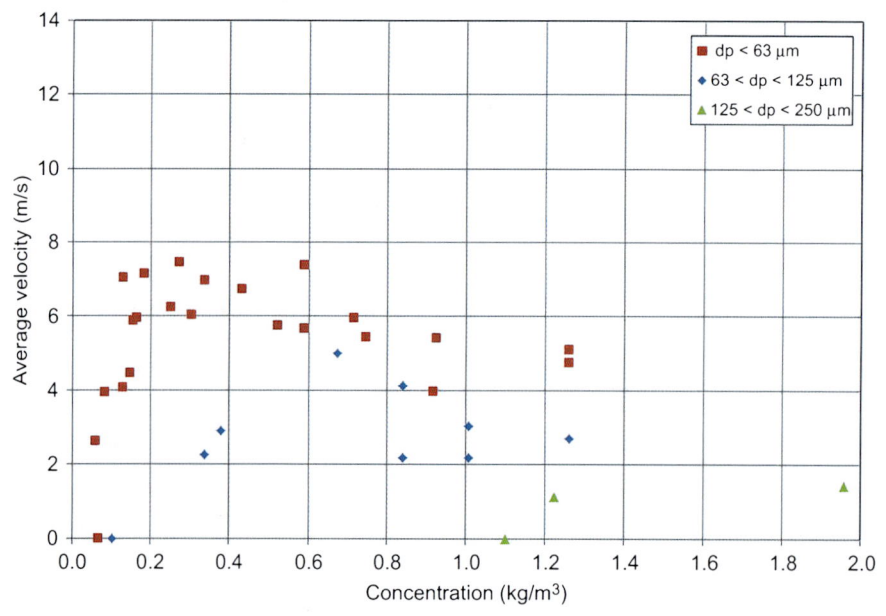

Figure 5.14 Explosion flame-front velocity and MEC of pulverised white wood pellets obtained through a disc impaction mill (dried at 75 °C).

5.8 Conclusions and future trends

Upon market implementation of thermally pre-treated biomass fuels, the need to convert or replace the handling and storage infrastructure to facilitate biomass co-firing will likely reduce. The recent examples in Canada, the UK and elsewhere in the world signify that the investment that is necessary to convert a pulverised coal-fired boiler to 100% white wood pellets can be large. By shifting towards pre-treated biomass, many safety aspects that are related to the perishable nature of biomass will become more straightforward to manage. An examples is the mechanical and biological degradation of white wood pellets, which can result in increased dust formation and combustible off-gases respectively. These require an elaborate and expensive monitoring and safety infrastructure, besides fully covered handling and storage. The nature of thermally pre-treated biomass fuels displays more resemblance with coal, which makes these fuels easier to implement. Industrial trials have demonstrated that handling and conversion of increased thermally pre-treated biomass co-firing shares is possible without any major hardware modifications. Technologies like torrefaction and steam explosion are on the verge of market breakthrough, and, once large-capacity plants are established, the production costs including overseas transport are expected to be similar to white wood pellets (Arpiainen and Wilen, 2014).

Beyond the nature of the fuel, the global expansion of biomass co-firing and re-powering efforts will also broaden experience. This will give rise to locally inspired innovative solutions and optimization approaches, as has been observed in the coal-fired power plants during each technological generation. This will also facilitate biomass co-firing and re-powering options to the fleet of new generation plants. New pulverised coal-fired plants can be incorporated with part of the critical biomass-related infrastructure, such as flex mills, reinforced pulverised-fuel feeding lines, better induced-draft fan controls, to be able to handle a flexible-fuel portfolio. The latter might reduce the large investment costs that are typically required to facilitate co-firing or re-powering of existing coal-fired power plants.

Nomenclature

ECN	Energy research Centre of the Netherlands
MIE	Minimum ignition energy
MEC	Minimum explosible concentration
PDI	Pellet durability index
vol%	Volume percentage
wt%	Weight percentage

Acknowledgement

Part of the work presented in this chapter was conducted as part of the SECTOR project, which has received funding from the European Union's Seventh Programme for research, technological development and demonstration under grant agreement no. 282826. Furthermore, the TKI Bio-based Economy (TKI BBE) and the Netherlands Enterprise Agency (RVO) are gratefully acknowledged for their financial support of the Pre-treatment/INVENT project under agreement no. TKIBE01011.

References

Arpiainen, V., Wilen, C., 2014. Deliverable D3.2: Report on Optimisation Opportunities by Integrating Torrefaction into Existing Industries. EU Sector Project.
Berry, R., 2015. Dealing with Dust. World Biomass Power Markets, Amsterdam.
Carbo, M., 2014. Characterisation of torrefied pellets. In: Workshop "Torrefaction of Biomass." University of Leeds, Leeds.
Dorp, E.v, 2015. Determining and Implementing the Critical Success Factors of the Process to Achieve the Safe, Efficient and Sustainable Co-firing of Biomass. World Biomass Power Markets, Amsterdam.
Eckhoff, R., 2003. Dust Explosions in the Process Industries, third ed. Gulf Professional Publishing, Burlington, MA.
EN13821, 2002. Potentially Explosive Atmospheres—Explosion Prevention and Protection—Determination of Minimum Ignition Energy of Dust/Air Mixtures.
EN15210-1, 2009. Solid Biofuels—Determination of Mechanical Durability of Pellets and Briquettes—Part 1: Pellets.
European Pellet Council, 2014. Results and Main Lines of Action, 2nd International Workshop on Pellet Safety. Fügen.
Griffin, M., 2011. The development of the 500 MW co-firing facility at Drax power station. In: 1st IEA Clean Coal Centre Workshop on Cofiring Biomass with Coal. Drax, United Kingdom.
Hedlund, F., Astad, J., 2013. Safety: a neglected issue when introducing solid biomass fuel in thermal power plants? Some evidence of an emerging risk. In: Huang, C., Kahraman, C. (Eds.), Intelligent Systems and Decision Making for Risk Analysis and Crisis Response. CRC Press Taylor and Francis Group, London.
Hedlund, F., Astad, J., Nichols, J., 2015. Inherent hazards, poor reporting and limited learning in the solid biomass energy sector: a case study of a wheel loader igniting wood dust, leading to fatal explosion at wood pellet manufacturer. Biomass Bioenergy 66, 450—459.
Holdich, R., 2002. Fundamentals of Particle Technology. Midland Information Technology and Publishing, Loughborough, United Kingdom.
Huéscar Medina, C., Phylaktou, H., Sattar, H., Andrews, G., Gibbs, B., 2013. The development of an experimental method for the determination of the minimum explosible concentration of biomass powders. Biomass Bioenergy 53, 95—104.
Khodayari, R., 2012. Vattenfall strategy and experiences on co-firing of biomass and coal. In: 2nd International Workshop on Cofiring Biomass with Coal. IEA Clean Coal Centre, Copenhagen, Denmark.
Marshall, L., 2015. OPG Thunder Bay Unit 3, the World's First Coal-biomass Conversion with Advanced Wood Pellets. World Biomass Power Markets, Amsterdam.

Padban, N., 2014. First experiences from large scale co-gasification tests with refined biomass fuels. In: Central European Biomass Conference. Graz, Austria.
Solids Online, 2015. Retrieved March 20, 2015, from: http://www.solidsonline.com.
SolidsWiki, 2015. Retrieved March 20, 2015, from: http://www.solidswiki.com.
Stelte, W., 2013. Steam Explosion for Biomass Pre-treatment. Danish Technological Institute, Taastrup, Denmark.
Storm, R., Reilly, T., 1987. Coal fired boiler performance improvement through combustion optimization. In: Joint ASME/IEEE Power Generation Conference. Miami, FL. ASME Paper 87-JPGC-PWR-6.
Verhoeff, F., Adell i Arnuelos, A., Boersma, A., Pels, J., Lensselink, J., Kiel, J., et al., 2011. TorTech: Torrefaction Technology for the Production of Solid Bioenergy Carriers from Biomass and Waste. ECN, Petten, the Netherlands.

Production of syngas, synfuel, bio-oils, and biogas from coal, biomass, and opportunity fuels

6

James G. Speight
CD&W Inc., Laramie, WY, USA

6.1 Introduction

Synthesis gas (also called *syngas*), a mixture produced by the gasification of carbonaceous material (e.g., coal, petroleum residua, biomass, and opportunity fuels such as industrial and municipal waste) composed primarily of carbon monoxide and hydrogen but also contain water, carbon dioxide, nitrogen, and methane, has been produced on a commercial scale. The process for producing synthesis gas comprises two major components: (1) synthesis gas generation and (2) gas processing. Within each of these systems, several options are available. For example, synthesis gas can be generated to yield a range of compositions ranging from high-purity hydrogen to high-purity carbon monoxide. Two major routes can be utilized for high-purity gas production: (i) pressure swing adsorption and (ii) utilization of a cold box in which separation is achieved by distillation at low temperatures. In fact, both processes can also be used in combination. However, to address these concerns, research and development is ongoing, and success can be measured by the demonstration and commercialization of technologies, such as a permeable membrane for the generation of high-purity hydrogen, which in itself can be used to adjust the H_2/CO ratio of the synthesis gas produced.

It is the purpose of this chapter to provide a general description of synthesis gas production from various carbonaceous feedstocks by means of available technologies and the potential of the process. In addition, the conversion of synthesis gas to hydrocarbon fuel is also described to place synthesis gas production in the proper perspective.

6.2 Gasification

Gasification as defined for the purposes of this chapter is the conversion of a carbonaceous feedstock (or a mixture of carbonaceous feedstocks) to produce a mixture of produced gases (of which synthesis gas is one) and process heat. Thus, the gasification of any carbonaceous feedstock or a derivative (i.e., char produced from the feedstock) is essentially conversion of the feedstock (by any one of a variety of processes) to produce combustible gases (Fryer and Speight, 1976; Radović et al., 1983; Radović and

Walker, 1984; Garcia and Radović, 1986; Calemma and Radović, 1991; Kristiansen, 1996; Speight, 2008, 2013a,b). With the rapid increase in the use of coal from the fifteenth century onward, it is not surprising that the concept of using coal, especially the use of water and hot coal, to produce a flammable gas became commonplace.

Gasification of coal is the oldest forms of synthesis gas production and offers one of the most versatile methods (with less environmental impact than combustion) to produce electricity, hydrogen, and other valuable energy products and is considered to be the prime example for the production of synthesis gas in this chapter. However, hydrogen and other gases can also be used to fuel power-generating turbines, or be used as the starting chemicals for a wide range of commercial products. Gasification is also an extremely flexible technology that can be used to produce clean-burning hydrogen for automobiles of the future as well as power-generating fuel cells.

6.2.1 Coal

Coal gasification—the oldest form of synthesis gas production—is a commercially available proven technology. The process was first used to produce gas for lighting and heat in the United Kingdom more than 200 years ago. In fact, gasification processes have been evolving since the early days of the nineteenth century when town gas became a common way of bringing illumination and heat not only to factories but also to the domestic consumer. This led to the successful development of three key process technologies: (1) the Lurgi fixed-bed gasifier, (2) the high-temperature Winkler fluidized-bed gasifier, and (3) the Koppers—Totzek entrained-flow gasifier. In each case, steam, air, and oxygen are passed through heated coal, which may either be a fixed bed, a fluidized bed, or entrained in the gas. Exit gas temperatures from each reactor are 500 °C (930 °F), 900–1100 °C (1650–2010 °F), and 1300–1600 °C (2370–2910 °F), respectively. In addition to the steam—air—oxygen mixture being used as the feed gases, steam—oxygen mixtures can also be used. As with combustion processes, coal characteristics such as rank, mineral matter, particle size, and reaction conditions are all recognized as having a bearing on the outcome of the gasification process, not only in terms of gas yields but also of gas properties (Massey, 1974; Hanson et al., 2002; Speight, 2013a,b).

6.2.1.1 Technologies

Coal gasification processes combust coal in a measured supply of air (or pure oxygen) to generate a variety of gaseous products that can then be used to generate electrical energy using high-efficiency gas turbines engineered to eliminate soot and minimize formation of nitrogen oxides (NO_x), precursors to ozone-related smog and acid rain. In addition, a coal-gasification power plant typically produces less solid waste than other coal-fired power plants.

Reactors may also be designed to operate either at atmospheric pressure or at high pressure. In the latter type of operation, the hydrogasification process is optimized, and the quality of the product gas (in terms of heat, or Btu, content) is improved. In addition, the reactor size may be reduced and the need to pressurize the gas before it is introduced into a pipeline is eliminated (if a high heat-content gas is the ultimate product). However, high-pressure systems may have problems associated with the introduction of the coal into the reactor.

There has been a general tendency to classify gasification processes by virtue of the heat content of the gas that is produced; it is also possible to classify gasification processes according to the type of reactor vessel and whether or not the system reacts under pressure. However, for the purposes of the present text, gasification processes are segregated according to the bed types, which differ in their ability to accept (and use) caking coals: (1) fixed-bed processes, (2) fluid-bed processes, (3) entrained-bed processes, and (4) molten salt processes.

In the *fixed-bed process*, the feedstock is supported by a grate and combustion gases (steam, air, oxygen, etc.) passed through the supported coal, whereupon the hot produced gases exit from the top of the reactor. Heat is supplied internally or from an outside source, but caking coals cannot be used in an unmodified fixed-bed reactor. On the other hand, the *fluid-bed process* uses finely sized coal particles, and the bed exhibits liquid-like characteristics when a gas flows upward through the bed. Gas flowing through the coal produces turbulent lifting and separation of particles, and the result is an expanded bed having greater coal surface area to promote the chemical reaction; however, such systems have only a limited ability to handle caking coals.

The fluidized-bed system requires the reactant gases to be introduced through a perforated deck near the bottom of the vessel. The volume rate of gas flow is such that its velocity (1—2 ft/s) is high enough to suspend the solids but not high enough to blow them out of the top of the vessel. The result is a violently boiling bed of solids (that simulates a boiling liquid) having very intimate contact with the upward-flowing gas. This gives a very uniform temperature distribution, and the gas flows uniformly upward with no possible countercurrent flow. If a degree of countercurrent flow is desired, two or more fluid-bed stages are placed one above the other. On the other hand, the entrained-flow reactor uses a still finer grind of coal (80% through 200 mesh) than the fluid-bed reactor, and the coal is conveyed pneumatically by the reactant gases. Velocity of the mixture must be about 20 ft/s (6.1 m/s) or higher depending upon the fineness of the coal. In this case, there is little or no mixing of the solids and gases, except when the gas initially meets the solids.

An *entrained-bed process* uses small-size coal particles that are introduced into the gas stream prior to entry into the reactor, and combustion occurs with the coal particles suspended in the gas phase; the entrained system is suitable for both caking and noncaking coals. The molten salt system employs a bath of molten salt to convert the charged coal (Howard-Smith and Werner, 1976).

Finally, *molten salt processes*, as the name implies, use a molten medium of an inorganic salt to generate the heat to decompose the coal into products. In molten-bath gasifiers, crushed coal, steam, air, and/or oxygen are injected into a bath of molten salt, iron, or coal ash. The coal appears to *dissolve* in the melt in which the volatiles crack and are converted into carbon monoxide and hydrogen. The fixed carbon reacts with oxygen and steam to produce carbon monoxide and hydrogen. Unreacted carbon and ash float on the surface from which they are discharged. High temperatures, around 900 °C (1650 °F) and above, depending on the nature of the melt, are required to maintain the bath molten. Such temperature levels favor high reaction rates and throughputs and low residence times. Consequently, tar and distillable oil are not produced in any great quantity, if at all. Gasification may be enhanced by the catalytic properties of the melt used. Molten salts, which are generally less corrosive and have lower melting points than molten metals, can strongly catalyze the steam−coal reaction and lead to very high conversion efficiencies.

Once the coal is fed into the reactor, contacting the solid coal with reactant gases to accomplish the required gasification is a second major mechanical problem (Bodle and Huebler, 1981; Probstein and Hicks, 1990; Speight, 2013a,b). Lumps of coal (ca. 1/8−1 in, 3−25 mm, diameter) are laid down at the top of a vessel, whereas reactant gases are introduced at the bottom of the vessel and flow at relatively low velocity upward through the interstices between the coal lumps. As the coal descends, it is reacted first by devolatilization using the sensible heat from the rising gas, then hydrogenated by the hydrogen in the reactant gas, and finally burned to an ash. The reactions are, therefore, carried out in a countercurrent fashion.

In addition, an extensive variety of individual gasifiers are being developed and are all influenced largely by mechanics. These include (1) introducing the solid feedstock into the gasifier, often at high pressure; (2) contacting the feedstock with reactant gases; (3) removing solid ash or semisolid slag; and (4) collecting the fine, partially reacted dust carried out of the reactor with the gaseous products.

In a gasifier, the coal particle is exposed to high temperatures generated from the partial oxidation of the carbon. As the particle is heated, any residual moisture (assuming that the coal has been pre-fried) is driven off, and further heating of the particle begins to drive off the volatile gases. Discharge of these volatiles will generate a wide spectrum of hydrocarbons ranging from carbon monoxide and methane to long-chain hydrocarbons comprising tars, creosote, and heavy oil. At temperatures above 500 °C (930 °F) the conversion of the coal to char and ash is complete. In most of the early gasification processes, this was the desired byproduct, but for gas generation the char provides the necessary energy to effect further heating and, typically, the char is contacted with air or oxygen and steam to generate the product gases.

The issues inherent in gasification processes include emissions of: (1) particulate matter, (2) sulfur oxides, (3) nitrogen oxides, (4) carbon dioxide, and (5) hazardous species such as mercury, which must be removed from the volatile products (Speight, 2008, 2013a,b; Chadeesingh, 2011). One of the major environmental advantages of the gasification process is the opportunity to remove impurities such as sulfur, mercury, and soot-generating constituents *before* burning the coal, using readily available process options. In addition, the ash produced is in a vitreous or glass-like state, which can be recycled as concrete aggregate, unlike pulverized coal-fired plants that generate ash that must be landfilled, potentially contaminating groundwater.

6.2.1.2 Product properties

The influence of physical process parameters and the effect of coal type on coal conversion are important parts of any process in which coal is used as a feedstock, especially with respect to gasification (Speight, 2013a,b). Thus, variations in coal quality can have an impact on the heating value of the syngas produced by the gasification process. However, a desired throughput can be selected, and then the size and number of gasifiers can be determined within the specific range of coal types considered.

The reactivity of coal generally decreases with increase in rank (from lignite to subbituminous coal to bituminous coal to anthracite). Furthermore, the smaller the particle size, the more contact area there is between the coal and the reaction gases causing faster reaction. For medium- and low-rank coals, reactivity increases with an increase in pore volume and surface area, but for coals having carbon content greater than 85% w/w, these factors have no effect on reactivity. In fact, in high-rank coals, pore sizes are so small that the reaction is diffusion controlled.

The volatile matter produced by the coal during thermal reactions varies widely for the four main coal ranks and is low for high-rank coals (such as anthracite) and higher for increasingly low-rank coals (such as lignite). The higher the volatile matter production, the more reactive a coal and the reactive coals can be more readily converted to gas while producing lower yields of char than a less-reactive coal. Thus, for high-rank coals, the utilization of char within the gasifier is much more of an issue than for lower-rank coal. However, the ease with which they are gasified leads to high levels of tar in the gaseous products, which makes gas cleanup more difficult (Mokhatab et al., 2006; Speight, 2013a, 2014).

The mineral matter content of the coal does not have much impact on the composition of the produced synthesis gas. Gasifiers may be designed to remove the produced ash in solid or liquid (slag) form. In fluid-bed or fixed-bed gasifiers, the ash is typically removed as a solid, which limits operational temperatures in the gasifier to well below the ash melting point. In other designs, particularly slagging gasifiers, the operational temperatures are designed to be above the ash melting temperature. The selection of the most appropriate gasifier is often dependent on the melting temperature and/or the softening temperature of the ash and the coal that is to be used at the facility.

In fact, coals that display caking or agglomerating characteristics (Speight, 2015) are usually not amenable to treatment by gasification processes employing fluidized-bed or moving-bed reactors; in fact, caked coal is difficult to handle in fixed-bed reactors. The pretreatment involves a mild oxidation treatment that destroys the caking characteristics of coals and usually consists of low-temperature heating of the coal in the presence of air or oxygen.

High-moisture content of the feedstock lowers internal gasifier temperatures through evaporation and the endothermic reaction of steam and char. Typically, a limit is set on the moisture content of coal supplied to the gasifier, which can be met by coal-drying operations if necessary. For a typical fixed-bed gasifier and moderate rank and ash content of the coal, this moisture limit in the coal limit is on the order of 35% w/w. Fluidized-bed and entrained-bed gasifiers have a lower tolerance for moisture, limiting the moisture content to approximately 5–10% w/w a similar coal feedstock. Oxygen supplied to the gasifiers must be increased with an increase in mineral matter content (ash production) or moisture content in the coal.

In regard to the maceral content, differences have been noted between the different maceral groups (Speight, 2013a,b). In terms of the character of the coal, gasification technologies generally require some initial processing of the coal feedstock with the type and degree of pretreatment a function of the process and/or the type of coal. For example, the Lurgi process will accept "lump" coal (1 in, 25 mm, to 28 mesh), but it must be noncaking coal with the fines removed. The caking, agglomerating coals tend to form a plastic mass at the bottom of a gasifier and subsequently plug up the system, thereby markedly reducing process efficiency. Thus, some attempt to reduce caking tendencies is necessary and can involve preliminary partial oxidation of the coal to destroy the caking properties.

Finally, depending on the type of coal being processed and the analysis of the gas product desired, pressure also plays a role in product definition. In fact, some (or all) of the following processing steps will be required: (1) pretreatment of the coal (if caking is a problem); (2) primary gasification of the coal; (3) secondary gasification of the carbonaceous residue from the primary gasifier; (4) removal of carbon dioxide, hydrogen sulfide, and other acid gases; (5) shift conversion for adjustment of the carbon monoxide/hydrogen mole ratio to the desired ratio; and (6) catalytic methanation of the carbon monoxide/hydrogen mixture to form methane. If high-heat content (high-BTU) gas is desired, all of these processing steps are required because coal gasifiers do not yield methane in the concentrations required (Mills, 1969; Cusumano et al., 1978).

The products from the gasification of coal may be of low-, medium-, or high-heat (high-BTU) content as dictated by the process as well as by the ultimate use for the gas (Fryer and Speight, 1976; Mahajan and Walker, 1978; Anderson and Tillman, 1979; Cavagnaro, 1980; Bodle and Huebler, 1981; Argonne, 1990; Baker and Rodriguez, 1990; Probstein and Hicks, 1990; Lahaye and Ehrburger, 1991; Speight, 2013a,b).

However, the quality of the gas generated in a system is influenced by coal characteristics, gasifier configuration, and the amount of air, oxygen, or steam introduced into the system. The output and quality of the gas produced is determined by the equilibrium established when the heat of oxidation (combustion) balances the heat of vaporization and volatilization plus the sensible heat (temperature rise) of the exhaust gases. The quality of the outlet gas (BTU/ft^3) is determined by the amount of volatile gases (such as hydrogen, carbon monoxide, water, carbon dioxide, and methane) in the gas stream.

As a very general *rule of thumb*, optimum gas yields and quality are obtained at operating temperatures of approximately 595–650 °C (1100–1200 °F). A gaseous product with a higher heat content (BTU/ft^3) can be obtained at lower system temperatures but the overall yield of gas (determined as the *fuel-to-gas ratio*) is reduced by the unburned char fraction. With some coal feedstocks, the higher the amounts of volatile matter produced in the early stages of the process, the higher the heat content of the product gas. In some cases, the highest gas quality may be produced at the lowest temperatures, but when the temperature is too low, char oxidation reaction is suppressed, and the overall heat content of the product gas is diminished.

6.2.2 Biomass

The search for alternative fuels (Rutz and Janssen, 2007; Fairbridge, 2013) has led to the acceptance of biomass as an alternative feedstock to conventional fossil fuel. Generally, most biomass materials are easier to gasify than coal because they are more reactive with higher ignition stability. This characteristic also makes them easier to process thermochemically into higher-value fuels such as methanol or hydrogen. The mineral matter content (therefore the ash-producing propensity) is typically lower than for most coal types, and sulfur content is much lower than for many fossil fuels. Unlike coal ash, which may contain toxic metals and other trace contaminants, biomass ash may be used as a soil amendment to help replenish nutrients removed by harvest. Some biomass feedstocks stand out for their peculiar properties, such as high silicon or alkali metal contents—these may require special precautions for harvesting, processing, and combustion equipment. Note also that mineral content can vary as a function of soil type and the timing of feedstock harvest. In contrast to their uniform physical properties, biomass fuels are rather heterogeneous with respect to their chemical elemental composition.

Biomass gasification has been a major subject of interest in recent years to estimate efficiency and performance of the process using various types of feedstocks. These include sugarcane residue (Gabra et al., 2001), rice hulls (Boateng et al., 1992), pine sawdust (Lv et al., 2004), almond shells (Rapagnà and Latif, 1997; Rapagnà et al., 2000), wheat straw (Ergudenler and Ghali, 1993), food waste (Ko et al., 2001), and wood biomass (Pakdel and Roy, 1991; Bhattacharaya et al., 1999; Chen et al., 1992; Hanaoka et al., 2005).

Recently, significant interest has been shown in the co-gasification of various biomass feedstocks with coal, such as Japanese cedar wood (Kamabe et al., 2007), sawdust (Vélez et al., 2009), pine chips (Pan et al., 2000), and birch wood (Collot et al., 1999; Brage et al., 2000). The process not only produces a low-carbon footprint on the environment, but also improves the hydrogen/carbon monoxide (H_2/CO) ratio in the produced gas, which is required for further use of the gas as a starting feedstock for the production of liquid fuel (Sjöström et al., 1999; Kumabe et al., 2007). In addition, inorganic matter present in biomass catalyzes the gasification of coal.

Feedstock combinations including Japanese cedar wood and coal (Kumabe et al., 2007), coal and saw dust (Vélez et al., 2009), coal and pine chips (Pan et al., 2000), coal and silver birch wood (Collot et al., 1999), and coal and birch wood (Brage et al., 2000) have been reported in gasification practice. Co-gasification of coal and biomass has some synergy—the process not only produces a low-carbon footprint on the environment, but also improves the H_2/CO ratio in the produced gas, which is required for liquid fuel synthesis (Sjöström et al., 1999; Kumabe et al., 2007). In addition, the inorganic matter present in biomass catalyzes the gasification of coal. However, co-gasification processes require custom fittings and optimized processes for the coal and region-specific wood residues.

Although co-gasification of coal and biomass is advantageous from a chemical viewpoint, some practical problems are present on upstream, gasification, and downstream processes. On the upstream side, the particle size of the coal and biomass is required to be uniform for optimum gasification. In addition, moisture content and pretreatment (torrefaction) are very important during upstream processing.

Although upstream processing is influential from a material handling point of view, the choice of gasifier operation parameters (temperature, gasifying agent, and catalysts) dictate the product gas composition and quality. Biomass decomposition occurs at a lower temperature than coal, and therefore different reactors compatible to the feedstock mixture are required (Speight, 2011a, 2013a,b; Brar et al., 2012). Furthermore, feedstock and gasifier type along with operating parameters not only decide product gas composition but also dictate the amount of impurities handled downstream.

Downstream processes need to be modified if coal is co-gasified with biomass. Heavy metal and impurities such as sulfur and mercury present in coal can make synthesis gas difficult to use and unhealthy for the environment. Alkali present in biomass can also cause corrosion problems and high temperatures in downstream pipes. An alternative option to downstream gas cleaning would be to process coal to remove mercury and sulfur prior to feeding into the gasifier.

However, first and foremost, coal and biomass require drying and size reduction before they can be fed into a gasifier. Size reduction is needed to obtain appropriate particle sizes; however, drying is required to achieve moisture content suitable for gasification operations. In addition, biomass densification may be conducted to prepare pellets and improve density and material flow in the feeder areas.

It is recommended that biomass moisture content should be less than 15% w/w prior to gasification. High-moisture content reduces the temperature achieved in the gasification zone, thus resulting in incomplete gasification. Forest residues or wood has a fiber saturation point at 30–31% moisture content (dry basis) (Brar et al., 2012). Compressive and shear strength of the wood increases with decreased moisture content below the fiber saturation point. In such a situation, water is removed from the cell wall leading to shrinkage. The long-chain molecule constituents of the cell wall move closer to each other and bind more tightly. A high level of moisture, usually injected in the form of steam in the gasification zone, favors formation of a water–gas shift reaction that increases hydrogen concentration in the resulting gas.

As a point of reference, biomass decomposition occurs at a lower temperature than coal, and therefore different reactors compatible to the feedstock mixture are required (Unruh et al., 2010; Brar et al., 2012). Furthermore, feedstock and gasifier type along with operating parameters not only decide product gas composition but also dictate the amount of impurities handled downstream. In addition, at high temperature, alkali present in biomass can cause corrosion problems in downstream pipes. Size reduction is needed to obtain appropriate particle sizes; however, drying is required to achieve moisture content suitable for gasification operations. In addition, densification of the biomass may be done to make pellets and improve density and material flow in the feeder areas.

It is recommended that biomass moisture content should be less than 15% w/w (in some cases, less than 15% w/w) prior to gasification. High-moisture content reduces the temperature achieved in the gasification zone, thus resulting in incomplete gasification. Forest residues or wood has a fiber saturation point at 30–31% moisture content (dry basis) (Brar et al., 2012). Compressive and shear strength of the wood increases with decreased moisture content below the fiber saturation point. In such a situation, water is removed from the cell wall, which causes shrinkage of the cell wall and which can cause more extreme conditions for decomposition.

The torrefaction process is a thermal treatment of biomass in the absence of oxygen, usually at 250–300 °C (290–570 °F) to drive off moisture, completely decompose hemicellulose, and partially decompose cellulose (Speight, 2008, 2011a). Torrefied biomass has reactive and unstable cellulose molecules with broken hydrogen bonds and not only retains 79–95% of feedstock energy but also produces a more reactive feedstock with lower atomic hydrogen–carbon and oxygen–carbon ratios than the original biomass. In addition, pretreatment of the feedstock by torrefaction results in higher yields of hydrogen and carbon monoxide in the gasification process.

Finally, the presence of mineral matter in the biomass feedstock is not always appropriate for fluidized-bed gasification. Low melting point of ash present in woody biomass leads to agglomeration, which causes defluidization of the ash and sintering, deposition, and corrosion of the gasifier construction-metal bed (Vélez et al., 2009). Biomass containing alkali oxides and salts with the propensity to produce yields of ash higher than 5% w/w of the feedstock causes clinkering/slagging problems (McKendry, 2002). Thus, it is imperative to be aware of the melting of biomass ash, its chemistry within the gasification bed (no bed, silica/sand bed, or calcium bed), and the fate of alkali metals when using fluidized-bed gasifiers.

Biomass fuel producers, coal producers, and waste companies have realized the benefits of co-gasification that are amenable to supplying co-gasification power plants and with alternative feedstock. The benefits of a co-gasification technology include use of a reliable coal supply, which allows the economies of scale from a larger plant than could be supplied just with waste and/or biomass. In addition, the technology offers a future option for refineries for hydrogen production and fuel development. In fact, oil refineries and petrochemical plants are opportunities for gasifiers when hydrogen is particularly valuable (Speight, 2011b).

6.2.2.1 Technologies

The fluidized-bed gasifier can operate in a highly back-mixed mode, thoroughly mixing the feedstock particles with those particles already undergoing gasification. Because of the highly back-mixed operation, the gasifier operates under isothermal conditions at a temperature below the ash fusion temperature of the coal, thus avoiding clinker formation and possible collapse of the bed. The low-temperature operation of this gasifier means that fluidized-bed gasifiers are best suited to relatively reactive feeds, such as biomass, or to lower quality feedstocks such as high mineral matter biomass or waste. This gives the gasifier the following characteristics: (1) can accept a wide range of solid feedstocks, including high-mineral feedstocks, including wood and solid waste, (2) uniform, moderate temperature, (3) moderate oxygen and steam requirements, and (4) char recycling.

Most small- to medium-sized biomass gasifiers are air blown and operate at atmospheric pressure and at temperatures in the range 800–100 °C (1470–2190 °F). They face very different challenges to large gasification plants—the use of small-scale air-separation plant should oxygen gasification be preferred. Pressurized operation, which eases gas cleaning, may not be practical. Fluidized-bed gasifiers can also convert biomass to a combustible gas that can be fired in a boiler, kiln, or other energy load. The gasifier can be installed as an add-on to a coal-fired power plant to provide a means to convert a portion of the fuel supply to clean, renewable biomass-based fuel.

6.2.2.2 Fuel properties

Size reduction is needed to obtain appropriate particle sizes; however, drying is required to achieve moisture content suitable for gasification operations. In addition, densification of the biomass may be done to make pellets and improve density and material flow in the feeder areas.

Although upstream processing is influential from a material-handling point of view, the choice of gasifier operation parameters (temperature, gasifying agent, and catalysts) decides product gas composition and quality. Biomass decomposition occurs at a lower temperature than coal, and therefore different reactors compatible to the feedstock mixture are required (Brar et al., 2012). Furthermore, feedstock and gasifier type along with operating parameters not only decide product gas composition but also dictate the amount of impurities to be handled downstream. Downstream processes need to be modified if coal is used with biomass in gasification. Heavy metal and impurities such as sulfur and mercury present in coal can make syngas difficult to use and unhealthy for the environment. In addition, at high temperature, alkali metals present in biomass can cause corrosion problems in downstream pipes.

It is recommended that moisture content of the biomass should be less than 15% w/w (in some cases, less than 15% w/w) prior to gasification. High-moisture content reduces the temperature achieved in the gasification zone, thus resulting in incomplete gasification. Forest residues or wood has a fiber saturation point at 30—31% moisture content (dry basis) (Brar et al., 2012). Compressive and shear strength of the wood increases with decreased moisture content below the fiber saturation point. In such a situation, water is removed from the cell wall, which causes shrinkage of the cell wall. The long-chain molecules, which make up the cell wall move closer to one another and bind more tightly.

The torrefaction process is a thermal treatment of biomass in the absence of oxygen, usually at 250—300 °C (482—572 °F) to drive off moisture, decompose hemicellulose completely, and partially decompose cellulose (Speight, 2011a). Torrefied biomass has reactive and unstable cellulose molecules with broken hydrogen bonds and not only retains 79—95% of feedstock energy but also produces a more reactive feedstock with lower atomic hydrogen—carbon and oxygen—carbon ratios than the original biomass. Torrefaction results in higher yields of hydrogen and carbon monoxide in the gasification process.

Finally, the presence of mineral matter in the coal—biomass feedstock is not appropriate for fluidized-bed gasification. Low melting point of ash present in woody biomass leads to agglomeration that causes defluidization of the ash, sintering, deposition, and corrosion of the gasifier construction-metal bed (Vélez et al., 2009). Biomass containing alkali oxides and salts with the propensity of produce yield higher than 5% w/w ash causes clinkering/slagging problems (McKendry, 2002). Thus, it is imperative to be aware of the melting of biomass ash, its chemistry within the gasification bed (no bed, silica/sand, or calcium bed), and the fate of alkali metals when using fluidized-bed gasifiers.

6.2.3 Opportunity fuels

For the purposes of this chapter, an opportunity fuel is any carbonaceous fuel that does not fall under the definition of coal or biomass or refinery residua—municipal solid waste (MSW) is such a fuel and often has minimal presorting. In addition, waste may also be refuse-derived fuel (RDF), which has had significant pretreatment, usually mechanical screening and shredding. Other more specific wastes, excluding hazardous waste, and possibly including petroleum coke, can provide niche opportunities for gasification and/or co-gasification.

Thus, waste may be MSW, which has had minimal presorting, or RDF, which has had significant pretreatment, usually mechanical screening and shredding. Other more specific wastes, possibly including petroleum coke, which is unsatisfactory for production of electrode carbon, may provide niche opportunities for co-utilization.

Co-utilization of waste and biomass with coal may provide economies of scale that help achieve the policy objectives identified above at an affordable cost. In some countries, governments propose co-gasification processes as being well suited for community-sized developments suggesting that waste should be dealt with in smaller plants serving towns and cities, rather than moved to large, central plants (satisfying the so-called *proximity principle*).

Use of waste materials as co-gasification feedstocks is starting to attract significant interest but the availability of sufficient fuel locally for an economic plant size is often a major issue, as is the reliability of the fuel supply. Use of more-predictably available coal alongside these fuels overcomes some of these difficulties and risks. In fact, coal could be regarded as the *flywheel* or *bread and butter feedstock*, which keeps the plant running when the waste feedstocks are not available in sufficient quantities.

Furthermore, as the disposal of municipal waste and industrial waste becomes a more urgent issue because the traditional means of disposal—the landfill—has become environmentally much less acceptable than previously. New, much stricter regulation of these disposal methods will make the economics of waste processing for resource recovery much more favorable (Gay et al., 1980).

The gasification of petroleum residua and petroleum coke to produce synthesis gas, hydrogen, and/or power may become an attractive option for refiners (Dickenson et al., 1997; Gross and Wolff, 2000; Speight, 2014). The premise that the gasification section of a refinery will be the *garbage can* for de-asphalter residues, high-sulfur coke, as well as other refinery wastes is worthy of consideration.

In summary, coal might be co-gasified with petroleum residua, waste, or biomass for environmental, technical, or commercial reasons (Speight, 2008, 2013a,b). It allows larger, more efficient plants than those sized for the biomass grown or waste arising within a reasonable transport distance; specific operating costs are likely to be lower; and fuel supply security is assured. Co-gasification technology varies and is usually site specific with high dependence on the feedstock. At the largest scale, the plant may include the well-proven fixed-bed and entrained-flow gasification processes. At smaller scales, emphasis is placed on technologies that appear closest to commercial operation. Pyrolysis and other advanced thermal conversion processes are included in which power generation is practical using the on-site feedstock produced. However, needing to be addressed are (1) the core fuel handling and gasification/pyrolysis technologies, (2) the fuel gas cleanup, and (3) the conversion of fuel gas to electric power (Ricketts et al., 2002).

6.2.3.1 Technologies

The gasification MSW is a promising candidate for both disposal of the waste and synthesis gas production, and two major process steps of thermal degradation have been observed (Kwon et al., 2009; Arena, 2012). The first thermal degradation step occurs at temperatures on the order of 280–350 °C (535–650 °F) and consists mainly of the decomposition of the biomass component into low molecular weight (methane, ethane, propane) volatile hydrocarbons. The second thermal degradation step occurs between 380 and 450 °C (715 and 840 °F) and is mainly attributed to the decomposition of polymer components, such as plastics and rubber, in waste feedstock.

Thus, in the past two decades there has been growing interest in the use of gasification technologies to treat solid waste. The concept is not a new one but the systems available are typically used for the gasification of coal. Early attempts to use municipal waste as a feedstock ran into problems when scaled up unless the input was suitably homogeneous. Nevertheless, with its lure of low emissions and a greatly reduced, environmentally sound residue, the story was not going to end there.

The main reactors used for gasification of MSW are fixed beds and fluidized beds. Larger capacity gasifiers are preferable for treatment of MSW because they allow for variable fuel feed, uniform process temperatures due to highly turbulent flow through the bed, good interaction between gases and solids, and high levels of carbon conversion.

Advanced gasification technologies, such as the Westinghouse advanced gasification technology, includes at least one continuously operating gasification reactor. Within the reactor, the charge material is gasified into synthesis gas, which exits the top of the reactor through two outlets. The feedstock (MSW, biomass, refuse-derived fuel, hazardous waste), flux, and bed materials are delivered to the plant receiving facility. The feed is metered onto a common charge conveyor, which transports the feed to the gasification reactor. The majority of the mineral matter in the feedstock material forms molten slag, which flows through the tap-holes at the bottom of the reactor. The slag is then quenched and granulated upon exiting the reactor. The resulting vitreous granules are conveyed and loaded onto trucks for export off site. The synthesis gas is the product generated from the feedstock and exits at the top of the reactor. The gas is cooled and sent through a series of gas cleaning processes to remove particulate, chlorine, sulfur, and mercury.

An alternate gasification option using plasma technology is also available and offers much promise for use with solid waste and other carbonaceous feedstocks

(Gomez et al., 2009). Briefly, plasma is a superheated column of electrically conductive gas and plasma torches that burn at temperatures approaching 5500 °C (10,000 °F), and, when utilized for waste treatment, plasma torches are very efficient at causing organic and carbonaceous materials to vaporize into gas. Inorganic materials are melted and cool into a vitrified glass. Because waste gasification typically operates at temperatures on the order of 1500 °C (2700 °F), the plasma-based process is ideally suited to generate the high temperatures needed for gasification.

Plasma arc processing has been used for years to treat hazardous waste, such as incinerator ash and chemical weapons, and convert them into nonhazardous slag. Thus, it is not surprising that plasma gasification is an emerging technology that can process landfill waste to extract commodity recyclables and convert carbon-based materials into fuels. It can form an integral component in a system to achieve zero waste and produce renewable fuels, while caring for the environment. However, utilizing this technology to convert MSW to energy is not yet fully mature but does have the great potential to operate more efficiently than other systems due to the high temperature, heat density, and nearly complete conversion of carbon-based materials to syngas, and inorganic components to slag.

Plasma gasification is a multistage process that starts with feedstocks ranging from waste to coal to plant matter, and MSW, as well as hazardous waste. The first step is to process the feedstock to make it uniform and dry, and sort out the valuable recyclables. The second step is gasification, in which heat from the plasma torches is applied inside a sealed, air-controlled reactor. During gasification, carbon-based materials break down into gases and the inorganic materials melt into liquid slag, which is poured off and cooled. The heat causes hazards and poisons to be destroyed. The third stage is gas cleanup and heat recovery, in which the gases are scrubbed of impurities to form clean fuel, and heat exchangers recycle the heat back into the system as steam. The final stage is fuel production and the output can range from electricity to a variety of fuels, as well as chemicals, hydrogen, and polymers.

6.2.3.2 Product properties

Although evaluating suitability of a gasification technology for waste processing, the degree of preprocessing required in conversion of MSW into a suitable feed material is a major criterion. Unsorted MSW is not suitable for most thermal technologies because of its varying composition and size of some of its constituent materials. It may also contain undesirable materials, which can reduce the process efficiency or have an adverse effect on emission control systems. In fact, the variable mixed character of MSW leads to major environmental concerns.

Thus, the application of waste gasification technologies requires detailed knowledge of the feedstock (as it does with any other gasification feedstock) and the unique features of MSW such as high-moisture content (Yun et al., 2003). Due to the heterogeneous MSW matrix only limited information on the gasification of MSW gasification is available even though the gasification process for several of the components of the waste have been investigated (Choy et al., 2004; Jung et al., 2005; Cheung et al., 2007; Kwon et al., 2009). In fact, MSW is composed of approximately 60% w/w biomass or biomass-derived components including vegetation trimmings, wood, food scraps, and paper, which can be an advantage in terms of carbon credit.

Because of the widely heterogeneous nature of MSW (and other wastes), the products from the gasification process may be of low-, medium-, or high-heat content (high-BTU) as dictated by the process as well as by the carbon content of the waste. However, the quality of the gas generated in the process is also influenced by the configuration of the gasifier and the amount of air, oxygen, or steam introduced into the system. As for coal gasification, the output and quality of the gas produced is determined by the equilibrium established when the heat of oxidation (combustion) balances the heat of vaporization and volatilization plus the sensible heat (temperature rise) of the exhaust gases. The quality of the outlet gas (BTU/ft^3) is determined by the amount of volatile gases (such as hydrogen, carbon monoxide, water, carbon dioxide, and methane) in the gas stream.

With some feedstocks, the higher the amounts of volatile matter produced in the early stages of the process, the higher the heat content of the product gas. In some cases, the highest gas quality may be produced at the lowest temperatures but when the temperature is too low, char oxidation reaction is suppressed, and the overall heat content of the product gas is diminished.

6.3 Biogas

The term *biogas* (also known as *swamp gas*, *marsh gas*, *landfill gas*, *digester gas*) includes a large variety of gases resulting from specific treatment processes, starting from various organic waste industries. Moreover, biogas is not synthesis gas, and the term typically refers to a gaseous mixture produced by the anaerobic decomposition of organic matter and that —if cleaned sufficiently through gas processing sequences (Mokhatab et al., 2006)—has similar characteristics to natural gas. Biogas can be produced from raw materials such as recycled organic waste. The gas can be combusted or oxidized with oxygen to produce water and carbon dioxide and thus can be used as a fuel—it can also be used in a gas engine to convert the energy in the gas into electricity and heat. Being predominantly methane, biogas can be compressed in the same way that natural gas (predominantly also methane) can be compressed (to CNG) and used to power vehicles.

Biogas production has usually been applied for waste treatment, mainly sewage sludge, agricultural waste (manure), and industrial organic waste streams (Hartmann and Ahring, 2005). The primary source that delivers the necessary microorganisms for biomass biodegradation and, as well, one of the largest single sources of biomass from food/feed industry is manure from animal production, mainly from cow and pig farms (Nielsen et al., 2007).

The process for producing biogas involves anaerobic digestion of organic material by anaerobic bacteria or fermentation of biodegradable materials such as municipal waste, manure, sewage, plant material, and crops. It is primarily methane (CH_4) with smaller amounts of carbon dioxide (CO_2), carbon monoxide (CO), as well as hydrogen sulfide (H_2S), and water (H_2O) (Table 6.1) depending on the origin of the

Table 6.1 Typical constituents and variable composition of biogas

Constituent	Formula	General range % v/v	Biogas (agricultural) % v/v	Sewage gas % v/v	Landfill gas % v/v
Methane	CH_4	50–75	55–75	55–65	40–55
Carbon dioxide	CO_2	25–50	25–45	30–40	35–50
Carbon monoxide	CO	Trace–5			
Hydrogen	H_2	0–2			
Hydrogen sulfide	H_2S	0–5	0–1.5	<200 ppmv	150–300 ppmv
Nitrogen	N_2	0–5	0–10	0–10	0–20
Oxygen	O_2	0–5			

anaerobic digestion process (Richards et al., 1991, 1994; Coelho et al., 2006; Ramroop Singh, 2011). For example, landfill gas typically has methane concentrations of approximately 50% v/v whereas gas treatment (gas cleaning) can increase the methane content to 55–75% v/v, which can be increased to 80–90% methane using more extensive gas purification (Mokhatab et al., 2006).

6.3.1 Technologies

Currently, biogas production is mainly based on the anaerobic digestion of single-energy crops. Maize, sunflower, grass, and sudangrass are the most commonly used energy crops. In the future, biogas production from energy crops will increase and requires being based on a wide range of energy crops that are grown in versatile, sustainable crop rotations (Bauer et al., 2007).

The process production occurs under anaerobic conditions and in different temperature regions. Typically, biogas is produced as landfill gas (LFG) or digester gas from anaerobic digesters. The feedstock to an anaerobic digester can be the so-called energy crops (such as maize) or biodegradable waste (such as sewage sludge and food waste). During the process, an airtight tank transforms biomass waste into methane, producing renewable energy that can be used for heating, electricity, and many other operations such as internal combustion engines and gas turbines that are well suited to the conversion of biogas into electricity and heat.

The two key processes for biogas production are (1) the mesophilic process and (2) the thermophilic process. The mesophilic digester (mesophilic biodigester) operates at temperatures in the range 20–40 °C (68–104 °F). A thermophilic digester (thermophilic biodigester) operates in temperatures in excess of 50 °C (122 °F) to produce the biogas.

6.3.2 Fuel analysis

Biogas is primarily a mixture of methane (CH_4) and carbon dioxide (CO_2), and different sources of production lead to different specific compositions (Table 6.1). The presence of hydrogen sulfide, carbon dioxide, and water make biogas very corrosive and require the use of adapted materials. In addition, the composition of a gas issued from a digester depends on the substrate, its organic matter load, and the feeding rate of the digester.

6.3.3 Quality control

The presence of hydrogen sulfide in biogas is a major factor in the quality of the gas. For further use, the gas must undergo a desulfurizing step (Mokhatab et al., 2006; Speight, 2014)—hydrogen sulfide is capable of causing damage due to corrosion effects to the downstream piping or to the co-generation engine or can cause serious damage to Fischer—Tropsch catalysts.

The presence of water, in the gaseous form of vapor, is inevitable in a biogas mixture due to the type of biochemical reactions that take place in anaerobic digestion. Like hydrogen sulfide, water is also undesirable in a biogas stream because it can contribute to corrosion effects caused by hydrogen sulfide. A high concentration of water can also cause the typically noncorrosive carbon dioxide into a corrosive compound due to formation of carbonic acid:

$$H_2O + CO_2 \rightarrow H_2CO_3$$

As a result, water removal from biogas is another necessary pretreatment step to mitigate the potential corrosion effects of other constituents of the gas.

6.4 Other methods for producing synthesis gas

A variety of other methods are available for producing gas from feedstocks. For the most part, the methods induced here are indirect methods insofar as the gaseous products are not prime products and, therefore, are byproducts that require use either as process fuel or for the production of useful products.

6.4.1 Liquefaction

The liquefaction process (coal was the feedstock of choice for many decades in the twentieth century) is a process used to convert a solid fuel into a substitute for liquid fuels such as diesel and gasoline. Coal liquefaction has historically been used in countries without a secure supply of petroleum, such as Germany (during World War II) and South Africa (since the early 1970s). The technology used in coal liquefaction is quite old, and was first implemented during the nineteenth century to provide gas for indoor lighting (Speight, 2013a). In the process, gases are also produced, and the gas mix can be used either as (1) fuel for process heat or (2) products that can be converted to synthesis gas and thence to liquid products. In fact, the concept is still often cited as a viable option for alleviating projected shortages of liquid fuels as well as offering some measure of energy independence for those countries with vast resources of coal who are also net importers of crude oil (Speight, 2011a,b,c).

Bio-oil (also known as pyrolysis oil, biocrude, or bio-oil) is a synthesis fuel that is produced from biomass by destructive distillation (with simultaneous removal of distillate) of dried biomass at a temperature on the order of 500 °C (930 °F). The oil must typically contain high levels of oxygen to be classed as hydrocarbon oil and, because of this, bio-oil is chemically different from typical petroleum products.

Any form of biomass can be considered for thermal decomposition of bio-oil and includes: wood, agricultural wastes, olive pits, nut shells, energy crops such as *Miscanthus* and sorghum, forestry wastes such as bark and thinnings, and other solid wastes, including sewage sludge and leather wastes, have also been studied. There is a variety of temperatures, heating rates, residence times, and feedstock varieties that make process generalizations almost impossible (Mohan et al., 2006; Speight, 2011a).

There is no question that the production of bio-oil is another means to produce potentially valuable liquids. In addition gases and oil of questionable value for fuel production may be suitable for the production of synthesis gas by gasification. Such an option may only be used as a clean-up option because the original feedstock could be gasified directly.

6.4.2 Carbonization

Carbonization is also an old technology insofar as it was used for the production of refinable tar products and coke (Speight, 2008, 2013a,b). More correctly, carbonization is the destructive distillation of organic substances in the absence of air accompanied by the production of carbon and liquid and gaseous products. Next to combustion, carbonization is usually achieved by the use of temperatures up to 1500 °C (2730 °F). The degradation of the coal (or any carbonaceous feedstock) is severe at these temperatures and produces (in addition to the desired coke) substantial amounts of gaseous products.

Again (as for the liquefaction process), there is no question that the carbonization process was a means of producing valuable products and that use of these products for synthesis gas production was of a secondary nature because the feedstock can be gasified directly to synthesis gas. Again, such an option may only be used as a clean-up option because the original feedstock could be gasified directly.

6.5 Syngas conversion to products

The Fischer–Tropsch reaction is the means by which a range of hydrocarbon products and alcohols can be produced from synthesis gas via the hydrogenation of carbon monoxide (Table 6.2). The major catalysts used industrially are iron-based and cobalt-based catalysts but rubidium-based and nickel-based catalysts are also used.

Table 6.2 **Carbon chain groups which can be produced as Fischer–Tropsch products**

Carbon number	Group name
C1–C2	SNG (synthetic natural gas)
C3–C4	LPG (liquefied petroleum gas)
C5–C7	Light petroleum
C8–C10	Heavy petroleum
C11–C20	Middle distillate
C11–C12	Kerosene
C13–C20	Diesel
C21–C30	Soft wax
C31–C60	Hard wax

Mechanistically, the reactions can be regarded as a carbon chain-building process in which methylene ($-CH_2-$) groups are attached sequentially in a carbon chain:

$$nCO + [n + m/2]H_2 \rightarrow C_nH_m + nH_2O \quad \Delta H: -ve$$

For example:

$$CO + 2H_2 \rightarrow -CH_2- + H_2O \quad \Delta H = -165 \text{ kJ/mol}$$

However, several other reactions also occur and, despite the volume of literature on the subject, the reaction mechanisms are still not well understood, and different reaction schemes are often proposed.

A common feature is the exothermic character of the reactions and, as a general rule of thumb, the reactions that produce water (H_2O) and carbon dioxide (CO_2) as products tend to be more exothermic on account of the very high heat of formation of these species. For example:

$$2CO + H_2 \rightarrow -CH_2- + CO_2 \quad \Delta H = -204 \text{ kJ/mol}$$
$$3CO + H_2 \rightarrow -CH_2- + 2CO_2 \quad \Delta H = -244 \text{ kJ/mol}$$
$$CO_2 + 3H_2 \rightarrow -CH_2- + 2H_2O \quad \Delta H = -125 \text{ kJ/mol}$$
$$CO + 2H_2 \rightarrow -CH_2- + H_2O \quad \Delta H = -165 \text{ kJ/mol}$$

In addition, the water–gas-shift reaction is also exothermic:

$$CO + H_2O \rightarrow H_2 + CO_2 \quad \Delta H = -39 \text{ kJ/mol}$$

Table 6.3 **Hydrogen−carbon monoxide ratios in synthesis gas from various processes**

Process	H_2/CO ratio
Steam−methane reforming (SMR)	3.0−5.0
SMR + oxygen secondary reforming (O2R)	2.5−4.0
Autothermal reforming (ATR)	1.6−2.65
Partial oxidation (POx)	1.6−1.9

Thus, because of the exothermic nature of the reactions an important issue is the need to avoid an increase in temperature—the need for mitigating a possible rise in temperature is of critical importance to: (1) maintain stable reaction conditions, (2) avoid the tendency to produce lighter hydrocarbons, and (3) prevent catalyst sintering and hence reduction in activity.

Products from processes that produce synthesis gas (Table 6.3) can range from (1) a range of hydrogen mixtures, (2) high-purity hydrogen, (3) high-purity carbon monoxide, and (4) high-purity carbon dioxide. In practice, however, the options are not limited to the ranges shown but rather even greater H_2/CO ratios, if adjustments are made like the inclusion of a shift converter to effect near-equilibrium water−gas shift conversion or by adjusting the amount of steam.

Catalysts for the Fischer−Tropsch (FT) synthesis are based on transition metals of iron, cobalt, nickel, and ruthenium. FT catalyst development has largely been focused on the preference for high molecular weight linear alkanes and diesel fuels production. Among these catalysts, it is generally known that: (1) nickel tends to promote methane formation, as in a methanation process and is generally not desirable, (2) iron has a higher water−gas-shift activity, and is therefore more suitable for a lower hydrogen/carbon monoxide ratio (H_2/CO) in the product synthesis gas, (3) cobalt is more active, and generally preferred over ruthenium because of the prohibitively high cost of ruthenium, and (4) in comparison to iron-based catalysts, cobalt-based catalysts tend to have a much lower water−gas-shift activity, and is much more costly.

Iron-based catalysts have been the commonly-used catalysts for converting coal-derived syngas into liquids and iron-based catalysts may be operated in both high-temperature regime (300−350 °C (570−660 °F)) and low-temperature regime (220−270 °C (430−520 °F)), whereas cobalt-based catalysts have been found to be more appropriate for use in the low-temperature range.

Although there are differences in the product distribution for use of cobalt-based catalysts and iron-based catalysts at similar temperatures and pressures—a cobalt-based catalyst has somewhat higher propensity to produce higher molecular weight hydrocarbons than an iron-based catalyst—the product distribution is primarily driven by the choice of operating temperature. For example, a higher temperature results in gasoline/diesel ratio on the order of 2:1 whereas a low-temperature results in gasoline/diesel ratio of 1:2, whether or not the catalyst iron-based or cobalt-based.

Higher temperatures shift selectivity toward (1) lower carbon number products, (2) more hydrogenated products, (3) increased branching in the products, and (4) an increase in byproducts such as ketones and aromatics also increases.

6.5.1 Technologies

Currently, four major reactor designs have been developed to maximize the efficiency of heat removal and enable optimal temperature control, namely: (1) the multi-tubular fixed-bed reactor—the ARGE reactor, (2) the fixed slurry-bed reactor, (3) the circulating fluidized-bed reactor—the Synthol reactor, and (4) the fixed fluidized-bed reactor—the Sasol advanced reactor (Chadeesingh, 2011). The issues that govern the design of reactors best suited to large-scale production of Fischer–Tropsch products are heat removal arising out of the exothermic reactions and temperature control—both issues are important to enable longer catalyst lifetimes and in obtaining optimal product selectivity.

The products of Fischer–Tropsch synthesis comprise a mixture of paraffins (C_nH_{2n+2}) and olefins (C_nH_{2n}). A reaction mechanism to explain the formation of this range of products must thus explain not only the formation of the different hydrocarbon functional groups, but also the build-up in carbon chain length. The mechanism proposed to explain this incorporates a chain initiation step. The idea of a theoretical optimal and stoichiometric chain growth can thus be deduced insofar as two hydrogen molecules are required for each molecule of carbon monoxide that becomes adsorbed on the surface of the catalyst. Thus, the optimal hydrogen/carbon monoxide ratio for the Fischer–Tropsch process is 2.0.

When the products desired are the shorter carbon chain lengths, for example, the light petroleum or gasoline fractions, the longer chain groups can be cracked accordingly. It would thus appear that Fischer–Tropsch synthesis conditions, which result in product distributions that provide longer carbon chains, are more amenable with a greater flexibility in choosing salable fractions of choice.

6.5.1.1 High-temperature and low-temperature Fischer–Tropsch

In practice, there are two Fischer–Tropsch process schemes—the low-temperature Fischer–Tropsch (LTFT) reaction and the high-temperature Fischer–Tropsch (HTFT) reaction. The efficiency (and progress) of each reaction is dictated by the type catalyst used and the temperature ranges utilized. If the catalyst used is iron-based, a temperature range of 300–350 °C (570–660 °F) is used and constitutes the high-temperature process (HTFT). On the other hand, if the catalyst selected is cobalt-based, the temperature range required is 200–240 °C (390–465 °F), which represents the low-temperature (LTFT) process but can, however, operate successfully using either cobalt-based or iron-based catalysts.

In principle, other catalysts can also be used in Fischer–Tropsch synthesis, especially those with rubidium or nickel active sites. In practice, however, because of the low availability of rubidium this type of catalyst has not managed to find a place in commercial-scale applications even though its activity is sufficient for a successful Fischer—Tropsch process. On the other hand, nickel-based catalysts although having high enough activities for commercial-scale application, suffer from the fact that they tend to produce too much methane. In addition, at high pressures the performance of nickel is considered poor due to the tendency for the production of volatile carbonyl compounds, that is, oxygenates. In actual practice, the two major catalysts used in industry remain those that are either iron-based or cobalt-based.

6.5.1.2 Steam-methane reforming

Steam reforming (sometimes referred to as steam-methane reforming, SMR) is carried out by passing a preheated mixture comprising essentially methane and steam through catalyst-filled tubes. Because the reaction is endothermic, heat must be provided to effect the conversion. The products of the process are a mixture of hydrogen, carbon monoxide, and carbon dioxide. To maximize the conversion of the methane feed, both a primary and secondary reformer are generally utilized. A *primary reformer* is used to effect a conversion of methane on the order of 90—92% v/v by partially reacting the feedstock with steam over a nickel-alumina catalyst to produce a synthesis gas ($H_2/CO = 3:1$). This is achieved using a fired tube furnace at 900 °C (1650 °F) at a pressure of 220—440 psi. Any unconverted methane is reacted with oxygen at the top of a *secondary autothermal reformer* containing nickel catalyst in the lower region of the vessel.

In autothermal reforming the organic feedstock (e.g., natural gas) and steam (and sometimes carbon dioxide) are mixed directly with oxygen and air in the reformer, which comprises a refractory lined vessel that contains the catalyst, together with an injector located at the top of the vessel. Partial oxidation reactions occur in a region of the reactor referred to as the combustion zone, and the product mixture from this zone flows through a catalyst bed in which the actual reforming reactions occur. Heat generated in the combustion zone from partial oxidation reactions is utilized in the reforming zone, so that in the ideal case, it is possible that the autothermal reformer can show a good heat balance.

When the autothermal reformer uses carbon dioxide, the H_2/CO ratio produced is 1:1 but when the autothermal reformer uses steam, the H_2/CO ratio produced is 2.5:1. Thus:

With carbon dioxide:

$$2CH_4 + O_2 + CO_2 \rightarrow 3H_2 + 3CO + H_2O + Heat$$

With steam:

$$4CH_4 + O_2 + 2H_2O \rightarrow 10H_2 + 4CO$$

The reactor itself consists of three zones: (1) the burner, in which the feed streams are mixed in a turbulent diffusion flame, (2) the combustion zone, in which partial oxidation reactions occurs to produce a mixture of carbon monoxide and hydrogen, and (3) the catalytic zone, in which the gases leaving the combustion zone reach thermodynamic equilibrium.

Combined reforming incorporates the combination of both steam reforming and autothermal reforming and, in such a configuration, the hydrocarbon (e.g., natural gas) is first only partially converted, under mild conditions, to syngas in a relatively small steam reformer. The off-gases from the steam reformer are sent to an oxygen-fired secondary reactor, the autothermal reformer in which the unreacted methane is converted to syngas by partial oxidation followed by steam reforming. Another configuration requires the hydrocarbon feedstock split into two streams that are then fed in parallel, to the steam reforming and autothermal reactors.

6.5.1.3 Water–gas shift

Two water–gas-shift (WGS) reactors are used downstream of the secondary reformer to adjust the hydrogen/carbon monoxide ratio, depending on the end use of the steam reformed products. The first of the two WGS reactors utilizes an iron-based catalyst, which is heated to approximately 400 °C (750 °F). The second WGS reactor operates at approximately 200 °C (390 °F) and is charged with a copper-based catalyst.

The deposition of carbon can be an acute problem with the use of Ni-based catalysts in the primary reformer (Alstrup, 1988; Rostrup-Neilsen, 2008). A successful technique is to use a steam/carbon ratio in the feed gas that does not allow the formation of carbon, but this method results in lowering the efficiency of the process. Another approach is to use sulfur passivation, which utilizes the principle that the reaction leading to the deposition of carbon requires a larger number of adjacent surface nickel atoms than does steam reforming (Udengaard et al., 1992; Rostrup-Neilsen, 2008). When a fraction of the surface atoms are covered by sulfur, the deposition of carbon is thus more greatly inhibited than steam reforming reactions. A third approach is to use Group VIII metals that do not form carbides, for example, platinum (Pt) but, as with rubidium, availability is an issue.

6.5.1.4 Partial oxidation

Noncatalytic partial oxidation (TPOX) and catalytic partial oxidation (CPOX) reactions occur when a substoichiometric fuel–air mixture is partially combusted in a reformer. The general reaction equation (without catalyst, TPOX) is of the form:

$$C_nH_m + (2n + m)/2\ O_2 \rightarrow nCO + (m/2)\ H_2O$$

For example:

$$C_{24}H_{12} + 12O_2 \rightarrow 24CO + 6H_2$$

The feedstock, which may include steam, is mixed directly with oxygen by an injector that is located near the top of the reaction vessel. Both partial oxidation reactions as well as reforming reactions occur in the combustion zone below the burner. The principal advantage of the partial oxidation process is its ability to process almost any feedstock, which can comprise very high molecular weight organic materials, such as petroleum coke (Speight, 2013a,b, 2014). A very high temperature, approximately

1300 °C (2370 °F), is required to achieve near-complete reaction. This necessitates the consumption of some of the hydrogen and a greater-than-stoichiometric consumption of oxygen, that is, oxygen-rich conditions. A possible means of improving the efficiency of synthesis gas production is via catalytic partial oxidation (CPOX) technology.

6.5.2 Product properties

The composition of synthesis gas is highly dependent upon the feedstock to the gasifier—several of the components of raw synthesis gas cause challenges that must be addressed at the outset, including tar constituents, hydrogen concentration, and moisture. The varying composition of the gas and the chemical behavior of the gas can place greater demands on the behavior of the gas in the Fischer–Tropsch process.

Hydrogen is produced from gasification of carbonaceous feedstock and although several gasifier types exist, entrained-flow gasifiers are considered most appropriate for producing both hydrogen and electricity from coal, because they operate at temperatures high enough (approximately 1500 °C, 2730 °F) to enable high-carbon conversion and prevent downstream fouling from tars and other high-boiling products. At the gasifier temperature, the ash and other coal mineral matter liquefies and exits at the bottom of the gasifier as slag, a sand-like inert material that can be sold as a coproduct to other industries (e.g., road building) (Speight, 2013a, 2014). The synthesis gas exits the gasifier at pressure and high temperature and must be cooled prior to the syngas cleaning stage.

The WGS reaction maximizes the hydrogen content of the synthesis gas, which consists primarily of hydrogen and carbon dioxide at this stage:

Water–Gas–Shift (WGS) Reaction:

$$CO + H_2O \rightarrow CO_2 + H_2$$

The synthesis gas is then scrubbed of particulate matter and sulfur is removed via physical absorption. The carbon dioxide is captured by physical absorption or a membrane and either vented or sequestered.

At this point, the hydrogen-rich synthesis gas is sufficiently pure for some stationary fuel cell applications and use in hydrogen internal combustion engines. However, for use in vehicles featuring proton exchange membrane fuel cells, the hydrogen must be purified to 99.999% using a pressure swing adsorption (PSA) unit. The high-purity hydrogen exits the PSA unit sufficiently compressed for pipeline transport to refueling stations. The purge gas from the PSA unit is compressed and directed to a combined cycle (gas and steam turbine) for co-production of electricity.

One other aspect of the Fischer–Tropsch reaction that must be given attention in the current context is the production of alcohols. In addition to alkane formation, competing reactions give small amounts of alcohols as well as alkenes and other oxygen-containing products. For example, in the presence of a cobalt catalyst at temperatures on the order of 180–200 °C (355–390 °F) and pressures from atmospheric to 1500 psi, the products, apart from water, are almost entirely hydrocarbon in nature.

The oxygen-containing substances (such as alcohols, aldehydes, ketones, acids) have been noted as byproducts—the result of side reaction and play no part in the main reaction mechanism. However, when iron catalysts are used in the synthesis at 1500–3000 psi, appreciable amounts of alcohols are produced. In addition, alkali metal-modified unsupported and supported cobalt catalysts can produce alcohols and also yield appreciable amounts of higher alcohols with more than four carbon atoms (C_{5+} alcohols), as much as 77% of the total alcohol distribution (Fiore et al., 2004; Fonseca et al., 2007; Xiang et al., 2008; Rabiu et al., 2012; Ishida et al., 2013).

The presence of alcohol products has raised the possibility that alcohols are a primary product of the Fischer–Tropsch synthesis. In fact, a decrease in the contact time leads to a significant increase in alcohol content, and the alcohol content of the product is appreciably higher at 1500 psi than at atmospheric pressure. The reaction mechanism is still being discussed, but some speculate that the hydrocarbon products are derived, not by synthesis from carbon monoxide and hydrogen, but by dehydration of the alcohols followed by reduction and hydrogenation cracking to form lower hydrocarbons, and by condensation reactions to form substances that are more complex.

6.6 Current status and future trends

The gasification-based refinery is another concept for the production of fuels, electricity, and chemical products (Speight, 2011b). Coal gasification has also been used for production of liquid fuels (Fischer–Tropsch diesel and methanol) via a catalytic conversion of synthesis gas into liquid hydrocarbons (Speight, 2008, 2013a,b; Chadeesingh, 2011).

6.6.1 Technical aspects

Combining biomass, refuse, and coal overcomes the potential unreliability of biomass, the potential longer-term changes in refuse, and the size limitation of a power plant using only waste and/or biomass. It also allows benefit from a premium electricity price for electricity from biomass and the gate fee associated with waste. If the power plant is gasification-based, rather than direct combustion-based, further benefits may be available. These include a premium price for the electricity from waste, the range of technologies available for the gas to electricity part of the process, gas cleaning prior to the main combustion stage instead of after combustion, and public image, which is currently generally better for gasification than for combustion. These considerations lead to the current study of co-gasification of wastes/biomass with coal (Speight, 2008, 2013a,b).

Use of waste materials as co-gasification feedstocks will attract significant attention as a means of reducing the space needed for landfill operations. Cleaner biomass materials are renewable fuels and may attract premium prices for the electricity generated. Availability of sufficient fuel locally for an economic plant size is often a major issue, as is the reliability of the fuel supply. Use of more-predictably available coal alongside these fuels overcomes some of these difficulties and risks. For the foreseeable future, the benefits of a co-gasification technology involving coal and biomass include use of a reliable coal supply with gate-fee waste and biomass, which allows the economies of scale from a larger plant than could be supplied just with waste and biomass. Furthermore, petroleum refineries and petrochemical plants are opportunities for gasifiers when hydrogen is particularly valuable (Speight, 2011a,b,c, 2014).

Electrical production or combined electricity and heat production remain the most likely area for the application of gasification or co-gasification. The lowest investment cost per unit of electricity generated is the use of the gas in an existing large power station. This has been done in several large utility boilers, often with the gas fired alongside the main fuel. This option allows a comparatively small thermal output of gas to be used with the same efficiency as the main fuel in the boiler as a large, efficient steam turbine can be used. It is anticipated that addition of gas from a biomass or wood gasifier into the natural gas feed to a gas turbine would be technically possible, but there will be concerns as to the balance of commercial risks to a large power plant and the benefits of using the gas from the gasifier.

The use of fuel cells with gasifiers is frequently discussed but the current cost of fuel cells is such that their use for mainstream electricity generation is uneconomic.

6.6.2 Environmental aspects

During the next five decades, fossil fuels will be the prominent options for producing liquid and gaseous fuels (Speight, 2007, 2008, 2009, 2011a,b, 2013a,b, 2014). However, the continued use of fossil fuels presents several serious environmental challenges, including significant air quality, climate change, and mining impacts. However, coal gasification technologies have been demonstrated that provide order-of-magnitude reductions in criteria pollutant emissions and, when coupled with carbon capture and sequestration (or storage) (CCS), the potential for significant reductions in carbon dioxide emissions. Therefore, although coal is a finite nonrenewable resource, coal-derived hydrogen with CCS can increase domestic energy independence, provide near-term carbon dioxide and criteria pollutant reduction benefits, and facilitate the transition to a more sustainable hydrogen-based transportation system. CCS is one of the critical enabling technologies that could lead to coal-based hydrogen production for use as a transportation fuel. However, other risks to the environment need to be addressed.

The increasing costs of conventional waste management and disposal options, and the desire in most developed countries to divert an increasing proportion of mixed organic waste materials from landfill disposal, for environmental reasons, will render the investment in energy from waste projects increasingly attractive. Most new projects involving the recovery of energy from municipal waste materials will involve the installation of new purpose-designed incineration plants with heat recovery and power generation. However, advanced thermal processes for MSW, which are based on pyrolysis or gasification processes, are also being introduced. These processes offer significant environmental and other attractions and will have an increasing role to play, but the rate of increase of use is difficult to predict.

Depending on subsequent processing and final use, various products and byproducts must be removed from the low- and medium-heat-content products that come from a gasifier. In all cases, hydrogen sulfide and other sulfur compounds must be removed, because (in addition to the environmental aspects of gas use) they can poison catalysts in subsequent processing. This may be essentially all of the cleanup that is necessary for low-heat-content gas destined for combustion, whereas gas that is to be methanated requires virtually complete removal of essentially all components except hydrogen and carbon monoxide.

The use of biomass and waste as feedstocks for (direct and indirect) energy production will require tar removal in the process. Catalytic cracking or thermal cracking, if they prove reliable, are generally regarded as the best processes as they retain much of the chemical energy of tars in the gas phase. However, experience to date on the reliability of tar cracking processes has been at best variable. Condensation and/or wet scrubbing are better proven than tar cracking processes. The collected tars are often toxic, carcinogenic, or difficult to break down even in combustion or oxygen gasification processes. Oxygen-blown entrained gasifiers are particularly good at breakdown of the most difficult tars, but it will be unlikely to find such a gasifier conveniently close to a smaller co-gasification unit.

However, at the current time, neither biomass nor wastes are produced, or naturally gathered, at sites in quantities sufficient to fuel a modern large and efficient power plant. The disruption, transport issues, fuel use, and public opinion all act against gathering such fuels at a single location. Biomass or waste-fired power plants are therefore inherently limited in size and hence in efficiency, labor costs per unit electricity produced, and in other economies of scale. The production rates of municipal refuse follow reasonably predictable patterns over periods of a few years. Recent experience with the very limited current *biomass for energy* harvesting has shown unpredictable variations in harvesting capability with long periods of zero production over large areas during wet weather.

References

Alstrup, I., 1988. On the kinetics of CO methanation on nickel surfaces. Journal of Catalysis 109, 241—251.

Anderson, L.L., Tillman, D.A., 1979. Synthetic Fuels from Coal: Overview and Assessment. John Wiley & Sons Inc, New York.

Arena, U., 2012. Process and Technological aspects of municipal solid waste gasification. A Review. Waste Management 32 (4), 625—639.

Argonne, 1990. Environmental Consequences of, and Control Processes for, Energy Technologies. Argonne National Laboratory. Pollution Technology Review No. 181. Noyes Data Corp., Park Ridge, New Jersey. Chapter 5.

Baker, R.T.K., Rodriguez, N.M., 1990. Coal—gasification. In: Fuel Science and Technology Handbook. Marcel Dekker Inc, New York. Chapter 22.

Bauer, A., Hrbek, R., Amon, B., Kryvoruchko, A.V., Machmüller, A., Hopfner-Sixt, K., Bodiroza, V., Wagentristl, H., Pötsch, E., Zollitsch, W., Amon, T., 2007. Potential of biogas production in sustainable biorefinery concepts. In: 5th Research and Development Conference of Central- and Eastern European Institutes of Agricultural Engineering, Kiev, June 20—24.

Bhattacharya, S., Siddique, A.H.Md M.R., Pham, H.-L., 1999. A study in Wood gasification on low tar production. Energy 24, 285–296.

Boateng, A.A., Walawender, W.P., Fan, L.T., Chee, C.S., 1992. Fluidized-bed steam gasification of Rice Hull. Bioresource Technology 40 (3), 235–239.

Bodle, W.W., Huebler, J., 1981. Coal gasification. Chapter 10. In: Meyers, R.A. (Ed.), Coal Handbook. Marcel Dekker Inc, New York, pp. 494–704.

Brage, C., Yu, Q., Chen, G., Sjöström, K., 2000. Tar evolution profiles obtained from gasification of biomass and coal. Biomass and Bioenergy 18 (1), 87–91.

Brar, J.S., Singh, K., Wang, J., Kumar, S., 2012. Cogasification of coal and biomass: a review. International Journal of Forestry Research 1–10.

Calemma, V., Radović, L.R., 1991. On the gasification reactivity of Italian Sulcis coal. Fuel 70, 1027.

Cavagnaro, D.M., 1980. Coal Gasification Technology. National Technical Information Service, Springfield, Virginia.

Chadeesingh, R., 2011. The Fischer-Tropsch process. Part 3, Chapter 5. In: Speight, J.G. (Ed.), The Biofuels Handbook. The Royal Society of Chemistry, London, United Kingdom, pp. 476–517.

Chen, G., Sjöström, K., Bjornbom, E., 1992. Pyrolysis/Gasification of wood in a pressurized fluidized bed reactor. Industrial and Engineering Chemistry Research 31 (12), 2764–2768.

Cheung, W.H., Lee, V.K.C., McKay, G., 2007. Minimizing dioxin emissions from integrated MSW thermal treatment. Environment and Science Technology 41 (6), 2001–2007.

Choy, K.K.H., Porter, J.F., Hui, C.W., McKay, G., 2004. Process design and feasibility study for small scale MSW gasification (Amsterdam, Netherlands). Chemical Engineering Journal 105 (1–2), 31–41.

Coelho, S.T., Velazquez, S.M.S.G., Pecora, V., Abreu, F.C., 2006. Energy generation with landfill biogas. In: Book of Proceedings of RIO6, World Climate & Energy Event, Rio de Janeiro, Brazil, November 17–18.

Collot, A.G., Zhuo, Y., Dugwell, D.R., Kandiyoti, R., 1999. Co-pyrolysis and cogasification of coal and biomass in bench-scale fixed-bed and fluidized bed reactors. Fuel 78, 667–679.

Cusumano, J.A., Dalla Betta, R.A., Levy, R.B., 1978. Catalysis in Coal Conversion. Academic Press Inc, New York.

Dickenson, R.L., Biasca, F.E., Schulman, B.L., Johnson, H.E., 1997. Refiner options for converting and utilizing heavy fuel oil. Hydrocarbon Processing 76 (2), 57.

Ergudenler, A., Ghaly, A.E., 1993. Agglomeration of alumina sand in a fluidized bed straw gasifier at elevated temperatures. Bioresource Technology 43 (3), 259–268.

Fairbridge, C., 2013. Conventional and Unconventional Fossil Fuel Supplies. CANMET Energy and Natural Resources Canada, Ottawa, Ontario, Canada.

Fiore, F., Lietti, L., Pederzani, G., Tronconi, E., Zennaro, R., Forzatti, P., 2004. Reactivity of paraffins, olefins and alcohols during Fischer-Tropsch synthesis on a Co/Al_2O_3 catalyst. Studies in Surface Science and Catalysis 147, 289–294.

Fonseca, Y.J., Fontal, B., Reyes, M., Suárez, T., Bellandi, F., Contreras, R.R., Cancines, P., Loaiza, A., Briceño, M., 2007. Hydrocarbon synthesis using iron and $Ruthenium/SiO_2$ with Fischer-Tropsch Catalysis. Avances en Química 2 (3), 15–21.

Fryer, J.F., Speight, J.G., 1976. Coal Gasification: Selected Abstract and Titles. Information Series No. 74. Alberta Research Council, Edmonton, Canada.

Gabra, M., Pettersson, E., Backman, R., Kjellström, B., 2001. Evaluation of cyclone gasifier performance for gasification of sugar cane residue–part 1: gasification of bagasse. Biomass and Bioenergy 21 (5), 351–369.

Garcia, X., Radović, L.R., 1986. Gasification reactivity of Chilean coals. Fuel 65, 292.

Gay, R.L., Barclay, K.M., Grantham, L.F., Yosim, S.J., 1980. Fuel production from solid waste (Chapter 17). In: Symposium on Thermal Conversion of Solid Waste and Biomass. Symposium Series No. 130. American Chemical Society, Washington, DC, pp. 227−236.

Gomez, E., Amutha, R.D., Cheeseman, C.R., Deegan, D., Wise, M., Boccaccini, A.R., 2009. Thermal plasma technology for the treatment of wastes: a critical review. Journal of Hazardous Materials 161 (2−3), 614−626.

Gross, M., Wolff, J., 2000. Gasification of residue as a source of hydrogen for the refining industry in India. In: Proceedings. Gasification Technologies Conference. San Francisco, California. October 8−11.

Hanaoka, T., Inoue, S., Uno, S., Ogi, T., Minowa, T., 2005. Effect of woody biomass components on air-steam gasification. Biomass and Bioenergy 28 (1), 69−76.

Hanson, S., Patrick, J.W., Walker, A., 2002. The effect of coal particle size on pyrolysis and steam gasification. Fuel 81, 531−537.

Hartmann, H., Ahring, B.K., 2005. The future of biogas production. In: Risø International Energy Conference on "Technologies for Sustainable Energy Development in the Long Term", Risø-r-1517(EN), May 23−25, pp. 163−172.

Howard-Smith, I., Werner, G.J., 1976. Coal Conversion Technology. Noyes Data Corp, Park Ridge, New Jersey.

Ishida, T., Yanagiharaa, T., Liu, X., Ohashi, H., Hamasaki, A., Honma, T., Oji, H., Yokoyama, T., Tokunaga, M., 2013. Synthesis of higher alcohols by Fischer−Tropsch synthesis over alkali metal-modified cobalt catalysts. Applied Catalysis A: General 458, 145−154.

Jung, C.H., Matsuto, T., Tanaka, N., 2005. Behavior of metals in ash melting and gasification-melting of municipal solid waste (MSW). Waste Management 25 (3), 301−310.

Ko, M.K., Lee, W.Y., Kim, S.B., Lee, K.W., Chun, H.S., 2001. Gasification of food waste with steam in fluidized bed. Korean Journal of Chemical Engineering 18 (6), 961−964.

Kristiansen, A., 1996. IEA Coal Research Report IEACR/86. Understanding Coal Gasification. International Energy Agency, London, United Kingdom.

Kumabe, K., Hanaoka, T., Fujimoto, S., Minowa, T., Sakanishi, K., 2007. Cogasification of woody biomass and coal with air and steam. Fuel 86, 684−689.

Kwon, E., Westby, K.J., Castaldi, M.J., 2009. An investigation into the syngas production from municipal solid waste (MSW) gasification under various pressures and CO_2 concentration atmospheres. Paper No. NAWTEC17-2351. In: Proceedings of the 17th Annual North American Waste-to-energy Conference. NAWTEC17. Chantilly, Virginia. May 18−20.

Lahaye, J., Ehrburger, P., 1991. Fundamental Issues in Control of Carbon Gasification Reactivity. Kluwer Academic Publishers, Dordrecht, Netherlands.

Lv, P.M., Xiong, Z.H., Chang, J., Wu, C.Z., Chen, Y., Zhu, J.X., 2004. An Experimental study on biomass air-steam gasification in a fluidized bed. Bioresource Technology 95 (1), 95−101.

Mahajan, O.P., Walker Jr., P.L., 1978. Reactivity of heat-treated coals. Chapter 32. In: Karr Jr., C. (Ed.), Analytical Methods for Coal and Coal Products, vol. II. Academic 6 Press Inc, New York, pp. 465−494.

Massey, L.G. (Ed.), 1974. Coal Gasification. Advances in Chemistry Series No. 131. American Chemical Society, Washington, DC.

McKendry, P., 2002. Energy production from biomass part 3: gasification technologies. Bioresource Technology 83 (1), 55−63.

Mills, G.A., 1969. Conversion of coal to gasoline. Industrial and Engineering Chemistry 61 (7), 6−17.

Mohan, D., Pittman Jr., C.U., Steele, P.H., 2006. Pyrolysis of wood/biomass for bio-oil: a critical review. Energy and Fuels 20 (3), 848−889.

Mokhatab, S., Poe, W.A., Speight, J.G., 2006. Handbook of Natural Gas Transmission and Processing. Elsevier, Amsterdam, Netherlands.

Nielsen, J.B.H., Oleskowicz-Popiel, P., Al Seadi, T., 2007. Energy crops potentials for bioenergy in EU-27. In: 15th European Biomass Conference & Exhibition. From Research to Market Deployment, Berlin, Germany, May 7–11.

Pakdel, H., Roy, C., 1991. Hydrocarbon content of liquid products and tar from pyrolysis and gasification of wood. Energy and Fuels 5, 427–436.

Pan, Y.G., Velo, E., Roca, X., Manyà, J.J., Puigjaner, L., 2000. Fluidized-bed cogasification of residual biomass/poor coal blends for fuel gas production. Fuel 79, 1317–1326.

Probstein, R.F., Hicks, R.E., 1990. Synthetic Fuels. pH Press, Cambridge, Massachusetts. Chapter 4.

Rabiu, A.M., Van Steen, E., Claeys, M., 2012. Further investigation into the formation of alcohol during Fischer Tropsch synthesis on Fe-based catalysts. Proceedings 2nd International Conference on Chemistry and Chemical Process (ICCCP 2012) May 5–6, 2012 APCBEE Procedia 3, 110–115.

Radović, L.R., Walker Jr., P.L., Jenkins, R.G., 1983. Importance of carbon active sites in the gasification of coal chars. Fuel 62, 849.

Radović, L.R., Walker Jr., P.L., 1984. Reactivities of chars obtained as residues in selected coal conversion processes. Fuel Processing Technology 8, 149.

Ramroop Singh, N., 2011. Biofuels. Part 1, Chapter 5. In: Speight, J.G. (Ed.), The Biofuels Handbook. The Royal Society of Chemistry, London, United Kingdom, pp. 160–200.

Rapagnà, Latif, A., 1997. Steam gasification of almond shells in a fluidized bed reactor: the influence of temperature and particle size on product yield and distribution. Biomass and Bioenergy 12 (4), 281–288.

Rapagnà, N.J., Kiennemann, A., Foscolo, P.U., 2000. Steam-gasification of biomass in a fluidized-bed of olivine particles. Biomass and Bioenergy 19 (3), 187–197.

Richards, B., Cummings, R.J., White, T.E., Jewell, W.J., 1991. Methods for kinetic analysis of methane fermentation in high solids biomass digesters. Biomass and Bioenergy 1 (2), 65–66.

Richards, B., Herndon, F.G., Jewell, W.J., Cummings, R.J., White, T.E., 1994. In Situ methane enrichment in methanogenic energy crop digesters. Biomass and Bioenergy 6 (4), 275–282.

Ricketts, B., Hotchkiss, R., Livingston, W., Hall, M., 2002. Technology status review of waste/biomass co-gasification with coal. In: Proceedings. Inst. Chem. Eng. Fifth European Gasification Conference. Noordwijk, The Netherlands. April 8–10.

Rutz, D., Janssen, R., 2007. Biofuel Technology Handbook. WIP Renewable Energies, Munich, Germany.

Rostrup-Neilsen, J.R., 2008. Steam Reforming–Energy Related Catalysis. John Wiley & Sons Inc, Hoboken, New Jersey.

Sjöström, K., Chen, G., Yu, Q., Brage, C., Rosén, C., 1999. Promoted reactivity of char in cogasification of biomass and coal: synergies in the thermochemical process. Fuel 78, 1189–1194.

Speight, J.G., 2007. Natural Gas: A Basic Handbook. GPC Books, Gulf Publishing Company, Houston, Texas.

Speight, J.G., 2008. Synthetic Fuels Handbook: Properties, Processes, and Performance. McGraw-Hill, New York.

Speight, J.G., 2009. Enhanced recovery methods for heavy oil and tar sands. Gulf Publishing Company, Houston, Texas.

Speight, J.G. (Ed.), 2011a. The Biofuels Handbook. Royal Society of Chemistry, London, United Kingdom.

Speight, J.G., 2011b. The Refinery of the Future. Gulf Professional Publishing, Elsevier, Oxford, United Kingdom.

Speight, J.G., 2011c. An Introduction to Petroleum Technology, Economics, and Politics. Scrivener Publishing, Salem, Massachusetts.

Speight, J.G., 2013a. The Chemistry and Technology of Coal, third ed. CRC Press, Taylor and Francis Group, Boca Raton, Florida.

Speight, J.G., 2013b. Coal-fired Power Generation Handbook. Scrivener Publishing, Salem, Massachusetts.

Speight, J.G., 2014. The Chemistry and Technology of Petroleum, fifth ed. CRC Press, Taylor and Francis Group, Boca Raton, Florida.

Speight, J.G., 2015. Handbook of Coal Analysis, second ed. John Wiley & Sons Inc, Hoboken New Jersey.

Udengaard, N.R., Hansen, J.H.B., Hanson, D.C., Stal, J.A., 1992. Sulfur-passivated reforming process lowers syngas H_2/CO ratio. Oil Gas Journal 90, 62–67.

Unruh, D., Pabst, K., Schaub, G., 2010. Fischer–Tropsch synfuels from biomass: maximizing carbon efficiency and hydrocarbon yield. Energy and Fuels 24, 2634–2641.

Vélez, J.F., Chejne, F., Valdés, C.F., Emery, E.J., Londoño, C.A., 2009. Cogasification of Colombian coal and biomass in a fluidized bed: an experimental study. Fuel 88, 424–430.

Xiang, M., Li, D., Xiao, H., Zhang, J., Qi, H., Li, W., Zhong, B., Sun, Y., 2008. Synthesis of higher alcohols from syngas over Fischer–Tropsch elements modified $K/\beta\text{-}Mo_2C$ catalysts. Fuel 87, 599–603.

Yun, Y., Chung, S.W., Yoo, Y.D., 2003. Syngas quality in gasification of high moisture municipal solid wastes. Preprints Division of Fuel Chemistry, American Chemical Society 48 (2), 823–824.

Part Three

Combustion and conversion technologies

Technology options for large-scale solid-fuel combustion   7

Markus Hurskainen, Pasi Vainikka
VTT Technical Research Centre of Finland Ltd, Jyväskylä, Finland

7.1 Introduction

The goal of CO_2 emission reduction and the subsequent renewable energy incentives have caused a switch from fossil fuels towards alternative fuels. The palette of biomass and waste derived fuels utilised in energy production has become more and more diverse and will continue to do so. Competitiveness of different fuels is highly dependent on legislation (taxes, subsidies), the changes of which can be difficult to predict. The availability of certain fuels can also change due to competition between different utilisers. Fuel flexibility is thus an important aspect of energy production. Plants having high fuel flexibility can adapt to the prevailing fuel market situation by changing their fuel mix to the most economical one.

In addition to enabling a wider range of fuels, co-firing offers the possibility to take advantage of synergy effects between various fuels. Combustion properties of biomass and especially waste-derived fuels limit the electrical efficiency of dedicated biomass or waste-to-energy plants. Biomass and waste-derived fuels are known to cause various ash-related operational problems such as slagging, fouling and high-temperature corrosion of heat-transfer surfaces. On the other hand, when these problematic fuels are co-fired with coal, the favourable properties of coal ash can prevent or largely reduce the aforementioned problems. At the same time, biomass ash may reduce acidic emissions from coal combustion. Thus, it is often more attractive to co-fire fuels rather than use dedicated boilers for each fuel type. However, co-firing might limit ash utilisation possibilities.

In this chapter, the technology options for biomass and waste fuel utilisation in large-scale combustion plants are presented with the focus on technologies allowing fuel flexibility. The three main combustion technologies for solid fuels are grate, fluidised-bed and pulverised-fuel combustion.

7.2 Combustion technologies for solid fuels

7.2.1 Grate combustion

7.2.1.1 Basics

Grate-fired boilers or stokers are the oldest method used for direct combustion of solid fuels. The main advantage of grate-fired boilers compared to their main rival, fluidised-bed boilers, is their ability to burn various fuels with little pre-handling. Grate-fired units are not sensitive to large metal, glass, stone impurities etc., and they do not suffer from bed agglomeration problems as the fluidised-bed boilers sometimes do. The disadvantages are lower combustion efficiency and higher emissions due to worse mixing of fuel and air, worse heat transfer characteristics and higher combustion temperature. As opposed to fluidised beds, the temperature in the combustion zone is not measured or controlled and it is typically in the range of 1200−1400 °C. In addition, for satisfactory combustion, grate-fired units usually need fuel with constant particle-size distribution and moisture content.

7.2.1.2 Modern stokers for biomass

From the early days, stokers have developed from inefficient and unstable boilers to a viable technology in certain cases. The main advancements have been the development of advanced over-fire air (OFA) systems and better distribution of the fuel, which have led to increased combustion efficiency, lower emissions and better stability of the combustion (Yin et al., 2008).

A modern stoker consists of:

- A fuel feeding system
- A grate for supporting the fuel and allowing under-grate (primary) air feeding
- An over-fuel air system for volatile combustion and minimising emissions
- An ash/residue discharge/re-injection system
- Flue gas treatment system

Depending on how the fuel is introduced to the boiler, stokers can be divided into underfeed and overfeed stokers. In the underfeed stokers both the fuel and combustion air are introduced from under the grate, whereas in overfeed stokers fuel is supplied from above the grate and air from below. Nowadays, overfeed stokers are far more common. They can be further divided into mass-feed and spreader stokers on the basis of fuel-feeding arrangements. In mass-feed stokers, fuel is continuously fed to one end of the grate after which it travels to the other end where the ash is then discarded. In spreader stokers, fuel is thrown (or spread) evenly to the whole grate area; the finer fuel particles burn in suspension and only the large particles drop onto the grate. For woody biomass and refuse-derived fuel (RDF), spreader stoker is the preferred choice. Mass feed has to be used for example for loose straw chaff, the density for which is too low for the spreader system and for untreated municipal solid waste (MSW) (Stultz and Kitto, 2005).

Grates come in various constructions. In modern spreader stokers designed for woody biomass, the grate is typically either a travelling grate or a water-cooled (or air-cooled) vibrating grate. Vibrating grates are often the preferred choice due to more robust structure. In Figure 7.1, a typical stoker for woody biomass with water-cooled vibrating grate is shown. For straw, mass-feed water-cooled vibrating grates are typically used to control slagging and ash melting. For untreated MSW, reciprocating and pusher-rod grates with mass feeding are common (waste incinerators).

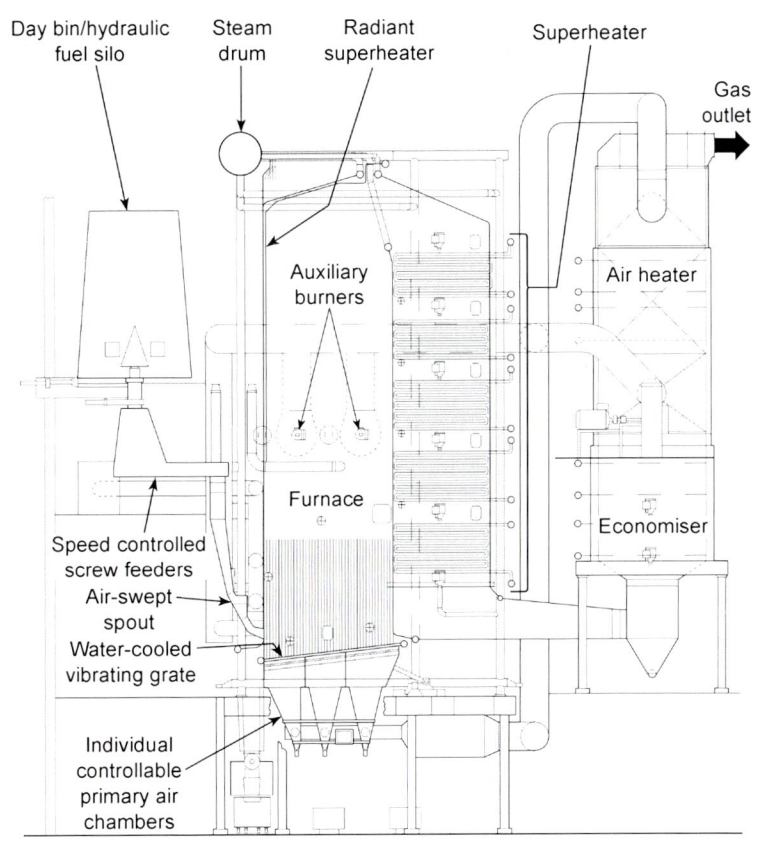

Figure 7.1 A typical biomass stoker with water-cooled vibrating grate for biomass. Courtesy of The Babcock & Wilcox Company, Stultz and Kitto (2005).

For fuels containing a high share of fines, a spreader is needed to avoid fuel segregation as the grate is typically suitable only for coarse particles. An example of a fuel feeder is shown in Figure 7.2. By using a spreader that throws the fuel onto the grate, finer particles have time to burn in suspension above the grate, whereas larger particles drop down to the grate and burn there (Yin et al., 2008). However, due to high shaft velocities in the lower furnace, this leads also to fuel 'carryover' out of the furnace: (partially) unburned fuel particles are entrained by the flue gases causing an efficiency loss of up to 4—6% (DeFusco et al., 2007). To avoid this marked loss, spreader stoker boilers can be equipped with carryover re-injection systems that recycle carryover particles to the furnace. In case of wood or bark boilers, a sand classifier is also needed before the re-injection to get rid of the highly abrasive sand particles. Re-injection systems are high-maintenance systems and according to The Babcock & Wilcox Company they have been shut down at many plants (DeFusco et al., 2007). A mechanical dust collector (cyclone) is also typically installed to prevent any heavy-particle

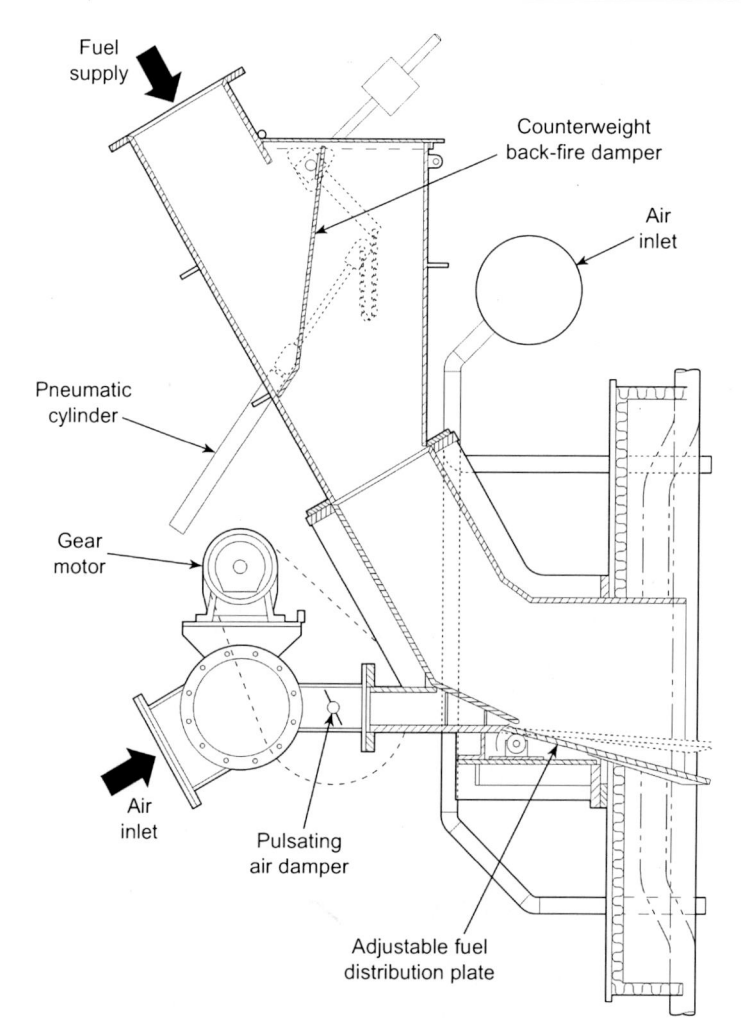

Figure 7.2 Air-swept spouts used for feeding of biomass fuels in stokers. Courtesy of The Babcock & Wilcox Company, Stultz and Kitto (2005).

carryover from reaching the main particulate reduction step which can be either electrostatic precipitator (ESP) or baghouse filter (BHF). In grate-fired units, ESPs are often preferred over BHF filters due to concerns on hot carryover particles possibly igniting the filter bags.

Even though grate boilers can utilise various fuels, they need a relatively constant fuel quality to be efficient: the moisture content and particle size should not vary too much. Thus, they are not very fuel flexible and are rarely used for co-firing purposes. The field in which grate combustion has found success is waste incineration, as no other traditional combustion technique can utilise untreated household or municipal waste.

7.2.2 Fluidised-bed combustion

7.2.2.1 Basics

In fluidised-bed combustion, fuels are burned in a bed of hot inert particles (typically sand) that are fluidised by the combustion air (primary air) fed from below the bed. Volatiles are then burned in the freeboard in which secondary and often tertiary air are introduced. The large mass of hot bed material is able to absorb fluctuations in fuel quality with little to no change in performance. Due to excellent mixing and heat-transfer characteristics, combustion efficiency of fluidised-bed boilers is superior to grate boilers. Fluidised boilers are suitable for co-firing fuels with different combustion properties, and they are not so sensitive to fluctuations in fuel moisture content or particle size.

One major difference of fluidised-bed combustion compared to grate and pulverised fuel (PF) combustion is the greatly lower combustion temperature, which is made possible by the efficient mixing and heat transfer due to the large mass of hot inert material. In fluidised-bed boilers bed temperature is measured and controlled to an optimal temperature that is typically in the range of 800–900 °C. The optimal bed temperature can be achieved by air staging (controlling the share of primary air), flue gas recirculation and fuel feeding. As the boiler operates at nearly a steady-state condition, a positive impact on boiler's emissions and performance is achieved. The lower temperature helps to reduce NO_x emissions which are controlled also by air staging and flue gas recirculation.

Another advantage of fluidised-bed boilers is the possibility to use in-furnace measures for SO_2 capture using limestone as an additive. Limestone is injected directly to furnace as solid particles, which means there is no need for a separate desulphurisation plant.

The main disadvantage of fluidised-bed combustion is the risk of bed agglomeration with certain fuels. The term bed agglomeration is used to describe the phenomenon in which separate bed particles adhere to each other to form larger particles. At some point, these particles are too large to stay in fluidised state any longer, and in severe cases this may result in total defluidisation and thus force a shutdown of the plant. Fuels that are problematic for fluidised beds are, for example, many agro-biomass fuels such as straw, olive residue and chicken litter which contain high amounts of alkalis. Alkalis are known to react with quartz (SiO_2), which is present in large amounts in typical bed sand, forming 'sticky' alkali-silicates that have a low melting point and thus cause a risk of bed agglomeration. In some cases bed agglomeration can be avoided using special bed materials (e.g., Silvennoinen, 2003) or additives such as kaolin (e.g., Öhman and Nordin, 2000).

The investment and operation and maintenance costs of fluidised-bed boilers are typically slightly higher than those of comparable grate or PF boilers. There are also some limitations on operation on low partial loads as fluidisation must be maintained at all times. In addition, the bed material (sand) and possible limestone in-furnace injection can affect the ash utilisation.

7.2.2.2 Bubbling fluidised-bed (BFB) and circulating fluidised-bed (CFB) technologies

Fluidised-bed combustion can be divided to bubbling fluidised-bed (BFB) and circulating fluidised-bed (CFB) combustion. In addition, there are turbulent, spouted bed and pressurised designs, but they have not been able to penetrate markers and are thus of lesser importance and will not be discussed here.

In BFB boilers, the bed material stays at the bottom part of the boiler as a dense bed (Figure 7.3). Typically, the bed height in the fluidised state is around 1 m. Typical fluidising velocities are in the range of 1–3 m/s and the average bed particle size is around 1 mm (Raiko et al., 2002).

The bed operates at reducing conditions: typically around 40% of the air is fed as primary air to minimise the NO_x emissions (Raiko et al., 2002). Volatiles are then burned using secondary and tertiary air in the freeboard area above the bed. In BFB boilers, the temperature tends to peak at the location where secondary air is introduced. The temperature increase can be some 200 °C. In some cases (e.g., Silvennoinen and Hedman, 2011) this can cause slagging problems.

By controlling the ratio of primary to secondary and tertiary air, it is possible to also control the bed temperature. The reducing conditions prevailing in the bed area means that all the oxygen introduced by the primary air gets consumed in combustion reactions generating heat. Thus, by increasing the share of primary air, more heat release

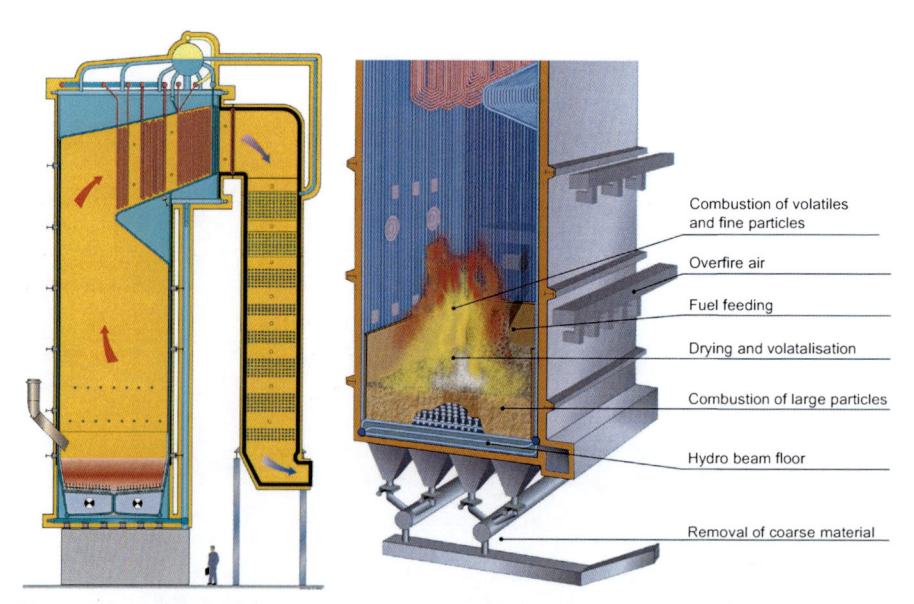

Figure 7.3 A typical BFB boiler cross-section and illustration of the bottom part of a BFB furnace.
Courtesy of Foster Wheeler (left) and Valmet Power (formerly Metso Power) (right).

takes place in the bed and subsequently the bed temperature increases. However, bed temperature responds quite slowly to the changes in primary airflow rate. According to Stultz and Kitto (2005), the increase of bed temperature by 15 °C and returning back to the original temperature can take 20 min or more. More rapid and accurate bed temperature control can be achieved by flue gas recirculation which at the same time reduces NO_x formation.

BFB boilers are especially well suited for highmoisture, low calorific value fuels with a high share of volatiles such as wood chips and residues. Typically, fuel particle size should be less than 80 mm. Two types of fuel feeders are typically used for biomass fuels: (1) chutes that direct fuel to small distinct areas and (2) air distribution feeders that distribute the fuel into larger areas. When fuel is fed into a small spot, the in-bed heat release is the lowest, and the highest in-bed release is achieved when the fuel is fed evenly into the whole bed area. By controlling the fuel distribution by adjusting the feeder airflow, this gives yet another possibility to control bed temperature, for example when fuel moisture content is changing.

In BFB boilers, the share of coal is limited to some 10–20%. Otherwise, char can accumulate to the bed and then cause sudden bed-temperature increase, when the primary air flow is increased for a load change. This can lead to bed agglomeration. BFB combustion technology requires that major part of the fuel energy is released above the bed.

Typical capacity range for BFB boilers is 20–200 MW_{fuel}. Boiler efficiencies are typically >90% on lower heating value (LHV) basis (Peña, 2011).

In CFB boilers, the fluidising velocity is so high that (the major) part of the bed material gets elutriated from the bed and has to be returned using a particle separator, which is almost always a cyclone (Figure 7.4). So-called U-beams are also used in some cases. Fluidising velocity is typically some 3–8 m/s and <0.5 mm bed particles are used (Raiko et al., 2002). In CFB boilers, the density of the bed decreases as function of the height so there is no distinct bed area. Due to better heat and mass transfer, the temperature profile of a CFB furnace is very even: the difference between the hottest and coldest point can be only some 30 °C. This has a favourable impact on emissions and the overall combustion process. Compared to BFB boilers, the process conditions (residence time, temperature, oxidising atmosphere) in CFB boilers are more favourable for SO_2 removal using in-furnace limestone injection.

CFB technology further broadens the co-firing possibilities compared to BFB boilers. With CFB boilers, it is possible to use 100% biowaste or 100% coal or any mixture of these making CFB combustion the most fuel-flexible combustion technology available. CFB boilers can also utilise low-quality coals with very high ash content. Recommended particle size of biomass fuels is less than 40 mm.

The excellent fuel flexibility of CFB boilers offers a possibility to control high-temperature corrosion by manipulating the ash chemistry by co-firing problematic fuels with coal or other fuels with suitable properties. In CFB boilers, it is also possible to place the final superheaters — which have the highest risk of high-temperature chlorine-induced corrosion — into the cyclone loop seal to mitigate corrosion. The idea behind the loop seal superheaters is that they are not in contact with the flue gases containing the corrosive alkali or heavy metal chloride vapours, and thus corrosion is reduced.

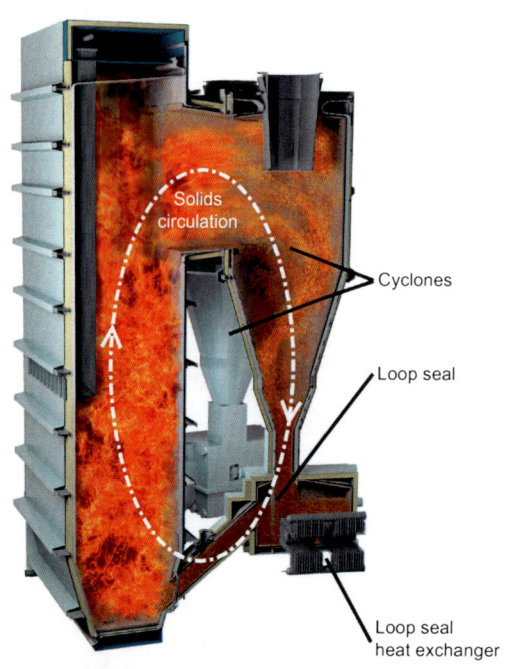

Figure 7.4 An illustration of a CFB boiler furnace and cyclones. Courtesy of Valmet Power.

Nowadays loop seal superheaters are widely used in CFB boilers and manufacturers have their own designs. Schematic of Foster Wheeler's loop seal heat exchanger (INTREX™) is shown in Figure 7.5. With INTREX™, the fluidising sand for the heat exchanger can be introduced from external circulation (from down comer) or from internal circulation (sand falling down the furnace walls). Sand is returned to the furnace via return channels. With Valmet Power's (formerly Metso Power) loop seal heat exchanger design, the amount of circulating material from external circulation that goes to the loop seal heat exchanger can be controlled by a bypass.

As there are no moving parts, the loop seal heat exchangers are reliable. Due to more efficient heat transfer, loop seal heat exchangers can be smaller than convective heat exchangers. In addition, the heat transfer can be controlled by the fluidising airflow adjustment. In addition, as fluidisation velocity can be kept low, erosion can be controlled. Due to special conditions, good knowledge of materials is, however, still needed especially when waste-derived fuels are used.

As CFB boilers are more complicated (=more expensive) than BFBs, CFB boilers are typically used in large-scale applications, when coal is also in the fuel mix or when the fuels are too dry for BFB boilers. Typical range for multi-fuel CFB boilers is some 50–500 MW$_{fuel}$ even though the largest ones are already close to 1000 MW$_{fuel}$. The smallest CFBs tend to be run mostly on waste-derived fuels and the largest ones mostly on coal. The boiler efficiency can be up to 95% LHV (Peña, 2011).

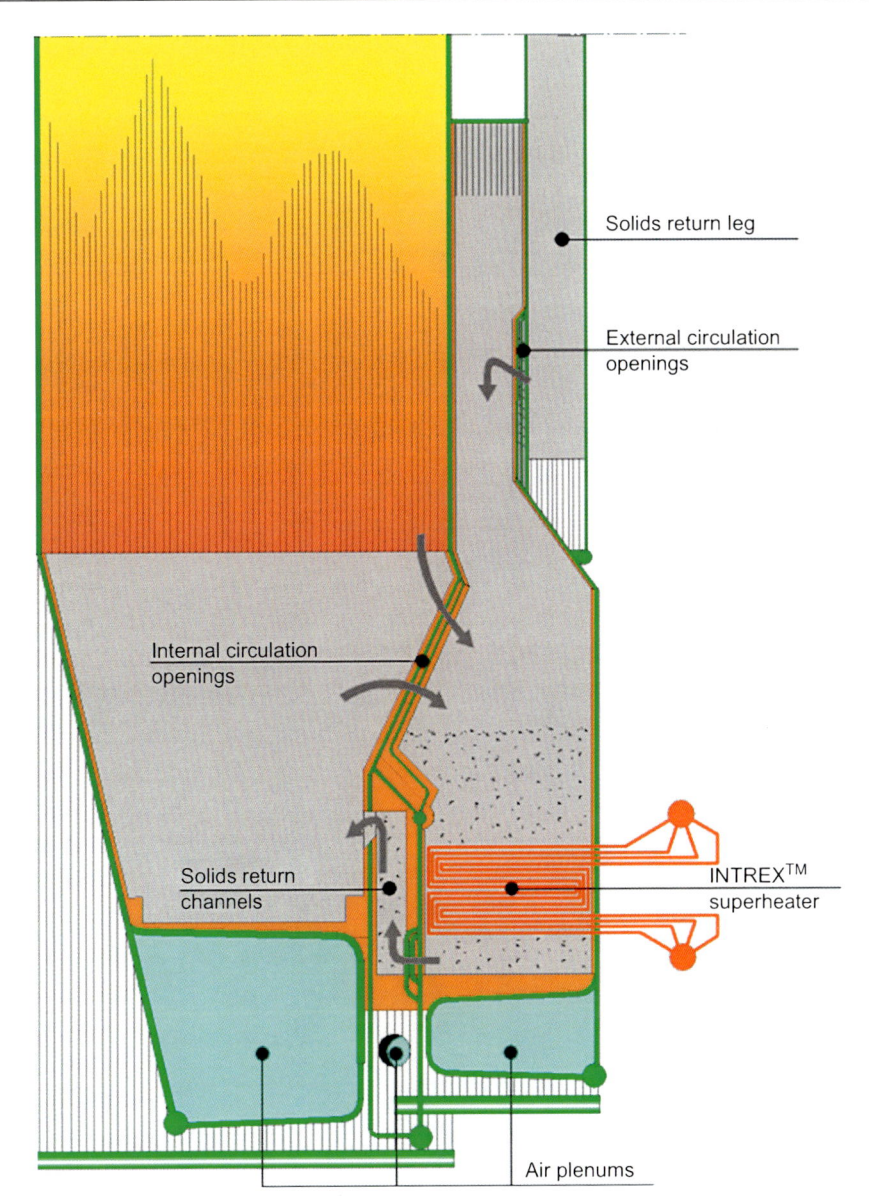

Technology options for large-scale solid-fuel combustion 185

Figure 7.5 INTREX™ loop seal heat exchanger.
Courtesy of Foster Wheeler.

7.2.2.3 Modern fluidised-bed boilers for biomass

Common design features of modern fluidised beds for biomass firing are:

- moderate fluidising velocity for reducing fouling and erosion
- wider superheater tube pitches compared to 'non-problematic' fuels or fuel mixes

- water- or steam-cooled solids separators (cyclones) in CFBs for less maintenance need (Figure 7.6)
- loop seal heat exchangers in CFBs
- optimised flue gas and superheater temperatures with respect to each other
- advanced coarse material removal

An example of a very large and fuel-flexible CFB boiler is Alholmens Kraft's AK2 boiler located in Pietarsaari, Finland. The plant produces electricity and supplies heat to an adjacent paper mill and district heating. The thermal power of the reheat boiler is 550 MW and the steam parameters are 165/38 bar and 545/545 °C. The plant can produce 240 MW power in condensing mode or 205 MW power, 100 MW process steam and 60 MW district heating. The boiler was supplied by Valmet Power and has been in operation since 2001.

The boiler was designed to use 100% peat, 100% coal, 100% biofuels or any mixture of those. Solid-recovered fuel (SRF) was also included in the fuel palette in 2008. In 2012, the fuel use was 53% wood, 20% peat, 18% coal, 9% SRF and 1% others on energy basis (Alholmens Kraft Website).

The positive effect of biomass co-firing on SO_2 emissions in the AK2 boiler is illustrated in Figure 7.7, in which the SO_2 reduction is shown as a function of biomass share in the fuel blend on full and partial loads. The reduction is caused by reaction between alkali and earth alkali metals present in biomass and SO_2 which is favoured in the CFB conditions. This is sometimes referred as 'SO_2 auto-reduction'.

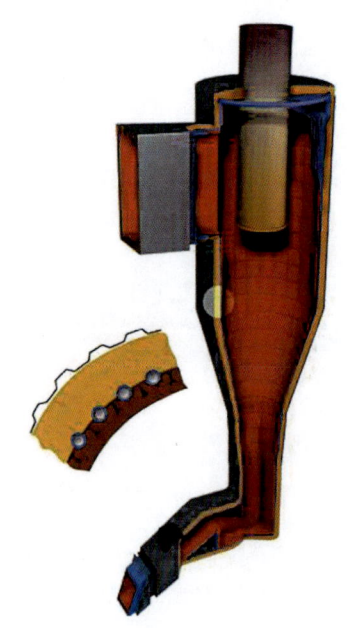

Figure 7.6 A modern water-cooled cyclone with light refractory coating. Courtesy of Valmet Power.

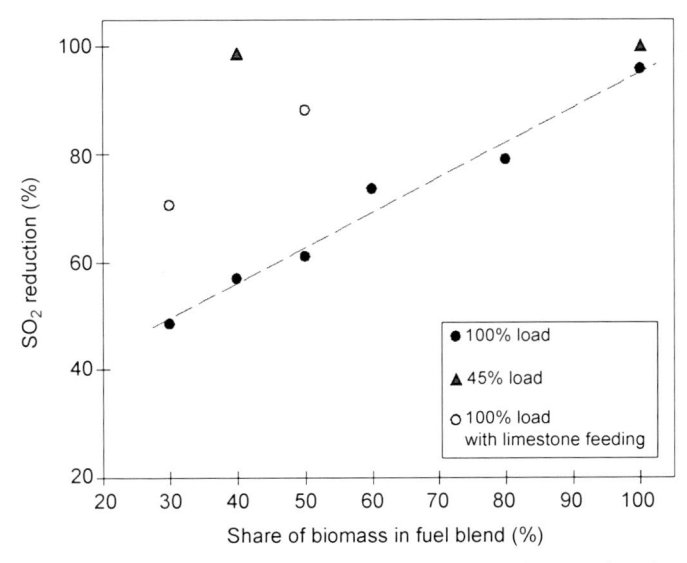

Figure 7.7 SO_2 reduction in the Alholmens Kraft's AK2 CFB boiler as a function of biomass energy share in the fuel mix. 'Auto-reduction' cases with black markers (VTT, Biomax: Maximum Biomass Use and Efficiency in Large-scale Cofiring—project).

7.2.3 Pulverised-fuel combustion

7.2.3.1 General

Pulverised coal (PC) combustion is the most widely used technology for utility-scale power generation in the world. In PC boilers, coal is ground into fine particles (~100 μm) and then injected with heated combustion air through a number of burners into the lower part of the furnace. Particles burn in suspension and release heat which is transferred into the steam cycle.

Based on burner arrangement, PC boilers can be divided into horizontally fired, tangentially fired and down-shot boilers. In horizontally fired boilers, burners are typically located on one (Figure 7.8) or two opposite walls and each burner produces an independent flame zone requiring a correct amount of secondary air supply for each burner. A high-turbulence swirl is created in the burners for efficient combustion.

In tangentially fired boilers, burners are located on the corners of the furnace. Together the flames produce a single cylindrical combustion zone in the centre of the furnace. Thus, the adjustment of secondary air for each burner separately is not so sensitive.

In down-shot boilers, coal is injected downwards into the refractory-lined lower furnace (Figure 7.9). This promotes complete combustion of low-volatile content coals such as anthracite. Nowadays, down-shot boilers are not common as boilers have become larger and other burner configurations can thus offer adequate flame lengths.

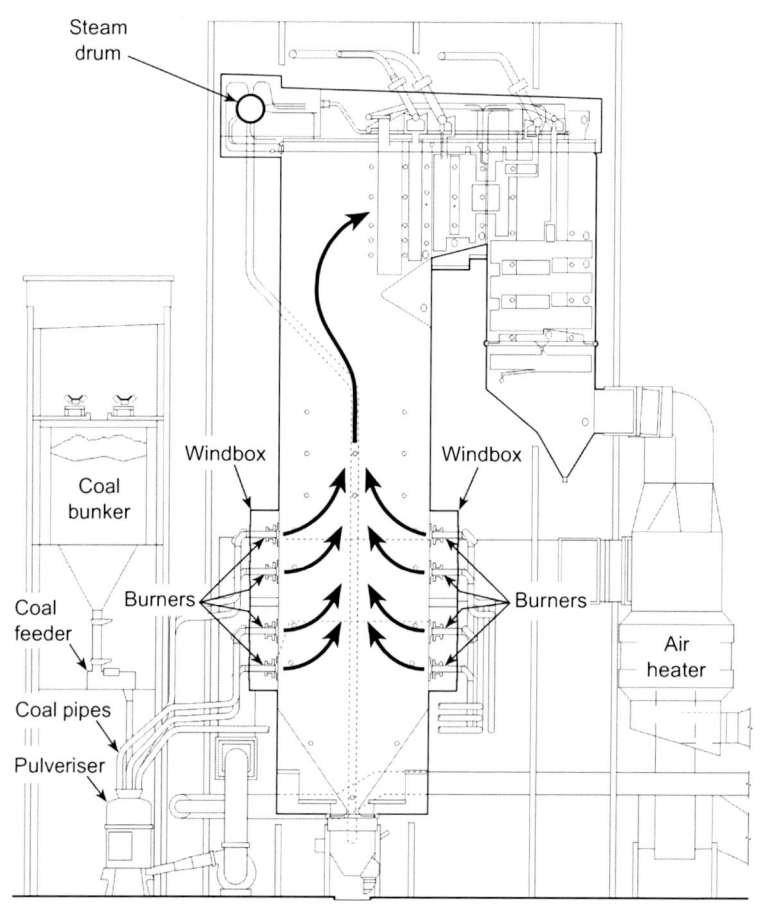

Figure 7.8 Schematic of a wall-fired PC boiler having burners on opposite walls. Courtesy of The Babcock & Wilcox Company, Stultz and Kitto (2005).

Compared to grate and especially fluidised-bed boilers, PC boilers have significantly stringent requirements for the fuels especially with respect to particle size and moisture content. Drying and milling consumes a lot of energy, reducing the overall efficiency of the plant.

The combustion temperature in PC boilers is high, some 1300–1700 °C and residence time is approximately 1–2 s. The high combustion temperature promotes good burnout but causes high NO and NO_2 emissions. So-called low-NO_x burners, which operate at 0.85–0.95 air ratio, and over-fire air supply are used for primary control of NO_x emissions. Reduction is limited by the high temperature needed to ensure ignition with low-NO_x burners (Jalovaara et al., 2003).

The main advantages of PC combustion are high efficiency and reliability, good load following capability and great upwards scalability. Units over 1000 MW_e (>2000 MW_{fuel}) are available.

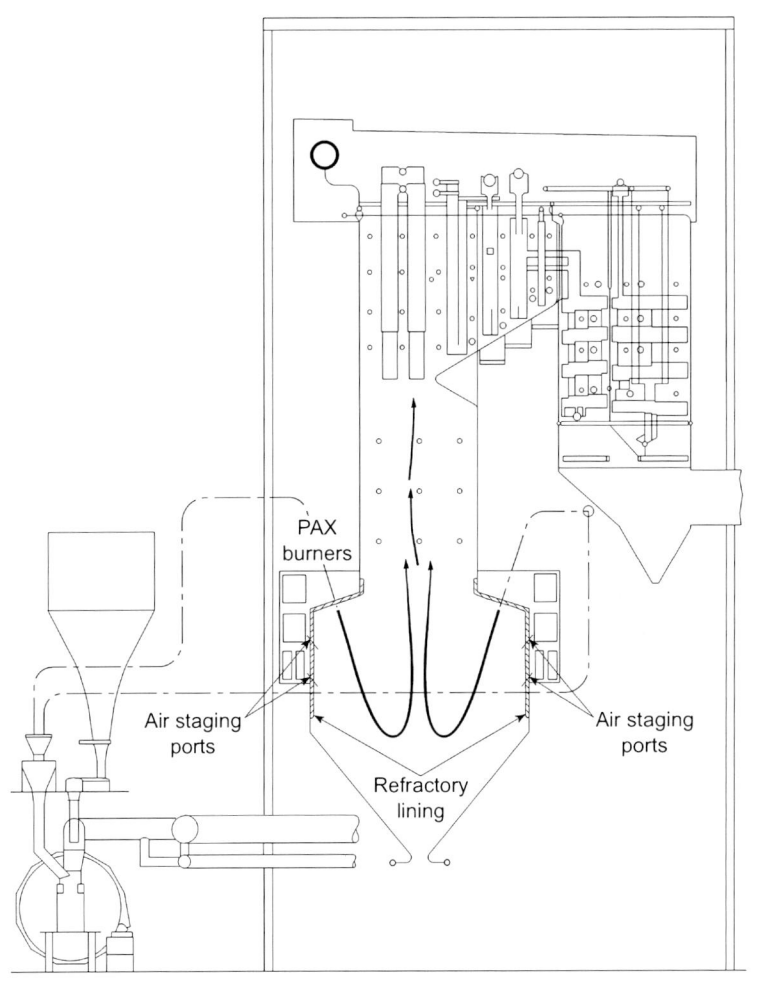

Figure 7.9 A down-shot PC furnace.
Courtesy of The Babcock & Wilcox Company, Stultz and Kitto (2005).

Almost all large-scale pulverised-fuel boilers were originally designed to use only coal but various ways of adopting biomass co-firing are available. These are discussed in the next chapter.

7.2.3.2   Technology options for co-firing in pulverised-coal plants
General
Co-firing biomass in traditional pulverised-coal (PC) boilers takes advantage of the existing infrastructure and the high efficiency of PC boilers requiring only very minor to moderate investments depending on the selected co-firing technology and the desired co-firing percentage.

Oftentimes (e.g. van Loo and Koppejan, 2002) the available co-firing technologies are divided into three groups:

1. Direct co-firing, in which biomass and coal are fired in the same furnace
2. Indirect co-firing, in which the solid biomass is first converted into gaseous or liquid form before firing in the same furnace
3. Parallel co-firing, in which biomass is fired on a separate boiler but the steam cycle of which is integrated to the coal boiler's steam cycle

In addition, a PC boiler can be converted into a BFB boiler after which it can be run on 100% biomass. Several subcategories can also be distinguished as shown in Figure 7.10. Next, these are discussed in more detail.

Direct co-firing in PC boilers

Direct co-firing is the most straightforward and the least capital-intensive method but it also has the most stringent requirements for fuel quality. In direct co-firing applications, biomass must be dried (to some <20 wt%) and comminuted to below some 2-mm particle size to achieve satisfactory combustion performance. As biomass particles tend to be long and narrow, the concept of particle size is a bit tricky. Acceptable length of the particles can be some 10–20 mm.

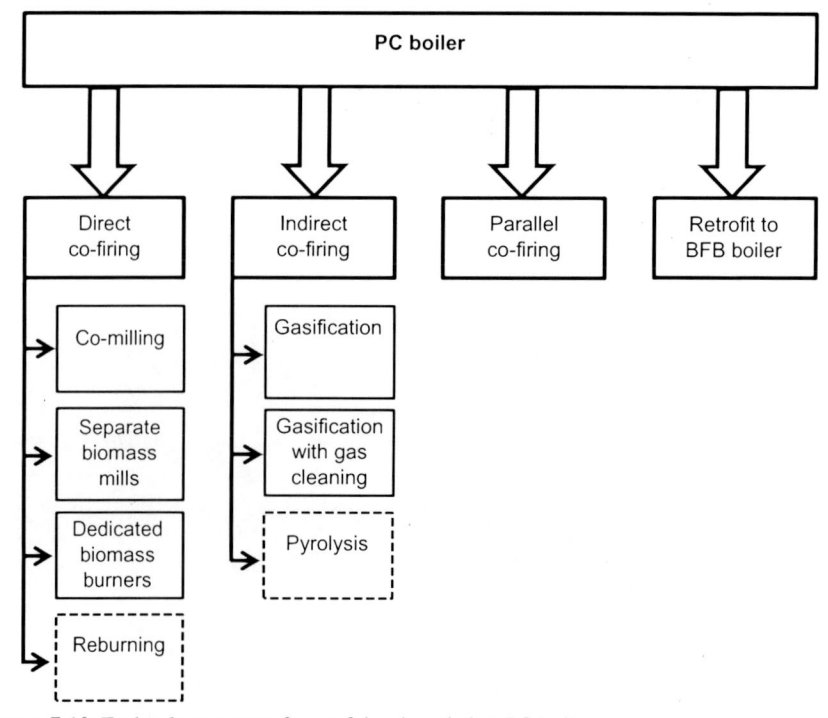

Figure 7.10 Technology routes for co-firing in existing PC boilers.

As shown in Figure 7.10, the direct co-firing technology options can be further divided into four classes based on the approach taken to pulverise and burn the biomass:

1. 'Co-milling': (pre-processed) biomass is mixed with coal before the coal mills
2. 'Separate milling': (pre-processed) biomass is pulverised on a separate mill and mixed with coal before the burners
3. 'Dedicated biomass burners': (pre-processed) biomass is pulverised on a separate mill and burned on dedicated biomass burners
4. 'Re-burning': (pre-processed) biomass is pulverised on a separate mill and is burned as re-burn fuel on dedicated burners located in the upper furnace for NO_x control

The most suitable method is dependent on various things such as the biomass type and quality, desired co-firing percentage, existing milling capacity and available space. Next, the pros and cons of these methods are described and examples are given of power plants, in which the methods have been adopted.

The main advantage of the co-milling option are the very low capital costs as investments are needed only for biomass receiving, storing and possibly pre-processing (drying, crushing/chipping). However, the risks for boiler operation are the highest and the maximum obtainable co-firing percentages and the range of usable fuels are the lowest. The main limitation in biomass share in this option is the milling capacity. The soft and fibrous structure of many biomasses makes their pulverisation in traditional coal mills ineffective. At the same time, the coal milling performance also decreases, which may cause unacceptable carbon losses due to incomplete combustion. This method is currently suitable practically only for sawdust and pellets. As pellets are composed of small particles, they are more readily pulverised in coal mills compared to untreated wood chips. Further improvement in grindability and other handling properties could be achieved through torrefaction or steam explosion of biomass before pelletisation. However, these technologies are not used at industrial scale today. For example, in Kärki et al. (2011) it was estimated that the maximum share for sawdust, traditional wood pellets and torrefied wood pellets are some 5, 15 and 50% (on energy), respectively.

One example of co-milling option is TSE's (Turun Seudun Energiantuotanto Oy) Naantali power plant located in Finland, where sawdust has been co-fired in PC boilers for many years on approximately 2% energy share. Sawdust is mixed with coal in the fuel yard and the mixture is fed via roller mills to the burners. In the tests carried out in 1999 and 2000, when the share of sawdust varied between 1 and 8% on energy, it was noticed that milling performance was decreased and that the roller mills mainly dried the larger sawdust particles rather than pulverising them. This caused smoking when the reject got into contact with primary air. The capacity of the mills restricted the share of sawdust that could be used without any problems to 4% on energy (Kostamo, 2000).

By utilising separate mills for biomass, the aforementioned limitations can be overcome. The milling performance of coal is not compromised and the separate biomass mills can be optimised for the specific properties of the used biomass type. This broadens the spectrum of usable biomasses and, for example, pre-dried wood chips can be used. Higher co-firing percentages are also achievable, but controlling of the burners can prove to be more difficult at high biomass shares. Disadvantages are the need for additional equipment such as the new biomass mills and biomass feeding lines for which there might be little or no space available. If there are surplus/backup coal mills available, they can be modified to biomass, which reduces the investment costs, but then the backup capacity is, of course, lost.

The separate milling option is in use in Essent 600 MW$_e$ Amer-9 power plant located in the Netherlands. The plant co-fires pelletised biomass on 27% energy share. Two of the coal mills were modified to pellets and after milling, biomass dust is directed to coal line. The tangentially fired PC boiler has seven rows of burners, six of which are for solid fuels and one for gasification gas. Gasification gas originates from the waste wood gasifier and its share is 5% on energy basis (IEA Bioenergy Task 32 Cofiring Database).

By using dedicated biomass burners at the end of the biomass line, the risks for boiler operation can be further reduced. The burners can be optimised to biomass properties offering better control and the operation of the coal burners is not compromised. The obvious disadvantage, compared to separate milling case with shared burners, is the additional investment needed for the biomass burners. Existing burners can be modified to biomass to reduce costs.

In unit 8 of the Amer power plant, the 'dedicated biomass burners' option has been adopted. For this tangentially fired 600 MW$_e$ PC boiler, two new mills were installed for pellets. The boiler has seven burner rows each having four burners. Coal is burned in four rows and the remaining two are for biomass. In this particular case the energy share of biomass is only 10–12% (IEA Bioenergy Task 32 Cofiring Database), but basically there is no limit for co-firing percentage as has been demonstrated, for example in Rodenhuise unit 4 located in Ghent, Belgium. This 240 MW$_e$ coal-fired boiler was converted to run on 100% wood pellets in 2011. After the conversion, the capacity is 180 MW$_e$. The boiler includes 24 biomass burners and four hammer mills. Similar conversions of PC boilers to 100% biomass have been carried out at Drax power plant in the United Kingdom.

The fourth option considered using biomass is re-burn fuel. Re-burning is an effective NO$_x$ reduction technology that utilises fuel injection above the main burner zone. Equipment-wise, the re-burn option is basically the same as the dedicated burner option, the only difference being the location of the burners. Small-scale experiments have shown that using coal and natural gas as re-burn fuels, NO$_x$ emissions can be reduced by more than 50–60% with about 15% of the heat input coming from the re-burn fuel, and the effect of wood has been found to be in the same range (Harding and Adams, 2000). However, unlike the first three options which can be considered mature technology, using biomass for re-burning is still in the development phase (van Loo and Koppejan, 2002).

Indirect co-firing technologies

Indirect co-firing is a process concept in which biomass or waste is first converted into gaseous (or liquid) fuel after which it is then fired together with the main fuel in the boiler. The most relevant indirect co-firing technology for PC boilers is atmospheric CFB gasification. CFB gasification has the same fuel flexibility-related advantages as CFB combustion and thus greatly reduces the fuel pre-treatment requirements for co-firing in a PC boiler. On the other hand, one could say that CFB gasification itself is the fuel pre-treatment step similar to biomass pulverisation indirect co-firing applications. The main drawbacks are the high investment cost which might reduce the willingness to invest if the biomass support mechanisms and policies are not clear in the future and the need of space.

The world's largest biomass gasifier is currently located at Vaskiluodon Voima Oy's combined heat and power (CHP) plant in Vaasa, Finland. For this 560 MW$_{th}$ (230 MW$_e$, 175 MW$_{dh}$) PC boiler, a 140 MW$_{fuel}$ CFB gasifier was supplied in 2012 by Valmet Power. The gasifier started commercial operation at the beginning of 2013 and uses mainly forest chips, but peat and small amounts of reed canary grass or straw can also be used. Moist fuels are dried to approximately 20 wt% moisture content in a wire-belt dryer before gasification. Some 20—40% of the coal can be replaced by biomass. By simply integrating the biomass gasification capability with the original coal-fired plant, the investment cost was said to be only about one-third of a similar-sized new biomass plant. Schematic of the concept is shown in Figure 7.11.

One additional benefit of the gasification route is the possibility to clean the product gas before the PC boiler, which further broadens the range of acceptable fuels and/or makes it possible to achieve higher co-firing percentages. By cleaning the gas, the PC boiler is not exposed to harmful compounds in biomass and waste fuels, namely alkali and heavy metal chlorides. In addition, less biomass/waste ash gets mixed with coal ash, so ash utilisation is not compromised. One drawback is that gasification ash from the filtration step of gas cleaning has to be handled. The unburned carbon content of ashes is typically high and can contain high amounts of heavy metals and water-soluble chlorine, depending on the fuels. In addition, there are still some technical challenges in gas-cleaning technologies (Simell et al., 2014).

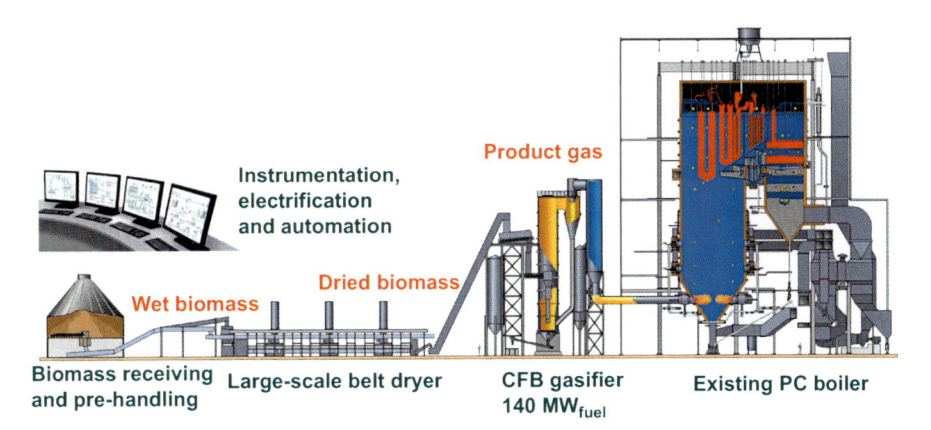

Figure 7.11 Vaskiluoto 560 MW$_{th}$ PC boiler equipped with a 140 MW$_{fuel}$ CFB gasifier. Courtesy of Valmet Power.

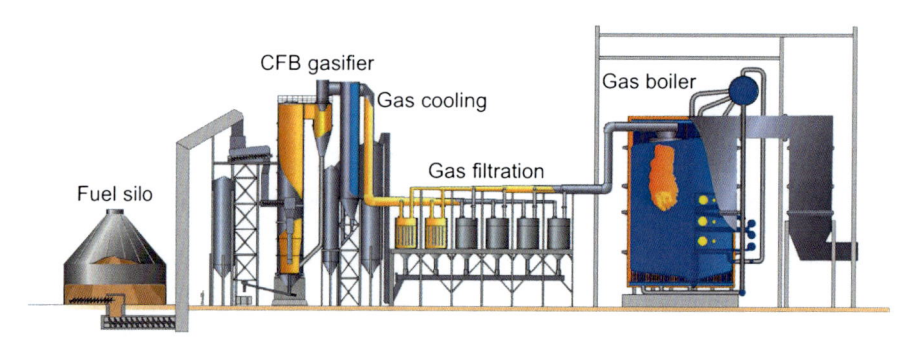

Figure 7.12 A schematic view on the Kymijärvi II unit.
Courtesy of Valmet Power.

Gas cleaning has been incorporated in the Lahti Energia's Kymijärvi II as combined heat and power unit located in Lahti, Finland. The plant produces 50 MW of electricity and 90 MW of district heat. The plant is based on CFB gasification of solid-recovered fuel, gas cooling and filtration followed by gas combustion as a single fuel in a specific gas boiler (Figure 7.12). Kymijärvi II is the first plant in the world to utilise such a concept. The unit has been in operation since 2012.

Kymijärvi II has two parallel gasification and gas-cleaning trains, having a combined capacity of 160 MW$_{fuel}$ and one gas boiler with four gas burners which can also utilise natural gas. Gasification takes place at 850–900 °C, after which the gas is cooled down to approximately 400 °C. This precipitates most of the harmful components — such as alkali chlorides — after which they can be removed in subsequent filter candles. Cooling the gas a bit further could be advantageous with respect to removal of all the harmful compounds, but it cannot be done as tar condensation must be avoided. Otherwise, tars would quickly clog the filters (Simell et al., 2014).

The steam parameters of the boiler are 540 °C/121 bar leading to approximately 31% electrical efficiency, which is significantly higher than in conventional waste incinerators. In condensing mode, >35% electric efficiency could be achieved with such steam parameters. As the boiler steels are not exposed to the corrosive environment otherwise prevailing in waste combustion, it has been possible to use less-expensive alloys resulting in reduced investment cost of the boiler.

Parallel co-firing

In the parallel co-firing concept, there are separate boilers for biomass and the main fuel(s), but the steam cycles are integrated. For example, steam can be generated in a biomass-fired boiler and superheated in a coal-fired boiler. This type of integration eliminates the risks related to deposit formation and corrosion caused by high steam parameters in a biomass-fired boiler.

Parallel co-firing has been adopted at the Avedøre 2 plant complex located in Denmark. The plant consists of (1) an ultra-supercritical (USC) main boiler, (2) a gas turbine unit and (3) a separate straw-fired boiler, the steam cycle of which is

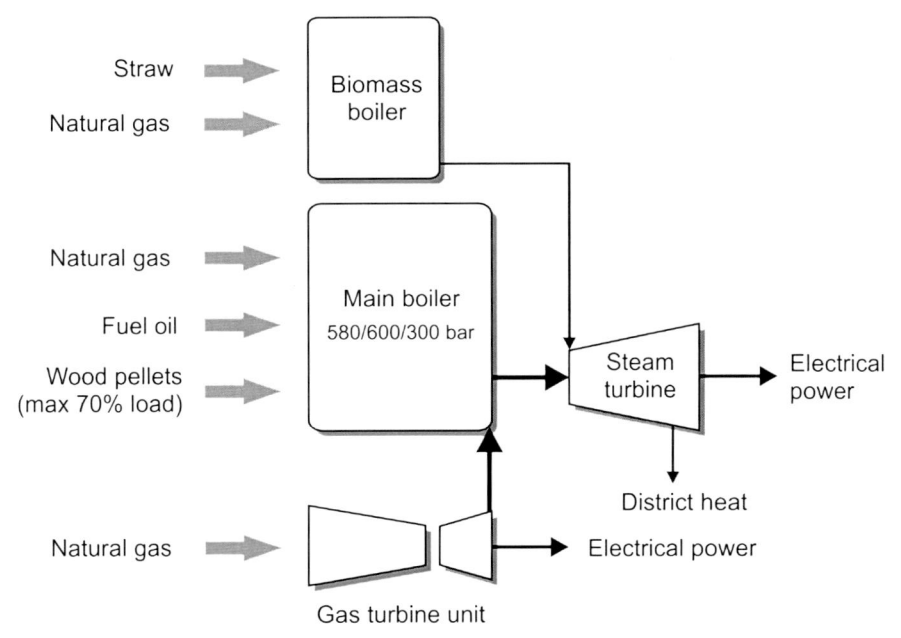

Figure 7.13 A simplified diagram of Avedøre 2.

combined to the steam cycle of the main boiler as shown in Figure 7.13. The Avedøre 2 began operation at the end of 2001.

The main boiler is an 800 MW$_{th}$ Benson-type boiler operating at steam parameters 582 °C/305 bar. It is capable of firing natural gas, oil and wood-pellet dust. The thermal power of the straw-fired grate boiler is 100 MW$_{th}$ and the live steam parameters are 545 °C and 310 bar. Using screw feeders, straw is fed onto a water-cooled vibration grate, in which up to 80% of the energy content is released by pyrolysis and gasification. The remaining straw/carbon will burn out on the grate. The boiler is equipped with a bag-filter system that removes >99% of the particulates from the flue gases. A submerged slag conveyor system is used to carry the slag and ash into the containers.

It should be noted that the steam parameters are exceptionally high for a straw-fired boiler. Typical Danish straw contains up to 0.4 wt% of chlorine (Montgomery et al., 2011), which means a very high risk of high-temperature chlorine-induced corrosion of superheaters. In addition to being highly corrosive, straw ash has also very high fouling and slagging propensities due to low ash-melting point. To cope with high-temperature corrosion and fouling, superheaters are used in so-called slagging mode (slag tap superheaters). Slagging-mode operation is based on a fact that the heat transfer from flue gases to the steam decreases when the deposit gets thicker. This in turn causes the temperature of the top layer of the deposit to increase. As the deposit temperature increases, there will be a point at which the rate of slag or melt falling off of the superheater is same as the ash-deposition rate. Thus, the deposit cannot grow infinitely. Then it follows that by leaving enough space between superheater pipes, it is possible to run the plant without having to soot blow the superheaters.

Converting a PC boiler into a BFB boiler

Another possibility for starting to utilise biomass in existing PC power plants is to convert the boiler into a BFB boiler. These PC-to-BFB retrofits are a routine practice for experienced fluidised-bed boiler technology suppliers. Several such conversions have been carried out especially in those European countries where the biomass utilisation in CHP has been promoted. For example, in Poland, at least eight of such conversions have been carried out since 2008. The capacities of the boilers that were retrofitted were in the range of 100–200 MW_{th} and after conversion they were run solely on wood and agro-biomass.

Conversions are typically carried out in relatively old boilers that would need modernisation in any case. The advantage of this approach is that the investment cost is approximately one-third to one-half of that of a green-field boiler plant. The main changes in PC-to-BFB conversion are modified boiler bottom and fuel feeding and air distribution systems and installation of start-up burners. Flue gas cleaning and heat-transfer surfaces may be renewed but depending on fuels this may not be necessary. Combustion air fans and air staging are modified to correspond to those in a BFB process enabling fluidisation and air staging to primary, secondary and tertiary air. It is essential that the fuel-feeding system from receiving all the way to the furnace wall is carefully designed so that the unit can achieve high availability and combustion stability.

Such retrofits have been carried out, for example, for a Polish power company Elektrociepłownia Białystok S.A. In 2011–2012 Valmet Power converted a 100 MW_{th} coal-fired PF boiler (type OP-140) to a biomass BFB boiler in Bialystok (Valmet Power, 2011). After the conversion the thermal power of the boiler was 75 MW_{th} and steam parameters were the same as before conversion, 540 °C/138 bar. The boiler's fuel mix consists of wood chips, forest residues, grain waste and willow. Valmet had carried out a similar conversion a few years earlier at the same site.

Summary of co-firing options in PC boilers

Figure 7.14 summarises the various technology options for co-firing biomass in PC boilers, and in Table 7.1 their main advantages and disadvantages are listed.

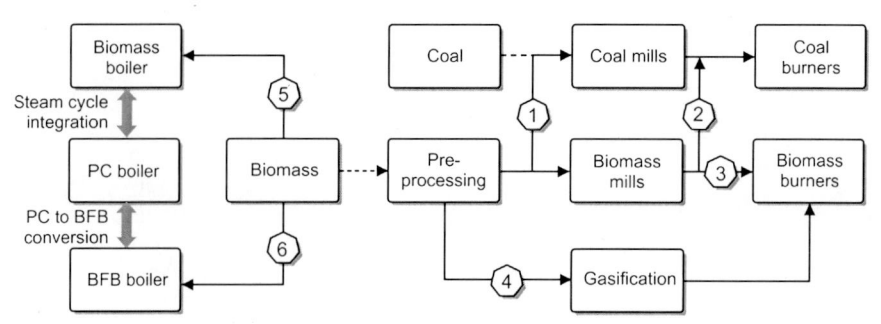

Figure 7.14 Technology options for biomass co-firing in PC boilers: (1) co-milling, (2) separate biomass mills, (3) dedicated biomass burners, (4) indirect co-firing via gasification, (5) parallel co-firing (integrated steam cycles) and (6) PC-to-BFB conversion.
Adapted from Gast et al. (2007).

Technology options for large-scale solid-fuel combustion

Table 7.1 **The main advantages and disadvantages of different co-firing technology routes for PC boilers**

Co-firing method	Advantages	Disadvantages
1. Co-milling	• Low investment cost	• Suitable to limited range of fuels (sawdust, pellets) only • Coal milling performance is decreased
2. Separate biomass mills	• Suitable to wider range of fuels • Optimised milling performance	• Need space for biomass lines inside the boiler house
3. Dedicated biomass burners	• Operation on coal not compromised • Optimised mills and burners	• Need space for biomass lines inside the boiler house
4. Indirect co-firing via CFB gasification	• High fuel flexibility • High co-firing ratio • Gas-cleaning option	• High investment cost • Filter ash disposal in gas-cleaning case • Challenges in filtering
5. Parallel co-firing (integrated steam cycles)	• Corrosion risks can be lower in the biomass boiler	• High investment cost • Complex operation
6. PC-to-BFB conversion	• Up to 100% biomass share • Low investment cost compared to greenfield biomass BFB boiler	• High investment cost compared to direct co-firing options

7.3 Summary

This chapter presented the characteristics of the most common solid-fuel combustion technologies: grate, fluidised-bed and pulverised-fuel combustion and discussed their suitability to co-firing and multi-fuel operation.

Grate boilers are rugged units capable of firing various fuels but have limited multi-fuel operation capability and are characterised by low efficiency and high emissions. Fluidised-bed combustion is the most fuel-flexible combustion technology and is also characterised by the lowest emissions. Pulverised-fuel combustion has the most stringent requirements for fuel quality but various methods are available to enable co-firing of various fuels even at high shares. More detailed summary of the advantages and disadvantages of each technology is given in Table 7.2.

Table 7.2 Summary of the advantages and disadvantages of the main solid fuel combustion technologies

Grate combustion (typical size range: <100 MW$_{fuel}$)	
• Rugged and reliable • Low investment and operation costs • Low erosion and dust load • Can handle various challenging fuels such as untreated MSW and straw	• Low efficiency • High emissions • Limited multi-fuel operation capability • Limited adaptability to changes in fuel quality
Bubbling fluidised bed (BFB) combustion (typical size range: 20–300 MW$_{fuel}$)	
• Suitability to various biomass fuels with varying particle size and moisture content • Combustion efficiency • Low NO$_x$ and SO$_2$ emissions • No moving mechanical parts in the hot region	• Risk of bed agglomeration with high-alkali fuels • Limited partial-load operation capability • Low maximum coal share • Erosion
Circulating fluidised-bed (CFB) combustion (typical size range: 50–800 MW$_{fuel}$)	
• Best fuel flexibility (0–100% biomass or coal) • Very high combustion efficiency • Lowest NO$_x$ and SO$_2$ emissions • Effective in-bed sulphur capture with limestone • Loop seal heat exchangers for corrosion mitigation • No moving mechanical parts in the hot region	• High in-house electricity consumption • High investment and operation cost • Risk of agglomeration in loop seal and dense bed • Limited partial-load operation capability • Erosion
Pulverised-fuel (PF) combustion (100–2000 MW$_{fuel}$)	
• High efficiency • Large unit sizes available • Good load-following capability • Various technology options available for co-firing available	• High NO$_x$ and SO$_2$ emissions • Stringent fuel quality requirements • Fuel flexibility

References

Alholmens Kraft Company website. Available from: http://www.alholmenskraft.com/ (accessed 11.11.13.).

DeFusco, J.P., McKenzie, P.A., Fick, M.D., 2007. Bubbling fluidized bed or stoker — which is the right choice for your renewable energy project? In: CIBO Fluid Bed Combustion XX Conference Lexington, Kentucky, USA.

Gast, C.H., Jelles, S.J., de Jong, M.P., Konings, A.J.A., Saraber, A.J., Vredenbregt, L.H.J., Witkamp, J.G., 2007. Technische Grenzen Maximaal Meestoken. KEMA Consulting, 07-2346.

Harding, N.S., Adams, B.R., 2000. Biomass as a reburning fuel: a specialized cofiring application. Biomass and Bioenergy 19, 429–445.

IEA Bioenergy Task 32 Cofiring Database. Available from: http://www.ieabcc.nl/database/cofiring.php (accessed 11.12.13.).

Jalovaara, J., Aho, J., Hietamäki, E., Hyytiä, H., 2003. Paras Käytettävissä Oleva Tekniikka (BAT) 5−50 MW:n Polttolaitoksissa Suomessa (Best Available Techniques (BAT) for 5−50 MW Combustion Plants in Finland). Suomen ympäristö 649, Suomen Ympäristökeskus, Helsinki, Finland (In Finnish).

Kostamo, J., 2000. Co-firing of sawdust in a coal-fired utility boiler. IFRF Combustion Journal. ISSN: 1562-479X. Article Number 200001.

Kärki, J., Flyktman, M., Hurskainen, M., Helynen, S., Sipilä, K., 2011. Replacing coal with biomass fuels in combined heat and power plants. Proceedings of the International Nordic Bioenergy 199−206.

van Loo, S., Koppejan, J. (Eds.), 2002. Handbook of Biomass Combustion and Co-firing. Twente University Press, Netherlands.

Montgomery, M., Jensen, S.A., Borg, U., Biede, O., Vilhelmsen, T., 2011. Experiences with high temperature corrosion at straw-firing power plants in Denmark. Materials and Corrosion 62, 593−605.

Öhman, M., Nordin, A., 2000. The Role of kaolin in prevention of bed agglomeration during fluidized bed combustion of biomass fuels. Energy and Fuels 14 (3), 618−624.

Peña, J.A.P., 2011. Bubbling Fluidized Bed (BFB) − when to Use This Technology? IFSA 2011. Industrial Fluidization South Africa, Johannesburg, South Africa.

Raiko, R., Saastamoinen, J., Hupa, M., Kurki-Suonio, I., 2002. Poltto Ja Palaminen (Combustion and Burning), second ed. Finland, International Flame Research Foundation - Finnish Flame Research Committee, Helsinki (In Finnish).

Silvennoinen, J., 2003. A new method to inhibit bed agglomeration problems in fluidized bed boilers. In: 17th International Conference on Fluidized Bed Combustion (FBC2003). Jacksonville, Florida, USA.

Silvennoinen, J., Hedman, M., 2011. Co-firing of agricultural fuels in a full-scale fluidized bed boiler. Fuel Processing Technology 105, 11−19.

Simell, P., Hannula, I., Tuomi, S., Nieminen, M., Kurkela, E., Hiltunen, I., Kaisalo, N., Kihlman, J., 2014. Clean syngas from biomass—process development and concept assessment. Biomass Conversion and Biorefinery 4, 357−370. http://dx.doi.org/10.1007/s13399-014-0121-y.

Stultz, S.C., Kitto, J.B., 2005. Steam: Its Generation and Use, 41st ed. The Babcock & Wilcox Company, Ohio, USA.

Valmet Power, 2011. Press Release on a Deal for a Conversion of Bialystok PC Boiler to BFB. Available from: http://www.metso.com/news/newsdocuments.nsf/web3newsdoc/968FFE156F4E00A3C22578FC00259EFD?OpenDocument&ch=ChMetsoWebEng&#. U2oPB1eLXDw (accessed 17.10.13.).

Yin, C., Rosendahl, L.A., Kær, S.K., 2008. Grate-firing of biomass for heat and power production. Progress in Energy and Combustion Science 34, 725−754.

Plant integrity in solid fuel-flexible power generation

Nigel J. Simms
Centre for Power Engineering, Cranfield University, Cranfield, Bedfordshire, UK

8.1 Introduction

The integrity of power-generation plants is critical for their continued safe and economic operation. Each power plant has thousands of components, many of which are not exposed to either fuels or their conversion products, and so are not directly affected by changes in fuel compositions. However, many components are affected by the fuels used, which include parts within:

- Fuel storage, preparation, handling and transport systems
- Combustion units
- Heat exchangers (including superheaters, reheaters, waterwalls, economisers, etc.)
- Ductwork
- Gas clean-up systems
- Chimneys

The operating environments for such fuel-path components (i.e. before and after combustion) within power plants are a result of a combination of the fuels used, the power plant design and the component operating conditions. The various interactions between these operating environments and the materials used for the component results in a wide range of potential degradation mechanisms. Some of these degradation mechanisms are relatively slow and will permit component lives of 10–30 years, whereas others are potentially very fast and can be component-life limiting. In these cases, the operating conditions/materials need to be optimised to permit viable component lives (usually 5–10 years, or at least longer than the plant inspection intervals).

This chapter describes potential solid fuels that can be considered, different types of power-generation systems and the variations in operating environments that are encountered with changes in fuels (and operating conditions) for selected components in such systems. The complex mechanisms that govern the environmentally induced degradation of these components are outlined, together with some of the available models of these processes. Monitoring of the degradation of these components is described, together with preventative measures. Finally, anticipated future challenges in plant integrity are considered in terms of likely future developments in fuels and operating conditions for solid fuel-fired power plants.

8.2 Potential solid fuels

A range of solid fuels can be combusted in power-generation systems. These can be classified into coals, biomass and waste products. Detailed descriptions of these general classes of fuels and the analytical methods used to characterise them are available elsewhere (e.g. Simms, 2011a; Speight, 1994), but for the purposes of this chapter the key characteristic properties of these fuels are summarised in the following sections.

8.2.1 Coals

Coals are solid fossil fuels derived from plant matter that has been saved by water and mud from oxidation and biodegradation and then subjected to high pressures and temperatures for prolonged periods; this process is described in detail elsewhere (e.g. Speight, 1994; Raask, 1985). Thus, coals can be classed as sedimentary organic rocks and are made up from several distinct parts (Figure 8.1). There are many variables that influence this process, including initial plant matter, pressure history, temperature history and time. Because of differences in such variables during the formation of coals, a wide range of coals can be formed. During the study of peats and coals, a number of different methods of classifying them have been developed. One commonly used method divides peats and coals into five broad types:

- 'peat'; material at an early stage in coal formation
- 'lignite' (or 'brown coal'); with a high moisture content
- 'sub-bituminous coal'
- 'bituminous coal'; a dense, usually black coal, frequently with a banded structure
- 'anthracite'; a glossy, hard, black coal, with a high carbon content and low in volatile matter

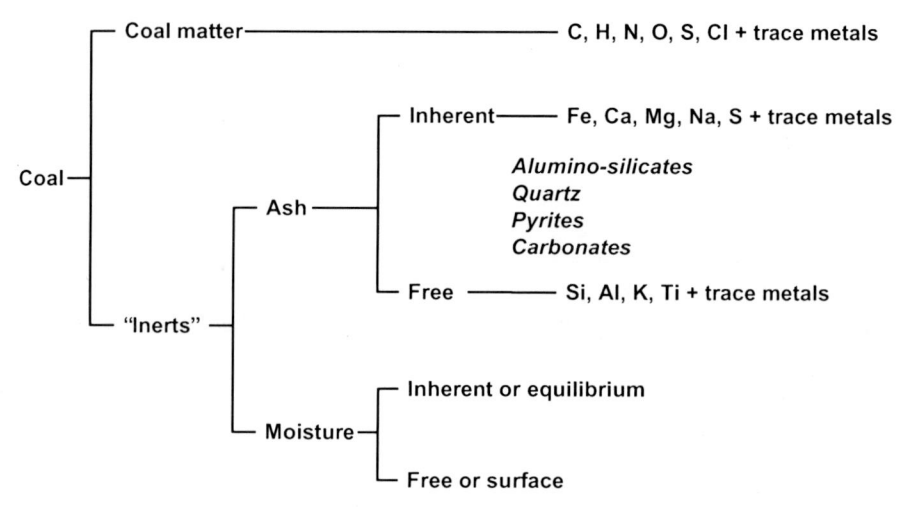

```
Coal ─┬─ Coal matter ──────────────────── C, H, N, O, S, Cl + trace metals
      │
      │            ┌─ Inherent ──── Fe, Ca, Mg, Na, S + trace metals
      │            │                Alumino-silicates
      │      ┌─ Ash ┤                Quartz
      │      │     │                Pyrites
      └─ "Inerts" ─┤                Carbonates
             │     └─ Free ──────── Si, Al, K, Ti + trace metals
             │
             │         ┌─ Inherent or equilibrium
             └─ Moisture ┤
                       └─ Free or surface
```

Figure 8.1 Breakdown of coal constituents (Simms, 2011a). Adapted from Jones (2005).

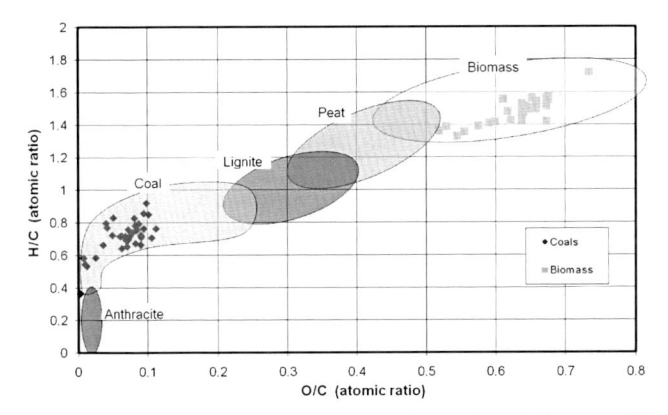

Figure 8.2 Relationship between H/C and O/C ratios for coals and biomass (Simms, 2011a).

From the point of view of using coals in power-generating systems, the properties of the coals progressively change through the various classification systems, with the anthracite coals having the highest calorific values and the lowest hydrogen/carbon (H/C) ratios, in contrast to the lignites, which have the lowest calorific values and the highest H/C ratios. One way of showing this progression in terms of fuel composition is using a van Krevelen diagram in which the H/C ratios of fuels are plotted as function of their oxygen/carbon (O/C) ratios (van Krevelen, 1950). Figure 8.2 includes various coal types in such a diagram and shows the position of biomass and peats (the data points on this diagram represent examples of individual coals and biomass used in power-generating systems).

In practice, for power-generation coals, it is necessary to have much more detailed analyses of coals covering major, minor and trace elements, as well as the energy content (Simms, 2011a; Speight, 1994; Raask, 1985; Clark and Sloss, 1992; Carpenter and Skorupska, 1993). These analyses are traditionally referred to as proximate, ultimate, ash, calorific value (CV) and trace metal. A wide range of standards have been developed over many years to enable their determination (summarised by, for example, Simms, 2011a; and Speight, 1994; standards available from American Society for Testing and Materials (ASTM), British Standards Institution (BSI), European Committee for Standardisation (CEN) etc.). Table 8.1 gives examples of some of these analyses for four example coals (from United Kingdom (UK), United States (USA), South America and South Africa) that are commonly used in power plants. During the last 20 years there has been a trend for some power-generating companies (depending on national policies/regulations) to move away from using locally mined coals towards using world-traded coals depending on fuel availability and cost.

A wide range of compositions can be found for coals mined in diverse locations around the world, because of differences in their formation and the local geology. Variations between coals mined in smaller geographic areas (though they are still influenced by the local geology) are less significant, but some differences even exist between coals produced from different seams in the same coal mine.

Although the form and quantities of the major elements present in coal are critical in combustion processes (as well as gasification and pyrolysis processes), the minor and

Table 8.1 **Typical coal analyses**

Parameter	Unit	UK (Thorseby)	USA (Pittsburgh#8)	South America (El Cerrejon)	South Africa (Koornfontein)
Moisture	%wt ar	4.8	2.6	7.0	3.8
Ash	%wt dry	11.8	9.4	9.0	13.9
CV (gross)	MJ/kg daf	34.1	34.4	33.1	34.1
CV (net)		32.9	33.2	31.8	32.9
C	%wt daf	84.3	84.6	79.9	84.5
H		4.6	5.06	5.3	5.2
O		7.9	7.6	12.21	8.8
N		1.8	1.7	1.7	2.1
S		2.13	2.7	0.73	0.6
Cl		0.67	0.05	0.03	0.1
Ash analysis (% on fuel ashing)
SiO_2		54.4	43.4	58.72	43.7
Al_2O_3		24.5	24.7	21.30	34.0
Fe_2O_3		10.7	12.9	7.19	3.0
CaO		2.36	6.4	2.20	7.2
MgO		1.62	1.5	2.81	2.2
K_2O		3.13	1.7	2.24	<0.5
Na_2O		1.88	0.5	1.03	0.4
TiO_2		1.07	1.1	0.89	1.7
BaO		0.11	n.d.	0.11	n.d.
Mn_3O_4		0.05	<0.1	0.06	n.d.
P_2O_5		0.15	1.0	0.21	1.0
SO_3		3.65	6.1	3.92	6.3

CV, calorific value; daf, dry ash free; wt%, weight%; ar, as received; n.d., not determined.
Adapted from Simms (2011a).

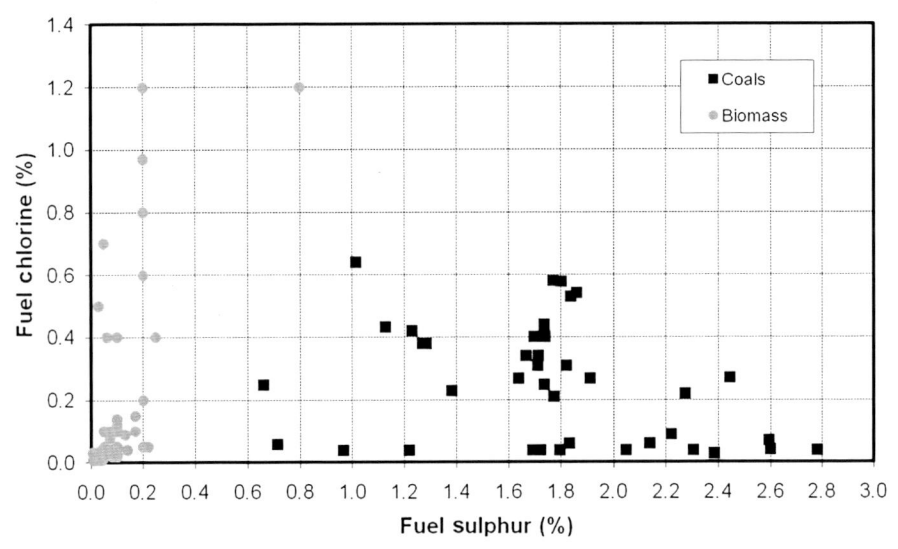

Figure 8.3 Sulphur and chlorine contents of selected biomass and coals (Simms, 2011a).

trace elements cause many of the operational challenges for practical power-generating systems (Raask, 1985). For example, these include emissions—gas cleaning system requirements for SO_x and NO_x and other species; fouling and slagging on the various heat exchanger surfaces; fireside corrosion; and dew point corrosion.

Figure 8.3 (Simms, 2011a) illustrates the variation in S and Cl contents for a wide range of example coals (and biomass). For coals in general, S contents can range from approximately zero up to ~4 weight% (dry ash-free basis), whereas the Cl content can range from approximately zero up to ~0.7 weight% (wt%) (dry ash-free basis). Higher S content coals require more gas cleaning to remove and control SO_x emissions. Higher S and Cl levels both play a role in the formation of slagging and fouling deposits, as well as the various corrosion processes (Raask, 1985).

The ash in the fuel analysis mostly arises from the mineral impurities that are found in coals (Francis and Peters, 1980; Raask, 1985; Speight, 1994). Table 8.2 lists the main mineral types found in coals and provides examples of minerals that are frequently found in coals (e.g. Raask, 1985; Stringer, 1995). It is worth noting that the results of standard ash analyses on coals do not give a good representation of the form of mineral-based elements in coal as they report each element in terms of its highest pure oxide (Stringer, 1995). The decomposition and interaction reactions of these minerals during combustion (or gasification—pyrolysis) produce most of the ash (bottom ash and fly ash) as well as the slagging and fouling deposits (Raask, 1985).

Coals also contain many trace elements, i.e. elements present at levels below 1000 ppm. In fact, most naturally occurring elements can be found in different coals (Clarke and Sloss, 1992). In practice, these trace elements can behave in a wide range of different ways during combustion processes (and differently during gasification and pyrolysis), with the result that they can be distributed (or partitioned) between bottom

Table 8.2 Common minerals found in coal

Mineral group	Mineral name	Chemical formula
Clay (alumina-silicates) and silicate minerals	Montmorillonite	$Al_2Si_4O_{10}(OH)_2 \cdot H_2O$
	Illite	$KAl_2(AlSi_3O_{10})(OH)_2$
	Kaolinite	$Al_4Si_4O_{10}(OH)_8$
	Muscovite	$K_2O \cdot 3Al_2O_3 \cdot 6SiO_2 \cdot 2H_2O$
	Quartz	SiO_2
	Albite	$NaAlSi_3O_8$
	Orthoclase	$KAlSi_3O_8$
	Fayalite	Fe_2SiO_4
	Anorthite	$CaO \cdot Al_2O_3 \cdot 2SiO_2$
	Chlorite	$Al_2O_3 \cdot 5(FeO,MgO) \cdot 3.5SiO_2 \cdot 7.5H_2O$
Oxide and hydrated oxide minerals	Haematite	Fe_2O_3
	Magnetite	Fe_3O_4
	Rutile	TiO_2
	Limonite	$Fe_2O_3 \cdot H_2O$
	Diaspore	$Al_2O_3 \cdot H_2O$
Sulphide minerals	Pyrite	FeS_2
	Marcasite	FeS_2
	Pyrrhotite	FeS_x
	Chalcopyrite	$CuFeS$
	Galena	PbS
	Sphalerite	ZnS
Sulphate minerals	Gypsum	$CaSO_4 \cdot 2H_2O$
	Anhydrite	$CaSO_4$
	Jarosite	$(Na,K)Fe_3(SO_4)_2(OH)_6$
	Kieserite	$MgSO_4 \cdot H_2O$
	Thenardite	Na_2SO_4
	Barytes	$BaSO_4$
Carbonate minerals	Calcite	$CaCO_3$
	Dolomite	$(Ca,Mg)CO_3$
	Siderite	$FeCO_3$
	Ankerite	$(Ca,Fe,Mg)CO_3$
Phosphate minerals	Apatitie	$Ca_{10}(PO_4)_6(Cl,F,OH)_2$
Chloride minerals	Halite	$NaCl$
	Sylvite	KCl

Adapted from Raask (1985) and Stringer (1995).

ash, fly ash, various deposits and the gas phase. For simplicity, this behaviour of elements can be grouped so that they are classed as having different volatilities during combustion and in combusted flue-gas streams:

- Group 1 (least volatile): Eu, Hf, La, Mn, Rb, Sc, Sm, Th and Zr
- Group 1–2: Ba, Be, Bi, Co, Cr, Cs, Cu, Mo, Ni, Sr, Ta, U, V and W
- Group 2: As, Cd, Ga, Ge, Pb, Sb, Sn, Te, Tl and Zn
- Group 2–3: B, Se and I
- Group 3 (most volatile): Br, Cl, F, Hg and Rn

Detailed reports are available from the International Energy Agency (IEA) Clean Coal Centre on many specific aspects of coal compositions and their impact on coal usage, for example: general impurity removal (Crouch, 1995), trace metals (Clarke and Sloss, 1992); halogens (Sloss, 1992); S and N species (Nalbandian, 2004).

Finally, it should be noted that, in practice, coals are often only partially analysed on a routine basis to determine just the parameters that are needed for pricing, quality control, plant operation, emissions and regulatory purposes (e.g. CV, moisture, ash, sulphur content).

8.2.2 Biomass fuels

A wide range of types of biomass could potentially be used in power-generation systems. However, the growth of biomass depends on local environments and soil conditions, so that the types that are available in practical terms vary between geographic regions. Biomass can be classed in different ways (van Loo and Koppeian, 2007; Livingston, 2009; White and Plaskett, 1981; Simms et al., 2007a), but one method uses the biomass-production route as the basis:

- Energy crops:
  - woods; e.g. coppiced willow, poplar, cottonwood
  - grasses; e.g. *Miscanthus*, reed canary grass, switch grass
- Agricultural and forestry residues:
  - straws, e.g. from wheat, barley, oats, rice, maize, oil seed rape
  - forest residues
- Processing residues from:
  - olives, almonds, palm nuts, sugarcane, rice
  - sawdust, bark, wood off-cuts
- Seaweeds: both naturally occurring and cultivated
- Animal wastes and sewage sludges

Some types of biomass can be supplied in a processed form as pellets, such as cereal co-products (CCP) or wood byproducts, which increase their energy density and make them easier to handle, transport and store. As a result of the low energy density of many types of biomass, many are only available to local markets, but processed biomass and those with higher energy densities (e.g. some forms of wood and pelletised products) are available on a world-traded basis (EUBIONET2, 2007; Simms et al., 2007a; Colechin, 2005).

To facilitate comparisons with other types of fuels used for thermal power generation, it is necessary to quantify biomass properties in terms of the traditional coal fuel parameters (i.e. proximate, ultimate, CV, ash, trace metal etc.). However, the differences in the structures of biomass and coals mean that to generate these fuel parameters for biomass it has been necessary to develop a new set of standard analytical methods (Simms, 2011a; CEN). In addition, other uses of biomass (e.g. anaerobic digestion) and crop optimisation require alternative types of biomass analyses. Thus, biomass can also be analysed in terms of its contents of lignin, hemi-cellulose, cellulose, starch, fats, proteins etc., for other applications (e.g. van Loos and Koppeian, 2007; White and Plaskett, 1981).

Table 8.3 gives examples of fuel analyses of selected biomasses. Fuel moisture has been omitted from this table as it is extremely variable and readily changes with fuel storage; for land-based harvested biomass, levels of 40—60% moisture may be found, but these can be reduced during storage to 15—20% and further with fuel processing (EUBIONET2, 2007). As such, the initial moisture contents are much higher than for traditional power-station coals. The ash contents of most biomass are generally lower than for coals, though some types of residual biomass can have relatively high ash contents (e.g. palm kernels in Table 8.3). For other traditional fuel parameters (e.g. CV, C, H and O contents), when expressed on a dry ash-free basis, they fall within a relatively narrow range of values, as illustrated in Table 8.3. Compared to traditional power station coals (e.g. Table 8.1), biomass have less C, but more H and O. This difference in basic fuel composition can also be seen in the trend from biomass through peat and coals to anthracite that was presented in Figure 8.2. The levels of other elements (including N, S, Cl, P and various metallic elements) vary widely between biomass types and the various parts of plants (e.g. heart wood compared to bark and leaves from the same tree). They also vary with growing conditions, the timing of harvesting, biomass storage etc. (EUBIONET2, 2007; van Loo and Koppeian, 2007; Doran, 2009; Livingston, 2009; White and Plaskett, 1981; Simms et al., 2007a). These are illustrated in Table 8.3 with an indication of the ranges found for some minor elements in particular types of biomass. Some useful trends in biomass compositions can be identified:

- S levels are low in biomass compared to most coals (Figure 8.2)
- the Cl contents of biomass (up to 2.5 wt% daf) span a wider range than for coals; some biomass types can have higher Cl contents than coals, but many do not; the faster growing biomass types (e.g. cereal crops) tend to have the higher Cl contents
- in biomass, the alkali metals are mostly present in a different chemical form than for coals (Livingston, 2009) and so they are much more readily released during combustion. A different balance also exists between the alkali metals, with much more K than Na; as with Cl, higher K levels are found in faster-growing biomass (Simms et al., 2007a)
- other elements can be found at high levels in selected biomass (e.g. N, P, Ca), which can lead to difficulties in using these potential fuels in some power systems (due to emissions, the formation of fouling deposits and/or corrosion issues). Figure 8.4 illustrates the relative abundance of selected elements in biomass compared with two typical bituminous power-station coals.

Table 8.3 **Typical biomass analyses**

Parameter	Unit	Short-rotation coppice (SRC) willow	Coniferous wood	Miscanthus	Straw (wheat, barley, rye)	Palm kernel	Olive residue
Ash	%wt dry	2	0.3	4	5	7.5	4.5
CV (gross)	MJ/kg daf	20.3	20.5	19.8	19.8	19.8	21.4
CV (net)		18.8	19.2	18.4	18.5	18.3	18.3
C	%wt daf	49	51	49	49	50.2	49
H		6.2	6.3	6.4	6.3	6.6	6.0
O		44	42	44	43	40	40
N		0.5	0.1	0.7	0.5	3.2	2.24
S		0.05	0.02	0.2	0.1	0.2	0.1
Cl		0.03	0.01	0.2	0.4	0.2	0.1
Elemental analysis (mg/kg dry basis)							
Al		3–1000	30–400	40–1400	50–700	470–1400	200–2700
Ca		2000–9000	500–1000	900–3000	2000–7000	2600–3800	5000–14,500
Fe		30–600	10–100	40–400	100–500	540–4300	350–2300
K		1700–4600	200–500	1000–16,000	2000–26,000	5900–6600	6600–34,000
Mg		200–800	100–200	300–900	400–1300	2700–3500	380–5100
Mn		80–160	n.a.	n.a.	n.a.	200–270	10–50
Na		10–450	10–50	200–500	500–3000	70–110	120–1750
P		500–1300	50–100	400–1200	300–2900	5700–7100	500–1600
Si		2–7200	100–200	2000–10,000	1000–20,000	2300–4800	500–1400

CV, calorific value; daf, dry ash free; wt%, weight%; n.a., not available.
Including some data from CEN/Technical Specification (TS) 14961:2005.

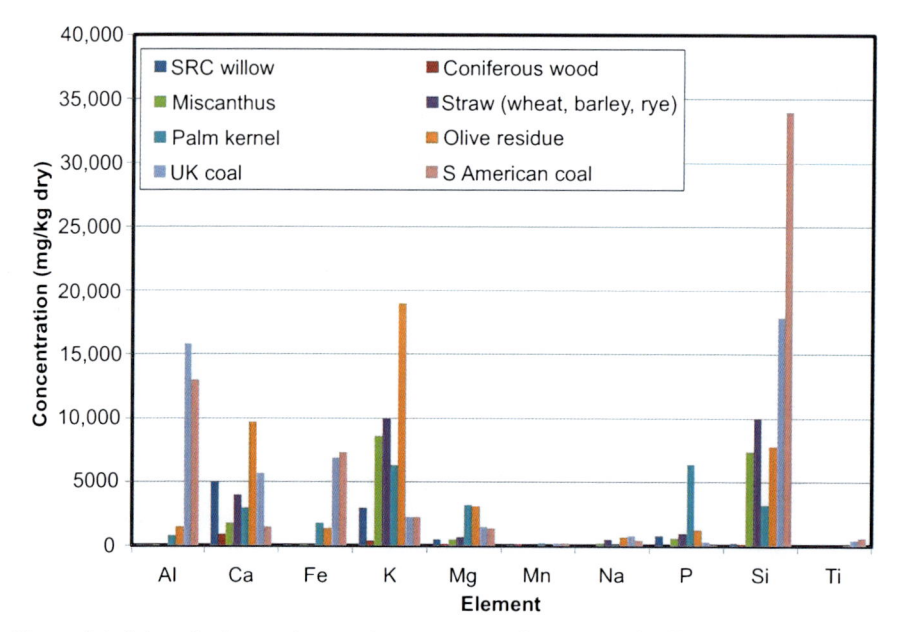

Figure 8.4 Selected minor and trace elements present in a range of different biomass.

The compositions of many types of biomass indicate that they have good potential to be used as fuels in combustion power systems, but their differences compared to coal highlight the need for care in choosing specific biomass types for use in particular power-generation systems (especially those that have been developed and designed for specific coals). The differences in the minor elements present in biomass can result in numerous issues related to emissions, deposition (fouling—slagging) and corrosion, causing both operational and maintenance issues, as well as restricting the efficiencies of the biomass-fired power-generation systems (compared to coal-fired systems) (Livingston, 2009; Simms et al., 2007a).

8.2.3 Waste-derived fuels

Wastes are usually classified in terms of their origins, but the classifications used vary around the world. As part of its activities to assess the emissions of greenhouse gases, the United Nations (UN) International Panel on Climate Change (IPCC) carried out an extensive assessment of the production and use of wastes around the world (Pipatti et al., 2006). It used the following categories for waste streams:

- municipal solid wastes (MSW)
- industrial wastes
- sludges
- other wastes

It was noted that many other waste classifications could be used and that some waste streams are allocated to different categories on a national or regional basis;

for example, commercial and demolition wood could be classified as industrial wastes, MSW or put into its own category depending on the location of the waste (Pipatti et al., 2006). For this study, MSWs were taken to include the following items (Pipatti et al., 2006): food waste; garden (yard) and park waste; paper and cardboard; wood; textiles; nappies (disposable diapers); rubber and leather; plastics; metal; glass (and pottery and china); other (e.g. ash, dirt, dust, soil, electronic waste).

Such surveys emphasise the extreme variability that can be found in waste streams. However, not all waste streams are destined to be considered as potential fuels, as with the currently favoured 'waste hierarchy' (EU Waste Framework Directive, 2008) favouring waste reduction, re-use and recycling over energy recovery. The materials that are left for use in energy recovery systems can be used as raw fuels or further processed to generate refuse-derived fuels (RDFs) or solid-recovered fuels (SRFs). In recent years to encourage the use of such fuels, there has been a move towards their standardisation by the CEN. As a result, a series of analytical standards are now available (CEN; Simms, 2011a) and a system for classifying SRFs in terms of their:

- mean net CV
- mean value of the chlorine content
- median and 80th percentile values of the mercury content (on an as-received basis)

Because of the variable waste feedstocks, the SRFs produced have variable properties, as illustrated in Table 8.4. Thus, the performance of such fuels cannot be generalised and has to be assessed on a site basis when the characteristics of the local fuel(s)

Table 8.4 Typical waste fuel analyses (Simms, 2011a)

Component	Units	Minimum value	Mean value	Maximum value
Water content	wt% as-received	2.9	14.6	38.7
Volatiles	wt% daf	74.6	88.7	99.4
Ash	wt% dry	4.4	17	44.2
CV (gross)	MJ/kg daf	13.130	24.597	44.029
CV (net)	MJ/kg daf	12.126	22.915	40.986
C	wt% daf	33.9	54.8	84.8
H	wt% daf	1.72	8.12	15.16
O	wt% daf	15.8	34	43.7
N	wt% daf	0.12	0.94	2.37
S	wt% daf	0.01	0.4	1.4
Cl	wt% daf	0.006	0.716	1.558

CV, calorific value; daf, dry ash free; wt%, weight%.

are known. However, the availability of the data required to do this in a standardised form enables the methods developed for use with coal and biomass fuel assessment to be translated (but these need to be applied with care).

8.3 Power plant types, component operating environments and fuel options

Traditionally, pulverised-coal combustion power plants have tended to be of large scale, with individual boilers of 500—800 MWe grouped together (to give total power plant capacities of 2000—4000 MWe). Plants that were built in the 1960 to 1970s often have steam systems that operate with maximum steam parameters of approximately 140—160 bars/540—560 °C and now operate with efficiencies of ~35—37% (following various upgrades and environmental protection measures that have respectively had the effect of increasing and decreasing system efficiencies over the years). New coal systems use individual boilers of similar sizes, but with steam systems with maximum operating parameters of approximately 290 bar/620 °C giving efficiencies of 46—47% (Farley, 2010).

In contrast, biomass combustion plants tend to be much smaller, with newly built power plants of up to ~44 MWe and with efficiencies of ~30% (e.g. Doran, 2009; Koppejan and van Loo, 2009). Waste-to-energy plants are also much smaller with electrical generating capacities of up to 30 MWe and efficiencies of up ~25%, although most are much less efficient (Prewin, 2011). However, the smaller scale and potential locations of biomass and waste generating plants lend themselves to possible use in combined heat and power applications (if appropriate local heat loads can be found), which would significantly increase the efficiency of energy use from these fuels.

In such systems, the hot gas streams can be produced by the combustion of a wide variety of fuels (solid, liquid or gaseous) that can contain a range of different impurities. As these hot combusted-gas streams pass through the boilers and over the various heat-exchanger surfaces, as well as transferring heat to the water—steam systems, they can interact to produce deposition, erosion and/or corrosion on the heat-exchanger surfaces. Both corrosion and erosion damage to the fireside surfaces of the heat exchangers cause metal losses and so reduce component lives (though there is usually a 'corrosion allowance' to enable design lives to be achieved). To avoid unexpected tube failures, which can result in costly plant outages, significant effort needs to be devoted to non-destructive examinations of heat-exchanger tubes during routine plant outages so that any excessively affected tubing can be identified and replaced. The replacement tubing can be the same (if the component life is acceptable), or a better material may be used (if one exists), or protective measures may be required (such as coatings, co-extruded tubes, bandages etc.). Alternatively, the boiler operating conditions could be changed to reduce the damage rates or the compositions of the fuels used in the boiler restricted. Deposit formation usually has the effect of reducing heat transfer from the hot gas stream to the water—steam system, which in turn reduces boiler efficiency. In addition, such deposits are involved in some of the corrosion damage mechanisms that have been found in boiler environments (Section 8.4.3).

8.3.1 Fuel preparation

Fuel preparation varies widely depending on the power plant requirements. For example, pulverised-coal plants require the use of finely ground fuels. These have traditionally been generated by routing the fuel-feed systems via ball mills to break up and grind the fuels just prior to their combustion; the technologies available for this are well established and rely on the physical properties of coals (especially the brittle nature of most power station coals). However, substitute and co-fired fuels also need to be finely ground to be used in these plants. This presents particular challenges for using alternative fuels. Biomass fuels have different physical properties to coal and also require different approaches to handling and storage (EUBIONET2, 2007; van Loo and Koppeian, 2007; Doran, 2009; Livingston, 2009; Francis and Peters, 1980; White and Plaskett, 1981; Simms et al., 2007a). Coal is traditionally stored in large heaps on the ground before being transferred to the power plants for use. However, biomass can absorb moisture, rot, generate odour, attract rodents etc., when left in the open; so dry storage for these fuels is required (but this limits the fuel reserve available on a power station site to relatively small quantities and creates the need for a robust supply chain with frequent deliveries). Ideally, biomass needs to be processed using appropriate cutting, shredding, milling, drying etc. technologies to produce the fuel in the form needed for its use (Doran, 2009). For small percentages of co-fired biomass (<5%), it has been found that it can be added to the coal feed just prior to grinding. However, for larger biomass use, dedicated hammer mills need to be used. Alternatively, the biomass can be prepared off-site (chopped, hammer milled, etc.) and made into pellets which are then transported to the pulverised-fuel power station and broken up in the coal mills prior to their use. Fuel preparation, storage, handling and supply chains are particularly critical to using biomass in large-scale pulverised-fuel power plants.

Other types of power plants have different fuel-preparation requirements:

- Fluidised-bed combustors can cope with fuels that are in suitable size ranges for their particular fuel-feed systems (e.g. conveying belts/bars, screws, pneumatic etc.). For biomass fuels, this could involve chopping the fuels down to <5 or <50 mm depending on the feed system design (Doran, 2009).
- Grate combustors need much less fuel preparation, but the fuel needs to be in a form that fits through the various meshes and grills and mechanical feeders in the fuel-supply systems, so waste fuels may need some sorting—mechanical shredding depending on their sources.

8.3.2 Superheaters—reheaters—waterwalls—etc.

The hot-gas paths of combustion systems contain a series of heat exchangers to generate high-temperature—high-pressure steam from water. Figure 8.5 shows a flow diagram for a typical water—steam system, with its series of heat exchangers; water flows into an economiser and then an evaporator after which the low temperature steam enters the superheater and then the high temperature steam enters the high-pressure steam turbine. In this system, the steam is reheated before entering the intermediate-pressure steam turbine. The highest steam temperatures in such a system are achieved in the final stages of the superheater and reheater.

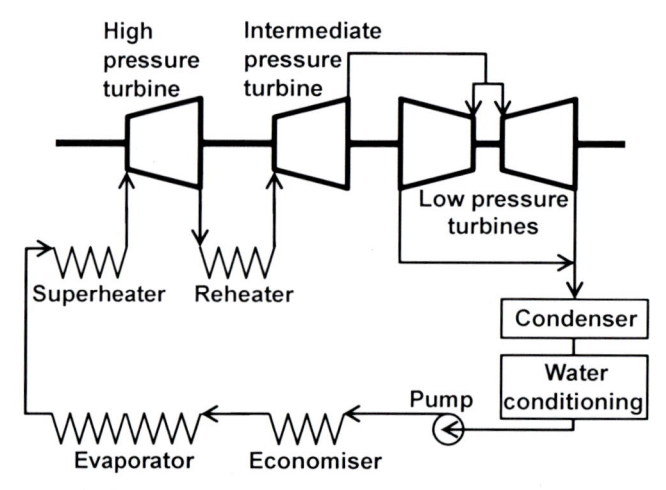

Figure 8.5 Schematic flow diagram for a power plant steam–water system showing the main component parts (Simms, 2011b).

Figure 8.6 illustrates the layout of these heat exchangers around a conventional pulverised fuel-fired combustion power plant; this shows that the combustion zone is surrounded by waterwalls and that the hot gases from the combustion process then flow past the various superheater and reheater stages before going through the economiser. In such a system, the waterwalls are relatively cool (up to 400 °C)

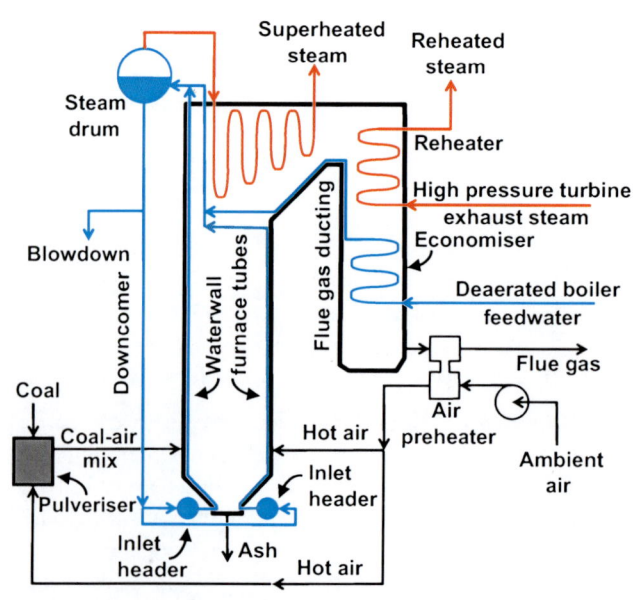

Figure 8.6 Schematic diagram of a pulverised-fuel power plant showing the position of the main heat exchangers.

despite enclosing the fuel burners and gases of up to 1600 °C, but have high heat fluxes (up to ~0.35 MW/m^2). The combustion gases have cooled to 1000 to 1200 °C by the time they pass through the superheaters and produce heat fluxes of ~0.2 MW/m^2; the steam temperatures exiting the superheaters can be ~540–620 °C depending on the age of the power plant. The combustion gases continue cooling through the superheaters, reheaters and economisers. The final stage of the reheaters is at similar steam and metal temperatures (but lower pressures) compared to the superheaters.

Figure 8.7 shows a different power plant configuration based on a circulating fluidised-bed combustion process. In this system, the combustion chamber is again surrounded by waterwalls, but superheaters are located in the gas pass after the cyclone. The flue-gas temperature approaching the superheaters is ~860–880 °C. The steam system operates with the final superheater output at 480 °C/80 bars.

Figure 8.8 shows a power plant configuration based on a grate-fired boiler, with this example based on a waste-to-energy process. In this system, the combustion gases initially pass through a chamber surrounded by waterwalls, but the superheaters are located in the third gas pass to overcome environmental degradation issues associated with such fuels (Section 8.4.3). In such systems, these issues limit superheated-steam temperatures to 360 °C (with pressures of ~33 bars) and such low-steam conditions restrict the generating efficiencies of this type of power plant.

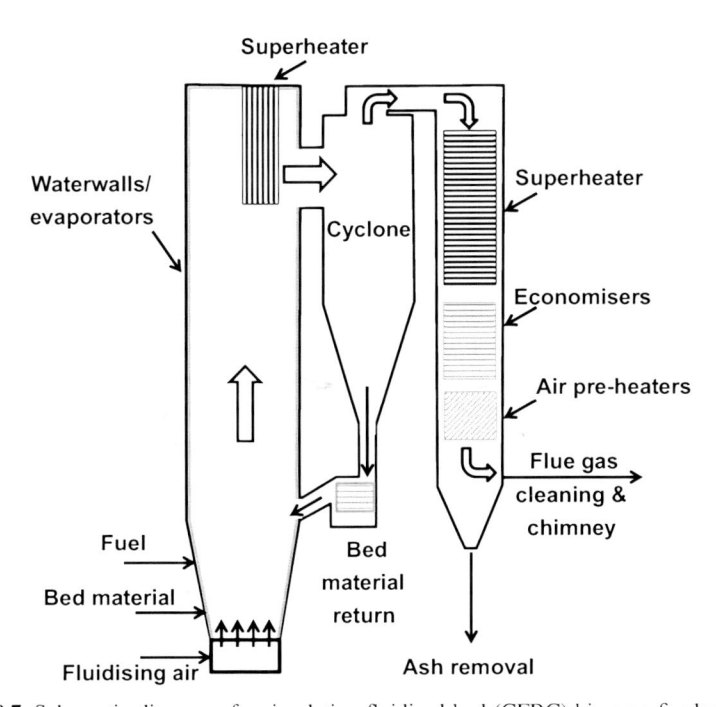

Figure 8.7 Schematic diagram of a circulating fluidised-bed (CFBC) biomass-fired unit.

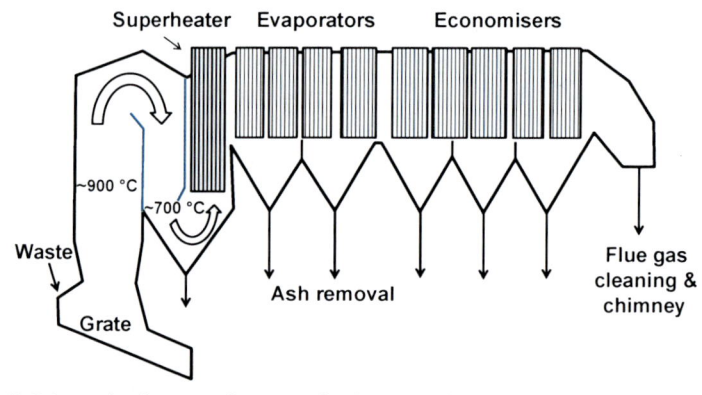

Figure 8.8 Schematic diagram of a waste-fired grate unit.

Within boilers, the environments around the heat exchangers depend on the chemical compositions of the fuels used, as well as the operating conditions used in the boilers. During the combustion processes, the fuels react with an oxidising gas stream, which is air in most current combustion plants, to produce a hot combusted-gas stream. Figure 8.9 illustrates this process for pulverised-coal combustion and shows the breakdown of the fuel in terms of the burnout of the combustible material, generation of ash particles and vapour-phase species (Simms, 2011b; Tomeczek and Palugniok, 2002). As a result of the complex reactions of the inorganic elements present in the fuels in various different forms (Section 8.2), minor and trace elements are partitioned between the coarse (bottom) ash, fly ash and gas—vapour phase. It is the fate of these elements as they pass through the hot-gas path of the power systems that is in a large part responsible for the environmental degradation of the heat exchangers. The particles produced can form deposits (Section 8.4.1) or cause erosion damage (Section 8.4.4);

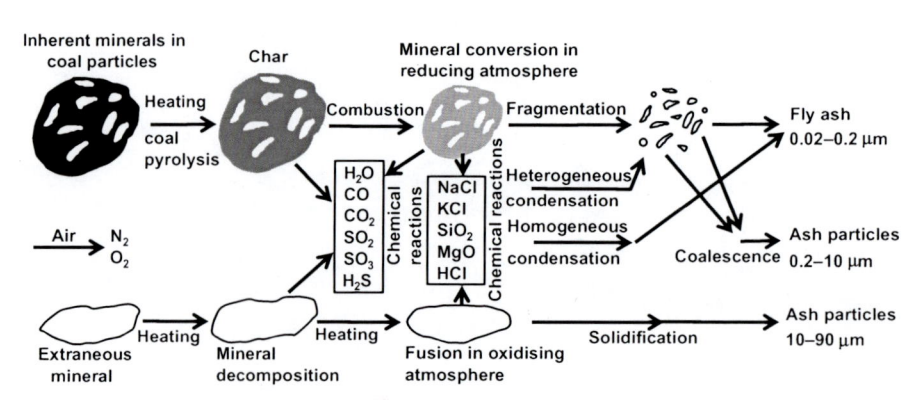

Figure 8.9 Schematic representation of fuel combustion (Simms, 2011b). Adapted from Tomeczek and Palugniok (2002).

the vapour-phase species can condense under particular conditions to become part of the deposits (Section 8.4.1); the gases, deposits and heat exchanger surfaces can react to cause accelerated corrosion damage (Section 8.4.3).

8.4 Degradation mechanisms and modelling

8.4.1 Deposition

The deposits that form around the fireside surfaces of heat exchanger tubes are created from the particles and vapours that pass through a boiler by the action of a number of different mechanisms which can occur in parallel in the local environments (Figure 8.10). For particles, the important potential deposition mechanisms are:

- Direct inertial impaction: This is the mechanism by which larger particles (usually >10 μm) deposit onto the surfaces of heat exchangers. This deposition mechanism is particularly important for the upstream surfaces of heat exchanger tubes, with the larger particles not being able to follow the gas-flow streamlines around the tubes (as a result of their having too high momentum). Particles hitting the tube surfaces may either rebound from the surfaces or stick to them, depending on their state (solid or sticky) and that of the tube surface (e.g. with solid, liquid or sticky deposits). Alternative approaches to calculating the deposition of particles by this mechanism are given by Zhou et al. (2007) and Tomeczek et al. (2004). Direct inertial impaction can also occur on the downstream

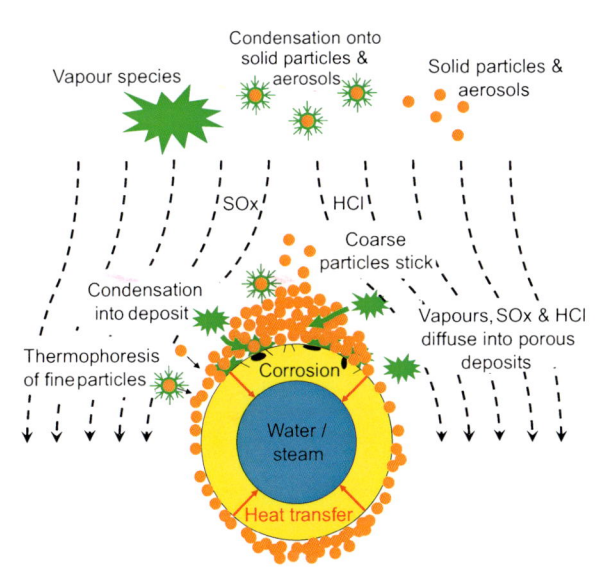

Figure 8.10 Schematic representation of interaction between superheater tube and its local environment.
Adapted from Simms et al. (2007b).

surfaces of tubes when particles that have passed by the tubes get caught up in turbulence and are then propelled towards a tube surface; an expression to calculate such deposition is given by Zhou et al. (2007).
- Thermophoresis, a process that results in the transport of smaller particles (typically <1 μm) through a gas along local temperature gradients (e.g. sub-micron particles from a hot-gas stream to a cooled heat exchanger tube). Jacobsen and Brock (1965) report a model for this process.
- Brownian–eddy diffusion; processes that enable the transport and deposition of sub-micron particles from turbulent gas streams. Wood (1981) provides a model for such deposition routes.

The latter two mechanisms are believed to play a relatively minor role in deposit formation in boilers.

For vapours, the deposition mechanisms are:

- Vapour condensation from hotter-gas streams onto cooler local surfaces; potential deposition fluxes via this mechanism may be calculated using expressions given in Tomeczek et al. (2004), Tomeczek and Wacławiak (2009) and Simms (2011b). Figure 8.11 illustrates the variation in the partial pressures of alkali species with temperature and shows the much higher levels for the chloride species (and so the relative stability of the sulphate compounds).
- Heterogeneous vapour condensation onto particles (which can then follow a particle deposition route dependent on the particle size).
- Homogeneous vapour condensation into aerosols (which then follow particle deposition routes appropriate to smaller particles).

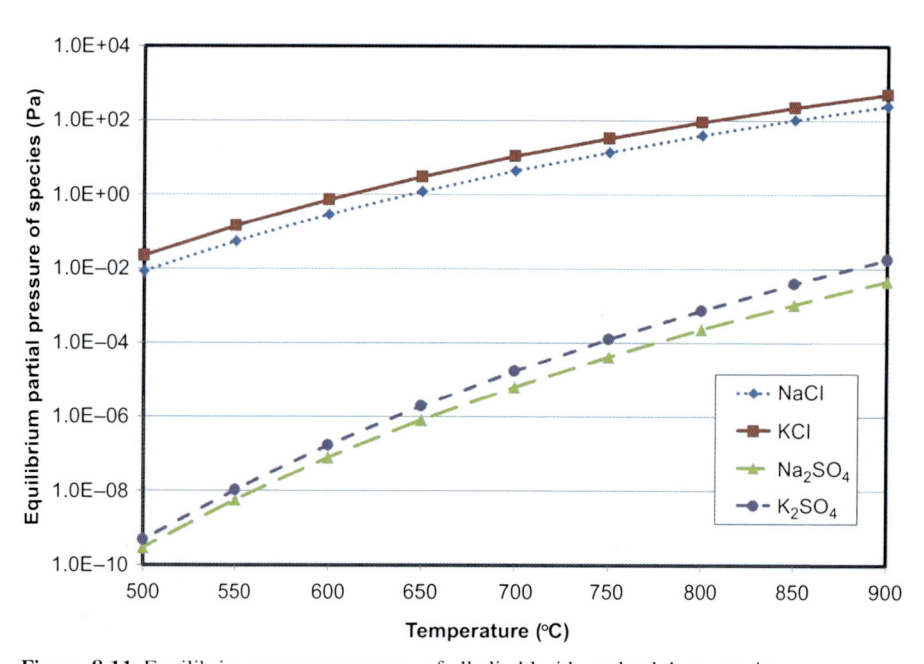

Figure 8.11 Equilibrium vapour pressures of alkali chloride and sulphate species.

Thus, the total deposit growth rate can be thought of as a sum of the various possible deposition mechanisms (Zhou et al., 2007):

$$dm(t,\theta)/dt = C(t,\theta) + TH(t,\theta) + BE(t,\theta) + I(t,\theta)$$

in which '$m(t,\theta)$' is the deposit weight, 't' is time and '$C(t,\theta)$', '$TH(t,\theta)$', '$BE(t,\theta)$' and '$I(t,\theta)$' represent the condensation, thermophoresis, Brownian and eddy diffusion and impaction rates, respectively, as a function of 't' and 'θ', the angle around the heat-exchanger tube.

8.4.1.1 Deposit compositions

The deposit compositions found on heat-exchanger surfaces are determined by the balance of the deposition processes occurring around that component and so can be boiler-design specific in addition to being dependant on the composition of the fuels and the element partitioning that happens during the combustion processes. There are many reports of different deposit compositions available in the literature; these just emphasise the sensitivity of the various depositional processes to the fuels, combustion processes and local component operating environments.

For superheater–reheater tubes in coal-fired systems, deposits can contain:

- Si–Al–O compounds
  - derived from alumina-silicate minerals
  - fixing Na, K if particle temperatures high enough
- Ca–Mg carbonates–sulphates–chlorides
- Na–K sulphates–chlorides
- Fe sulphates–chlorides–oxides–sulphides

An example of the type of deposit that is often found on superheater tubes in coal-fired power plants is given in Figure 8.12 (after Wright and Shingledecker, 2015; Simms, 2011b; Syrett, 1987; Nelson and Cain, 1960). This type of deposit often develops a layered structure that as it grows, with its shape gradually developing and its surface temperature increasing.

For biomass-fired systems, a similar range of compounds can be found in deposits, but the balance between elements varies because of the different fuel compositions and element partitioning. In addition, changes in the balance between the deposition processes are due to differences in the particle-size distributions produced and the availability of vapour-phase species (especially derived from K). One classification of deposits found in biomass systems produces three main groupings (Livingston, 2010):

- High silica–high K–low Ca ashes – low melting points
  - from agricultural residues (straws, etc.)
- Low silica–low K–high Ca ashes – high melting points
  - from woody materials
- High K–high P – low melting points
  - from animal wastes, etc.

Co-firing coal and biomass produces deposits that are a blend of those expected from the fuels when fired alone. However, there is a complex interaction due to the

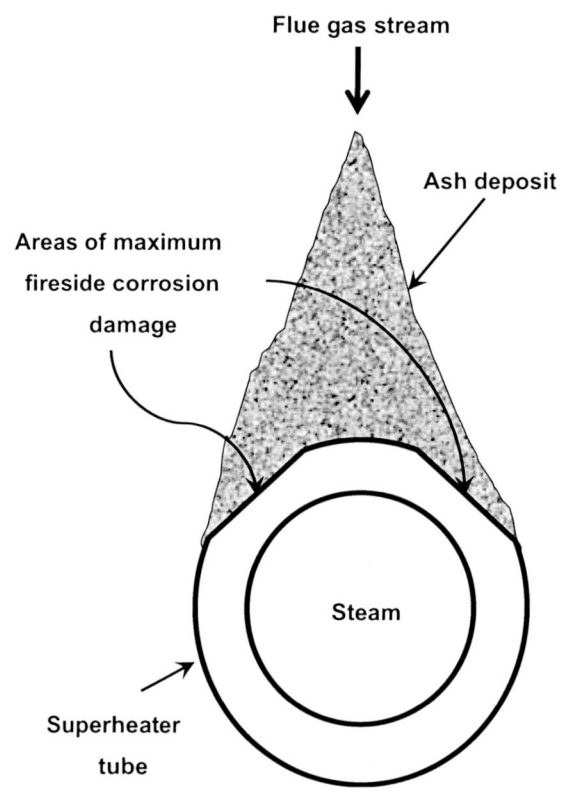

Figure 8.12 Example of the structure of a deposit observed on superheater tube in a coal-fired boiler.
Adapted from Wright and Shingledecker (2015), Simms (2011b), Syrett (1987), and Nelson and Cain (1960).

different size distributions of particles produced by the mixed fuels and vapour-phase species. These change the balance of the deposition mechanisms, and the resulting deposition fluxes and compositions depend on which specific fuels have been used.

For waste-fired systems, there are a much wider range of fuel compositions and so a correspondingly wider range of potential deposit compositions. For many waste streams there is a particular concern over the levels of heavy metals in the wastes as high levels of some of these (e.g. Pb, Zn, Cd, Sn) can produce vapour-phase species which condense as low melting-point chlorides on the surfaces of superheater tubes (and then cause rapid corrosion damage; Section 8.4.3).

Deposit formation in power plants is often referred to as slagging or fouling:

- slagging is the formation of molten deposits during plant operations at high temperatures and usually occurs on components in the combustion chamber.
- fouling is the formation of particulate-based deposits (or 'dry' deposits) and usually occurs further along on the hot-gas path. It is possible to move from fouling to slagging on a component if the local gas temperatures are hot enough and deposits are allowed to thicken sufficiently.

Because of fouling and slagging, a range of techniques have been developed to try to remove deposits from the surfaces of heat exchangers during operations (i.e. online cleaning) or periods of maintenance, including (Livingston, 2009; Stam et al., 2010; Vassilev et al., 2013):

- mechanically hitting (or rapping) the tube surfaces;
- soot blowing using compressed air, steam or water jets;
- sonic waves;
- explosive charges;
- water washing.

8.4.2 Oxidation

During the course of operation, the minimum chemical degradation that will be experienced by heat-exchanger tubes in combustion systems is oxidation. In this process, the tube materials react with oxygen (or oxygen-containing species) to generate surface-oxide scales. In its simplest form this can be represented as:

$$M<s> + \frac{1}{2}O_2<g> = MO<s>$$

in which M represents a metal and MO represents a metal oxide. The continued oxidation of this metal depends on the transport of metal ions and/or oxidant species through the metal-oxide scale. The most protective oxides that can grow are those that form dense, even scales and only permit a slow transport of metal and oxidant through them. The relative growth rates of different oxides are illustrated in Figure 8.13. The oxidation of metals has been thoroughly studied, and there are many textbooks that describe these processes well (e.g. Birks et al., 2006; Young, 2008; Kofstad, 1988).

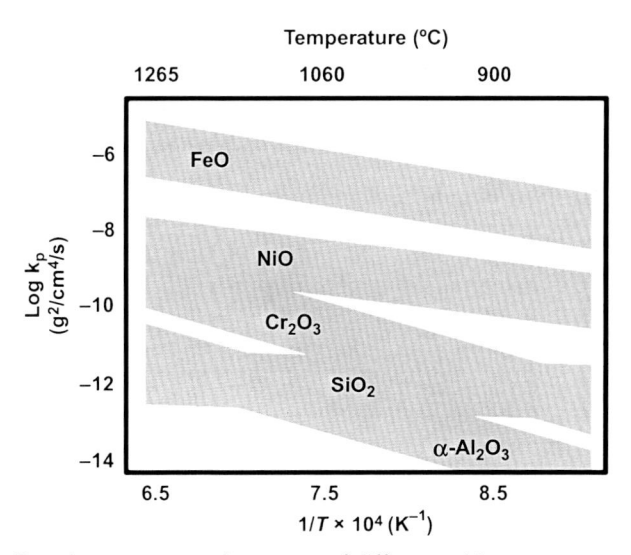

Figure 8.13 Growth rate constants for a range of different oxides.

Heat-exchanger materials are usually manufactured from various grades of steels, ranging from low-alloy ferritic steels through ferritic–martensitic varieties to austenitic stainless steels; examples of nominal compositions are given in Table 8.5. In some situations (Section 8.6.2), higher-alloyed materials are used as coatings and nickel-based alloys are also now being considered for use in future power plants with higher-temperature steam systems (Section 8.7).

The oxides that form on low-alloy steels ($< \sim 10-12$ wt% Cr) under oxidising conditions are multi-layered. At lower temperatures (below 570 °C for pure iron, but increasing with Cr content), the inner oxide layer is an inward-growing spinel $(Fe,Cr)_3O_4$, the central layer magnetite (Fe_3O_4) and the outer layer haematite (Fe_2O_3). At higher temperatures, wustite (FeO) forms an inner oxide layer beneath magnetite and haematite layers. However, the formation of wustite permits much higher oxidation rates, and so this provides one of the upper limits to the temperatures that such alloys can be used in practical systems (Stringer and Wright, 1995).

On higher-alloy materials (stainless steels, nickel-based alloys and coatings), oxide-scale growth is dominated by the formation of chromia (Cr_2O_3), which can be a relatively slow-growing protective oxide. In situations in which this layer breaks down, various mixed-oxide scales can also form (such as $(Cr,Fe)_2O_3$ and spinels $(Fe,Ni,Cr)_3O_4$) to produce multi-layered scale structures or scales with internal oxidation beneath them (Young, 2008; Birks et al., 2006; Bradford, 1987).

8.4.3 Fireside corrosion

Several different types of fireside corrosion have been found in boilers over the years. Despite many extensive studies, the detailed mechanisms of these degradation processes have proved to be difficult to fully define (e.g. Stringer and Wright, 1995; Syrett, 1987). However, the key factors causing such damage have been identified, with mechanistic and/or empirically based models being developed to describe the effects of some of these corrosion processes under certain conditions; those for superheater and reheater tubes have recently been reviewed by Wright and Shingledecker (2015).

8.4.3.1 Waterwall corrosion

In pulverised fuel-fired boilers, it is sometimes found that there are areas of high metal wastage on the waterwalls. These are usually associated with flame impingement or a failure to establish a combustion zone towards the centre of the furnace. As a result, areas of the waterwall can experience significant exposure periods under reducing conditions, as well as the more usual oxidising regime. The surface scales produced during this type of corrosion are generally based on magnetite, but can contain FeS lamella close to the metal surface, with FeS islands, fly ash spheres and unburnt carbon particles closer to the scale surface (Simms, 2011b; Stringer and Wright, 1995). Because of the operating conditions of the waterwall, the metal surface temperatures may only be in the range 300–400 °C, but the deposit surface may be molten. Many causes of this form of damage have been suggested over the years (Stringer and Wright, 1995; Syrett, 1987), but it is now believed that periods in the oxidising and reducing environments coupled with the presence of sulphur (and possibly carbon) are responsible.

Table 8.5 Nominal compositions of heat-exchanger tube materials

Element (weight%)

Alloy	Fe	Ni	Cr	Co	Mo	C	Si	Mn	S	P	Al	Ti	Nb	Cu	N	W	Others
1Cr	Bal		1.0–1.5		0.44–0.65	0.15	0.5–1.0	0.3–0.6	<0.025	<0.025							
T22	Bal		1.9–2.6		0.87–1.13	0.15	0.5	0.3–0.6	<0.025	<0.025							
T23	Bal		1.9–2.6		0.05–0.3	0.04–0.1	0.5	0.1–0.6	<0.01	<0.03			0.02–0.08		<0.03	1.45–1.75	$V = 0.2$–0.3; $B = 0.0005$–0.006
T24	Bal		2.2–2.6		0.9–1.1	0.05–0.15	0.15–0.45	0.3–0.7	<0.01	<0.02	<0.2	0.06–0.1			<0.012		$V = 0.2$–0.3; $B = 0.0015$–0.007
T91	Bal	0.4	8.0–9.5		0.85–1.05	0.08–0.12	0.2–0.5	0.3–0.6	<0.01	<0.02			0.06–0.10		0.03–0.07		$V = 0.18$–0.25; $Al < 0.04$
T92	Bal	<0.4	8–9.5		0.3–0.6	0.07–0.13	<0.5	0.3–0.6	<0.01	<0.02	<0.04		0.04–0.09		0.03–0.07	1.5–2	$V = 0.15$–0.25; $B = 0.001$–0.006
X20CrMoV121 (X20)	Bal	0.3–0.8	10–12.5		0.8–1.2			<1									$V = 0.25$–0.35
AISI316L	Bal	10–14	16.5–18.5		2.0–2.5	<0.03	<1.0	<2	<0.03	<0.04							
AISI347HFG	Bal	9–13	17–20			0.06–0.10	<0.75	<2									
304H	Bal	8–10.5	18–20			0.04–0.10	<0.75	<2	<0.045	<0.045							
Sanicro 25	Bal	25	22.5	1.5		<0.1	0.2	0.5	<0.015	<0.025			0.5	3.0	0.23	3.6	$8*C < Nb + Ta < 1.00$
HR3C	Bal	17–23	24–26			0.04–0.10	<0.75	<2	<0.03	<0.03			*		0.15–0.35		*Nb + Ta = 0.4
Alloy 625	<5	Bal	20–23	<1	8–10	<0.1	<0.5	<0.5	<0.015	<0.015	<0.4	<0.4	#				#Nb + Ta = 3.7
Alloy 263	0.7	Bal	19–21	19–21	5.6–6.1	0.04–0.08	0.4	0.6			0.6	1.9–2.4		<0.2			
Alloy 617 (mod)	3	Bal	20–24	10–15	8–10	0.05–0.15	1	1			0.8–1.9	0.6		0.5			
Alloy 740	0.7	Bal	25	20	0.5	0.03	0.5	0.3			0.9	1.8	2				

A predictive model for this type of damage has been suggested by Davis (2010):

$$M = C \times \left[(t_o \times K_{po})^{0.5} + (t_r \times K_{pr})^{0.5}\right] + \left[\frac{t_r \times ACR}{10^3}\right]$$

in which, K_{po} and t_o are the rate constant and the time, respectively, under oxidising conditions; K_{pr} and t_r are the rate constant and the time, respectively, under reducing conditions; C is a constant;

$$ACR = \left[(1425 \times \%Cl) \times (HF)^m \times e^{-\left[\frac{Q_{Cl}}{RT}\right]} - P\right]$$

where %Cl is the % of chlorine in the coal (by weight); HF is the heat flux; Q is an activation energy; R is the gas constant; m and P are other constants; and, T is the metal surface temperature (in Kelvin).

Solutions to this type of corrosion damage include:

- modifying the combustion environment (using improved burner designs, finer fuel particles or fuel mixes)
- providing an 'air curtain' in front of the waterwall to keep the environment oxidising
- replacing tubes with co-extruded tubes (an outer highly corrosion-resistant material outside waterwall tube material) (Meadowcroft and Manning, 1983)
- coating the fireside surfaces of the tubes (Section 8.6.2)

Another form of waterwall damage has been reported when the corroded tubes show deep parallel groves normal to the tube axis with spacings of the order of 1 mm (Stringer and Wright, 1995). These have been given a variety of descriptive names: for example, circumferential cracking, horizontal cracking, elephant hiding, alligator-skin cracking. The suggested cause is a combination of (1) thermally induced alternating stresses, (2) the fireside corrosion environment in areas of particularly high heat fluxes and (3) a rippled magnetite layer on the inside (waterside surface) of the tubes. Solutions include oxygenated-water treatment to reduce magnetite deposition on the waterside surfaces and methods for ensuring a more even heat-flux distribution (e.g. by improved control of slag removal from the fireside surface).

8.4.3.2 Superheater–reheater corrosion

There has been a long history of investigating the causes of fireside corrosion damage on superheater–reheater tubes in coal-fired power plants (described in detail by Stringer and Wright, 1995). However, it is now believed that the main cause of this form of accelerated damage is the presence of molten deposits on the surfaces of the tubes. These deposits can form as solid species and then become molten as a result of their reactions with other species in the deposit and the gas stream flowing around the tubes. In addition, corrosion products from initial reactions with the surfaces of the tube materials can also take part in further corrosion processes. Thus, in assessing this form of degradation it is necessary to consider both the immediate results of the deposition processes and the many potential further reactions that can take place.

Compounds that have been identified as having the potential to form in deposits and cause fireside corrosion damage include:

- Sulphate deposits:
  - Pyro-sulphates; for example, $(Na,K)_2S_2O_7$
  - Alkali—iron trisulphates; for example $(Na,K)_3Fe(SO_4)_3$
  - Mixed sulphate eutectics; for example, $Na_2SO_4-Fe_2(SO_4)_3$, or more generally mixes of $(K,Na,Fe)_XSO_4$
- Chloride deposits; with mixed compositions including Na, K, Fe, Ca, Mg and other metal elements depending on the fuel used
- Carbonates; with mixed compositions including Na, K, Fe, Ca, Mg and other metal elements depending on the fuel used
- Sulphate—chloride—carbonate 'soup' containing all the compounds above

In considering the potential for such compounds to both form in deposits and cause corrosion damage, it is necessary to assess the melting points of the compounds (with some examples given in Table 8.6) and the conditions necessary for their formation. However, this is only part of the story as deposited species can interact with the surrounding gaseous environment; for example, both alkali pyro-sulphates and alkali—iron trisulphates need sufficient vapour pressures of SO_3 around them for their stability to be maintained, with the alkali pyro-sulphates needing the higher levels. Mixed alkali-pyrosulphates have melting points down to ~345 °C (Table 8.6; Lindberg et al., 2006), whereas mixed alkali—iron trisulphates have melting points down to ~560 °C (Table 8.6; Cain and Nelson, 1961).

For biomass-fired systems which have deposits that contain higher levels of potassium chlorides as well as sulphates, the alkali chloride—sulphate system needs to be considered, with the lowest melting point of a mixed-alkali chloride—sulphate being ~517 °C (Table 8.6; Lindberg et al., 2007). In addition, for biomass-fired systems the effects of other compounds, such as carbonates, need to be considered (Blomberg, 2008).

For waste-fired systems, heavy-metal chloride compounds need to be considered in detail, as these can cause deposits that have much lower melting points (Spiegel, 2010); potential mixtures contain combinations of alkali metals, Fe, Pb, Zn, Cd and Sn as oxides, chlorides, sulphates and carbonates. For example, in the $ZnCl_2-KCl$ system the minimum melting point is ~240 °C (Table 8.6; Hack and Jantzen, 2008).

With increasing metal temperatures, such compounds can become unstable for a variety of different reasons, including (Nicholls and Simms, 2010):

- vapour condensation dew points being exceeded
- insufficient SO_3 being available to stabilise some sulphate phases (e.g. alkali pyro-sulphates, alkali—iron trisulphates or mixed sulphates); as the SO_3-SO_2 balance favours SO_3 at lower temperatures
- other phases becoming more stable with a change in temperature

The result of this is a 'bell-shaped' curve in materials corrosion (Figure 8.14). In this the lower limit is set by the melting point of a compound in the deposit and the increase in corrosion rate is dependent on the sensitivity of the corrosion reaction to temperature and the availability of reactants (deposition fluxes, gas partial pressures etc.) in the local

Table 8.6 Melting points of selected potential deposit constituents

Compound	Melting point (°C)	Compound mixtures (selected only)	Minimum melting point (°C)
Sulphates		**Mixed sulphates**	
Na_2SO_4	884	$Na_2SO_4-K_2SO_4$	834
K_2SO_4	1069	$Na_2S_2O_7-K_2S_2O_7$	~345
$PbSO_4$	1170	$(Na/K)_3Fe(SO_4)_3$	~560
$ZnSO_4$	600 (Decomposes)	$Na_2SO_4-Fe_2(SO_4)_3$	~620
$CaSO_4$	1450	**Mixed chlorides**	
$Fe_2(SO_4)_3$	480	$KCl-NaCl$	657
$FeSO_4$	680 (Decomposes)	$PbCl_2-FeCl_3$	175
$SnSO_4$	378	$KCl-PbCl_2$	411
Chlorides		$PbCl_2-FeCl_2$	421
$NaCl$	801	$KCl-FeCl_2$	355
KCl	770	$KCl-ZnCl_2$	~240
$PbCl_2$	501	$NaCl-FeCl_2$	372
$ZnCl_2$	283	$NaCl-ZnCl_2$	250
$CaCl_2$	782	$NaCl-PbCl_2$	~400
$FeCl_2$	~670	**Mixed sulphates–chlorides**	
$FeCl_3$	306	$KCl-NaCl-Na_2SO_4-K_2SO_4$	517
$SnCl_2$	246	$KCl-K_2SO_4$	690

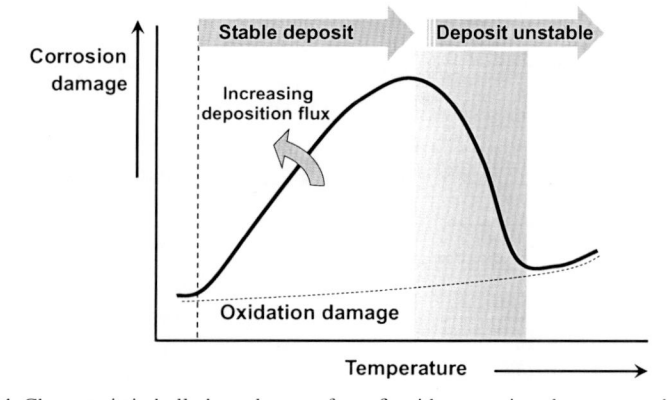

Figure 8.14 Characteristic bell-shaped curve for a fireside corrosion damage mechanism.

environment. The position of the peak (and downwards slope) depends on the cause(s) of the deposit instability (as listed above). Thus, the position of the peak and downwards slope of the bell-shaped curve can be influenced by the exposure environment, and so can be set differently in laboratory and plant exposures. Figure 8.15 shows multiple 'bell-shaped' corrosion peaks attributed to different compounds forming in deposits (Simms, 2011b; Natesan et al., 2003).

Heat-exchanger tube materials will respond in different ways to the aggressive deposits on their surfaces. For some lower-alloyed materials, the fluxing reactions in the molten deposits result in rapid corrosion, whereas for more highly alloyed materials, the chromia scale formed is more protective and can provide some protection against such deposits. However, much higher levels of chromium are needed in the alloy to provide significant resistance against fireside corrosion than used in standard stainless steels (hence the high chromium contents of materials selected as potential protective coatings for heat-exchanger surfaces in Section 8.6.2).

The relative corrosiveness of the superheater environments produced by one biomass (wheat straw) and coals are illustrated in Figure 8.16. This figure shows the results of corrosion damage measurements carried out on the stainless steels American Iron and Steel Institute (AISI) 347 and 347HFG. These were exposed in combustion units as part of the European Cooperation in Science and Technology (EU COST) 522 and 538 programmes (e.g. Henderson et al., 2002) and in subsequent EU research programmes (e.g. Stam, 2013), with the materials being exposed as parts of superheaters, reheaters and on cooled probes. The results show that wheat straw induces a higher range of corrosion damage rates than the coal or co-fired conditions used (<10% biomass by mass) at the same metal temperature. The range of corrosion

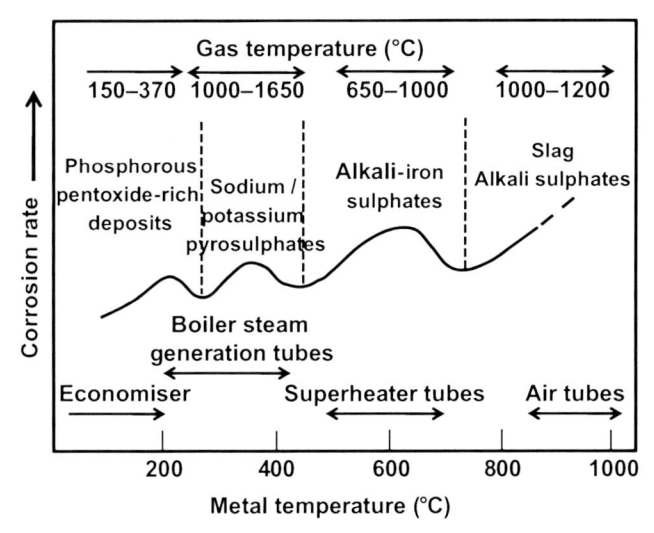

Figure 8.15 Effect of metal temperatures on corrosion rates in conventional pulverised fuel-fired power systems.
Adapted from Simms (2011b) and Natesan et al. (2003).

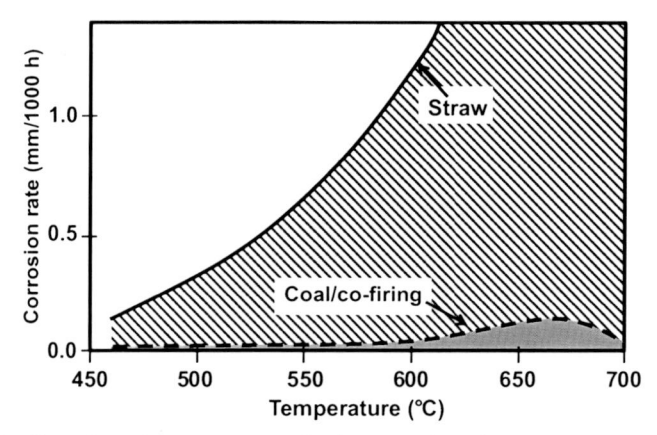

Figure 8.16 Effect of metal temperature on fireside corrosion of AISI347 and AISI347HFG for combustion systems fired on wheat straw, coals and co-firing.

damage rates observed reflects the known variability of wheat straw compositions and variations in local exposure conditions. These activities have been part of the driving force in developing aspirational targets for coal, biomass and waste-fired power plants (summarised in Table 8.7), which illustrate the generally increasing aggressiveness of fireside corrosion that has been found for coal, biomass and waste fuels.

As more data have been generated for fireside corrosion of superheater–reheater tubes, there has been a desire to generate mathematical models to represent such forms of corrosion damage. Different approaches have been developed ranging from mechanistic modelling through empirical curve fitting to neural networks

Table 8.7 **Aspirational targets for superheater/reheater tube lives, steam temperatures and degradation rates in coal, biomass and waste systems**

Fuel	Target lifetime (h)	Desired maximum steam temperature (°C)	Metal temperature (°C)	Maximum acceptable corrosion rate (μm/1000 h)
Coal	100,000	760	790–810	20
Straw	20,000	580	610–630	100
Wood	40,000	580	610–630	50
Waste	40,000	500	530–550	50

Adapted from EU COST522/538 targets as a result of subsequent EU research programmes.

(Saunders et al., 2002). A relatively simple model was generated for UK coals (James and Pinder, 1997):

$$\text{Corrosion rate} = \text{LE} \cdot A \cdot (T_g)^B \cdot (T_m - C)^D \cdot (\text{Cl}_{\%\text{fuel}} - E)$$

in which $A-E$ are constants, LE represents the leading edge of a tube bundle, T_g is the average gas temperature, T_m is an average metal temperature and $\text{Cl}_{\%\ \text{fuel}}$ represents the average fuel composition.

Alternative approaches have been recently reviewed by Wright and Shingledecker (2015) and include one proposed by Larson and Montgomery (2006), Simms et al. (2007b), Simms and Fry (2010), Heikinheimo et al. (2008), Linjewile et al. (2003) and Lant et al. (2011). All of these approaches have different benefits and limitations, and are still at various stages of development.

8.4.4 Erosion–abrasion–wear

Erosion is a damage mechanism that causes metal loss because of particles impacting on a surface (Finnie, 1995). Erosion damage has been found to vary with particle size, particle velocity, particle hardness, tube-surface hardness and impact angle. Brittle and ductile damage regimes have been identified depending on the impact conditions.

For heat-exchanger tubing operating at higher temperatures, interaction with the oxide scales that form result in what has been termed 'erosion/corrosion' damage (Stack et al., 1995). In this case, the impacting particles can interact with either a surface oxide layer or the underlying alloy depending on the exact exposure conditions. As a result, a number of different erosion–corrosion regimes have been identified ranging from pure erosion, through oxidation-enhanced erosion and erosion-enhanced oxidation to modified oxidation depending on the impact conditions and temperatures.

In pulverised-fuel boilers, erosion damage can occur to the waterwalls, superheaters–reheaters and the economiser (Stringer, 1995; Foster et al., 2004), with fly ash particles either directly eroding the tube material or the surface oxide (for tube surfaces >425 °C). For example, erosion–corrosion damage can be found in the superheater–reheater platens under conditions in which deposit blockages have built up between some tubes in these platens and so have caused locally higher gas velocities elsewhere in the platens.

An alternative cause of erosion damage to heat exchangers is caused by ash removal (or 'soot blowing') using steam or compressed air, in which ash becomes entrained in the high-velocity gas streams and impacts on the tube surfaces.

Erosion–corrosion conditions have been a particular challenge in fluidised-bed combustion systems, both for waterwalls and in-bed heat-exchanger tubing (Stringer, 1995). Because of many years of investigating cases of such damage, these damage modes can now usually be avoided by careful engineering design.

Another location for abrasion–wear issues that becomes important for fuel-flexible pulverised-fuel power plants is in the fuel-preparation equipment, such as ball mills, hammer mills etc. (Doran, 2009; Raask, 1985; Foster et al., 2004).

8.5 Flexible fuel use

The use of alternative solid fuels either alone (i.e. substitution) or by co-firing is influenced by many factors, including:

- fuel costs—subsidies
- emission penalties
- fuel availability
- suitability of alternative fuel for existing process
- fuel storage—handling—preparation
- power plant efficiencies and scale
- need for local use of heat—power
- need for additional gas cleaning facilities to meet environmental regulations

The compositions of the fuels particularly influence points 3—5 (e.g. Maciejewska et al., 2006). However, it has been found that the certain combinations of fuels are not desirable when co-firing due to their tendencies to increase deposition fluxes in different locations along the combustion system hot-gas paths and to increase corrosion damage to heat exchangers.

8.5.1 Fuel substitution

The availability of biomass and waste fuels, as well as the significant differences in both their physical and chemical properties, has guided their use as single fuels towards dedicated power plants. Many new biomass and waste-to-energy power plants have been built during the last decade and the numbers of these plants are expected to increase significantly in the immediate future as increasing emphasis is placed on switching to renewable and more sustainable fuels.

However, for old coal-fired power stations that are being decommissioned as part of current environmental initiatives, one alternative that has been investigated in the UK is to adapt them in such a way that 100% biomass can be fired in the boilers (e.g. at RWE npower's Tilbury power station and E.ON UK's Ironbridge power station). To make such schemes viable, several issues need to be addressed, including:

- sourcing very large quantities of easily transportable biomass
- using a power station suitably located to receive large quantities of biomass
- replacement of fuel handling, storage and preparation systems
- new fuel burners
- downrating the boiler (to use lower final steam temperatures and/or to cope with the large volumes of biomass required)
- installation of appropriate gas-cleaning systems
- long-term economic viability of such a scheme (given the unreliability of regulations and subsidies for sustained periods)

There is a need to improve the efficiencies of power plants firing biomass and waste fuels. One of the key limiting factors in restricting their efficiencies is deposit formation and corrosion on the final superheater (Sections 8.4.1 and 8.4.3). New boilers are designed to try to partially counter these effects, but one new technology developed by

Vattenfall is targeted at altering the environments generated in the boilers using these fuels. The 'ChlorOut' process sprays a sulphur-rich compound (ammonium sulphate) into the gas stream and controls this on the basis of minimising the alkali-metal chlorides that are present in the gas stream (Vattenfall, 2005). Initial trials have shown this effective in reducing both deposition and fireside corrosion with fuels rich in alkali chlorides (Vattenfall, 2005).

8.5.2 Co-firing fuels

Co-firing of biomass fuels in previously coal-fired power plants has proved to be a successful route to introduce significant quantities of biomass into the power-generation market. In the UK, the levels of biomass co-firing steadily increased during 2000—2010 up to 10% (on an energy basis) for some biomass—coal mixtures, with the use of still higher levels being actively investigated. The use of 10% biomass in a 2000 MWe power plant represents 200 MWe of biomass-derived power (and should be compared to the ~40—50 MWe output of a new 100% biomass-fired plant operating at a lower efficiency).

The coal—biomass mixes that can be used in such systems are limited by a number of factors:

- fuel transport—handling—storage systems originally designed for coal: ships, trains or lorries for fuel transport; external storage for coal or internal storage for biomass fuels
- fuel-preparation systems: coal-grinding mills can tolerate a few % biomass in a mixed-fuel feed, but at higher levels separate dedicated biomass mills are required, and then the two prepared fuel streams need to be blended
- combustion systems: adding biomass to existing coal feeds and then burning the mixed fuels in existing burners (designed for coal use only) is one alternative; another is to not mix the fuel streams and then use separate biomass and coal burners distributed evenly around the boiler
- slagging—fouling—corrosion: differences in fuel compositions can cause increased rates of deposition and different deposit compositions with some biomass—coal mixtures; such differences can result in reduced heat transfer/more frequent cleaning requirements, increased corrosion damage and ultimately shorter component lives coupled with reduced boiler reliability (Davis and Pinder, 2004; Livingston, 2010; Simms et al., 2007b).

To minimise the risks associated with the introduction of co-firing, technology developments have focused on all of these topics (Davis and Pinder, 2004; Livingston, 2010; Simms et al., 2007b, Waldron, 2010; Overgaard et al., 2004). For activities related to combustion and slagging—fouling—corrosion, these have included trials on both pilot plants and power station boilers, which have focused on specific coal—biomass combinations and included thorough monitoring of the power plants (e.g. Henderson et al., 2002a). In addition, additional fundamental supporting research has been carried out to gain a better understanding of the processes involved with multiple fuels, development of predictive models and discovery of approaches that can be used in controlling them (covered in Section 8.4). In particular, opportunities exist for minimising the risks involved by the careful selection of combinations of coal and biomass fuels.

8.6 Quantification of damage and protective measures

8.6.1 Component- and material-monitoring methods

The environmental degradation of heat-exchanger tubes has been studied in plants, pilot plants and laboratory tests. Each of these types of environment has its own particular benefits and limitation in terms of materials monitoring.

In plant environments, the traditional monitoring of heat-exchanger tube materials is carried out using a mixture of visual inspection, dimensional metrology and ultrasonic inspections. The data generated are used in combination with an assessment of the remaining life of the tubes to determine the risk of component failure before the next scheduled maintenance, and so whether any tubes need to be removed from service. Tubes removed from boilers (during plant maintenance or outages) can be destructively examined using standard laboratory techniques, including optical and electron microscopy, energy dispersive X-ray (EDX) analysis, X-ray diffraction, etc. to further investigate their performance (or determine the cause of failure). However, the data generated from plant heat exchanger tubing can be difficult to interpret as a result of changes in fuels (e.g. different coals or biomass, fuel preparation etc.) and operating conditions (e.g. gas temperatures, metal temperatures, air−fuel ratios, etc.), as well as a lack of monitoring data.

Specific materials evaluation programmes for extended periods (thousands or tens of thousands of hours) can be carried out in plants by several methods, including:

- Installing candidate materials within the heat exchangers; for example, as short lengths within a superheater−reheater, or as a small panel in a waterwall (Stam, 2013; Lant et al., 2011; Henderson et al., 2002a). These pieces then need retrieving at appropriate plant outages.
- Installing materials in separate water−steam-cooled loops within a boiler (Larson and Montgomery, 2006; Henderson et al., 2002a). These loops then need removal at appropriate plant outages.
- Exposing materials on cooled probes (using air, water or steam cooling) in the boiler environment (Stam, 2013; Lant et al., 2011; Henderson et al., 2002a). These can usually be retrieved during plant operation.
- Using online monitoring methods; these are at the relatively early stages of development, but are being used by research activities to try to assess corrosion rates, for example by using electrochemical noise (ECN) and/or linear polarisation resistance (LPR) (Linjewile et al, 2003; McGhee, 2009), and deposition rates.

In all these cases, it is also necessary to arrange for appropriate gas and temperature monitoring around the materials. Following their removal from the plant, the materials can be destructively examined to evaluate their performance.

The use of pilot plants offers an alternative approach to full-scale plant exposures with the advantages of still using real fuels, but allowing much easier and more extensive monitoring of the material's exposure conditions. However, pilot plants are expensive to operate for extended periods and so this usually limits the lengths of such exposures to tens or perhaps hundreds of hours at most.

Laboratory exposures can be carried out under much more controlled exposure conditions. However, these conditions are simulations of what happens in plants and are generally difficult to set up, with considerable care needed to relate them to the plant environments. The data generated have usually been reported in terms of mass change, but this has often proved misleading. Much more useful dimensional data can be used to generate datasets on metal losses, and this is now being increasingly generated following a draft EU standard method for corrosion testing (EC Project SMT4-CT95-2001, 2000), which has recently been modified into a series of International Organisation for Standardisation (ISO) standards (e.g. British Standards (BS ISO 26146, 2012, BS ISO 14802, 2012)). These tests are particularly valuable in enabling the effects of the different exposure variables to be separated out and their sensitivities determined (Saunders et al., 2002; Simms et al., 2007b).

Thus, investigations of materials performance in plant, pilot plant and laboratory environments all have roles to play in determining the environmental degradation of heat-exchanger materials, with each contributing to different aspects of generating the data required to understand the various processes involved.

8.6.2 Protective coatings

Traditionally, coatings have been used on the heat exchangers in power stations as one method of protecting the tubes from particularly challenging cases of environmentally induced degradation. A variation on the coating approach has been to use co-extruded tubes (Meadowcroft and Manning, 1983), with a highly alloyed material extruded around the outside of a standard low-alloy boiler tube material. As the thickness of the coatings that can be applied to heat-exchanger tubing is limited (usually less than 2 mm), it is necessary for the coating materials to be highly resistant to the exposure environment for them to have reasonable lives (at least enabling the component to continue in service until the next scheduled major overhaul). Another consideration is the cost of using coatings, with prices currently of the order of 1000–2000 euros/m^2.

To protect against fireside corrosion on waterwalls or superheaters—reheaters, it has been found that highly alloyed coatings are required; in UK, pulverised-fuel systems, coatings of IN671 or Ni-50% Cr have been used successfully, as have highly alloyed co-extruded tubes, such as American Iron and Steel Institute (AISI) 310 (Syrett, 1987). However, such coatings are only applied to limited areas of tubing to protect against specific localised environments that have been found to be causing accelerated damage. In contrast, waste-fired boilers have such aggressive conditions on heat exchanger surfaces that alloy 625 (Ni, 20–23%; Cr, 8–10%; Mo, <5%; Fe, 3.15–4.15%; Nb + Ta) is often now applied to large areas of these tubes (to prevent otherwise-frequent changes of the heat exchangers).

For all of these coating systems, it has been found that the coating quality is critical in providing adequate component life, as defects in the coatings can result in their rapid loss. The production of relatively smooth, defect-free coatings is easier in a production environment than in a power plant, although it is necessary to be able to apply such coatings in both types of locations. Weld overlay, high-velocity oxygen fuel thermal spraying (HVOF) and laser-cladding processes have all been used successfully in applying coatings to heat-exchanger tubing.

8.7 Future trends

There are currently many different pressures on the power-generation industry which will influence how it develops in the near future, including:

- a need to generate far more power worldwide to meet the needs of a growing world population and the expected economic development of many countries; IEA projections suggest 40% more energy will be needed by 2030 (IEA, 2009)
- concern over the emissions of greenhouse gases (GHG) leading to global warming and the varying desire to reduce such emissions around the world; current EU policy is for a 20% reduction (relative to 1990 levels) by 2020, 40% by 2030 and a 80–95% reduction by 2050 (European Community (EC), 2015)
- a desire to use more sustainable fuels; increasing the levels of biomass and waste used for power generation
- fuel supplies; in terms of total availability, geographic distribution and costs
- a wide range of alternative methods for generating power and heat that are at different stages of development and with varying possibilities for successful application (including wind, wave, solar and nuclear power systems)
- financial viability
- political policies and regulatory regimes that develop over time and vary with location

As a result, the nature of future power systems is currently considerably uncertain, but it is clear that they will have to be much more efficient than current systems, generate far fewer CO_2 emissions and be more fuel flexible. Specific fuel flexibility-related topics that will affect plant integrity are:

- higher steam system temperatures
  Proposed increases for pulverised-fuel power systems to ~760 °C/300 bar steam systems (from the state-of-the-art systems of ~600 to 620 °C/240 bar and most current systems of ~540 to 560 °C/160 bar) to increase efficiencies (and counter the efficiency penalty of CCS systems) present numerous material challenges in terms of component creep, fatigue, fireside corrosion, steam oxidation etc. These will require the increased use of stainless steels, nickel-based alloys and protective coatings (e.g. Shingledecker and Wright, 2006).
- increased biomass levels in co-firing
  The use of higher levels of biomass would further reduce the net CO_2 emissions, but many challenges are associated with this, particularly in terms of fuel handling, storage, combustion as well as economic viability (and subsidy levels). In terms of the environmental degradation of heat exchangers, the main concern is for the superheaters—reheaters, in which increasing the levels of biomass in a coal—biomass mix can increase the chance of aggressive chloride deposits forming and causing rapid corrosion damage (as observed in some biomass-fired power plants). To prevent this, it is necessary to carefully assess the particular combinations of specific coals and biomass that could be used to minimise the risks involved, or consider the use of fuel additives.
- Fuel switching within existing power plants
  It has been shown that pulverised-coal plants can be successfully converted to 100% biomass firing for limited periods. Given appropriate regulation—subsidy levels, this could become more widespread and would maximise the use of current power-generating capacity, but

would require modifications to enable biomass storage, handling, preparation, combustion, etc. Such systems could try to minimise the effects of the biomass compositions on deposition and corrosion by blending two (or more) types of biomass, or by using appropriate chemical additives.
• carbon capture systems
  These are beyond the scope of this chapter, but their implementation for coal-fired power generation (and associated impact on power-generation costs) would be expected to encourage the use of biomass and waste fuels, though this would be dependent on the regulation and subsidy levels applied to the various different technologies. The use of biomass co-firing with some types of CCS systems (e.g. oxy-firing) would need careful assessment in terms of its potential impact on heat-exchanger lives. The effect of using some types of oxy-fired pulverised-fuel systems (e.g. with hot flue-gas recycling upstream of a flue-gas desulphurisation unit) is to significantly change the flue-gas composition, with the levels of species such as SO_x and HCl increased by up to five times compared to air-fired systems (Figure 8.17), resulting in changes to the fireside corrosion environment (Bordenet and Kluger, 2008; Simms et al., 2007b).
• improved modelling for better predictions of potential lives and the effects of fuels
• improved online corrosion–deposition monitoring to more quickly measure the effects of different fuels and plant operating conditions to allow better optimisation of plant operations
• increased use of protective coatings on components (once adequately demonstrated)

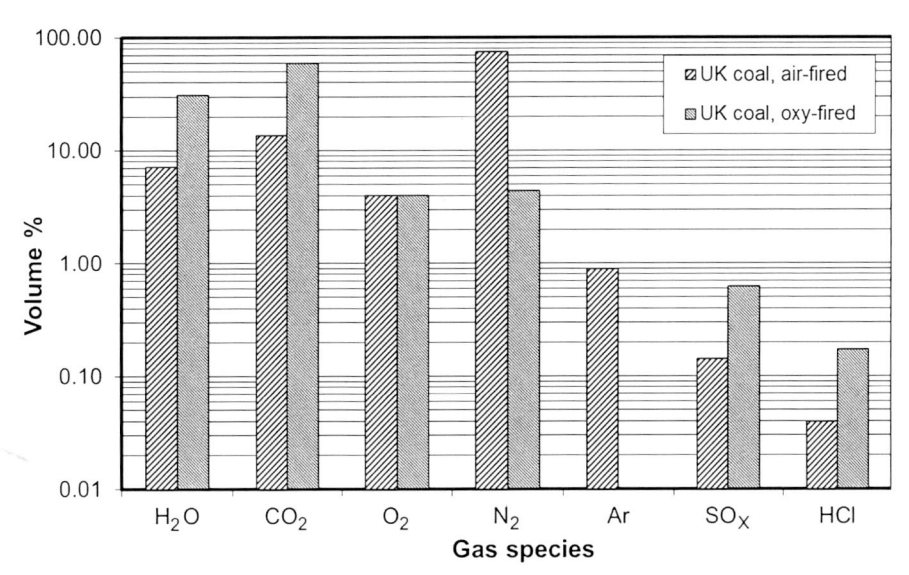

Figure 8.17 Comparison between gas compositions produced in pulverised-fuel systems fired with air and oxygen (with hot-gas recycle prior to flue-gas desulphurisation). Adapted from Simms et al. (2007a).

Sources of further information

Apart from the specific references provided throughout this chapter, further information on materials performance is available from the following references, conference series and websites:

- Failures in pulverised-coal boilers: French (1993).
- General oxidation and corrosion: Birks et al. (2006), Young (2008) and Kofstad (1988).
- Fouling, slagging and corrosion: Livingstone (2009), Zhou et al. (2007) and Tomeczek and Wacławiak (2009).
- Erosion: Stringer (1995) and Finnie (1995).
- Performance of materials in power plant environments:
  - Conference series: Materials for Advanced Power Engineering 1990, 1994, 1998, etc.
  - European Federation of Corrosion Book Series: Numbers 14, 34, 47
  - Parsons Conference series: 1984, 1988, 1995, 1997, 2000, 2003, 2007, 2011

For specific fuels, several books and websites provide far more detailed information than was possible in the sections of this chapter. These include:

- Speight, J.G., 2005. Handbook of Coal Analyses, John Wiley.
- van Loo, S., Koppeian, J., 2007. The Handbook of Biomass Combustion and Co-firing, Earthscan.
- British Coal Utilisation Research Association (BCURA): http://www.bcura.org/
- World Coal Institute: http://www.worldcoal.org/
- IEA Clean Coal Centre: http://www.iea-coal.org.uk/site/ieacoal/home

For fuel compositions, several databases can be accessed via the internet, for example:

- Energy Research Centre of the Netherlands: Phyllis, database for biomass and waste: www.ecn.nl/phyllis2/
- U.S. Department of Energy, Energy Efficiency and Renewable Energy, Biomass Program, Biomass Feedstock Composition and Property Database: www1.eere.energy.gov/biomass/feedstock_databases.html
- BIOBIB: A Database for Biofuels, University of Technology Vienna: www.vt.tuwien.ac.at/biobib/biobib.html
- IEA Bioenergy Task 32, Biomass Combustion and Co-firing: www.ieabcc.nl/

Fuel standards can be obtained from national and international standard institutions, for example:

- ASTM: www.astm.org/index.shtml
- BSI: www.bsigroup.com/en/
- CEN: www.cen.eu/cen/Pages/default.aspx

References

Birks, N., Meier, G.H., Pettit, F.S., 2006. High-Temperature Oxidation of Metals. Cambridge University Press.
Blomberg, T., 2008. What are the right test conditions for the simulation of high temperature alkali corrosion in biomass combustion. In: Schütze, M., Quadakkers, W.J. (Eds.), Novel

Approaches to Improving High Temperature Corrosion Resistance. European Federation of Corrosion Publications Number 47, Woodhead Publishing, pp. 501–513.

Bordenet, B., Kluger, F., 2008. Thermodynamic modelling of the corrosive deposits in oxy-fuel fired boiler. Materials Science Forum 595–598, 261.

Bradford, S.A., 1987. 'Fundamentals of corrosion in gases' metals handbook In: Corrosion, ninth ed., vol. 13. ASM International.

BS ISO 14802, 2012. Corrosion of Metals and Alloys—Guidelines for Applying Statistics to Analysis of Corrosion Data.

BS ISO 26146, 2012. Corrosion of Metals and Alloys—Method for Metallographic Examination of Samples after Exposure to High Temperature Corrosive Environments.

Cain, C., Nelson, W., 1961. Trans ASME. Journal of Engineering for Power 83 (Series A), 468.

Carpenter, A.M., Skorupska, N.M., 1993. Coal Combustion – Analysis and Testing. IEA Coal Research. IEACR/64.

Clarke, L.B., Sloss, L.L., 1992. Trace Elements – Emissions from Coal Combustion and Gasification. IEA Coal Research. IEACR/49.

Colechin, M., 2005. Best Practice Brochure: Co-firing of Biomass (Main Report). DTI Report No. COAL R287, DTI Pub URN 05/1160.

Commission of the European Communities, COM, 2005. 628 Final, Biomass Action Plan.

Couch, G.R., 1995. Power from Coal – Where to Remove Impurities? IEA Coal Research. IEACR/82.

Davis, C., 2010. Impact of Oxy-fuel Operation on Corrosion in Coal Fired Power Plant. http://www.specialmetalsforum.com/uploads/docs/12754067434.
EONEngineeringOxyfuelCorrosion.pdf (accessed 04.11.10.).

Davis, C.J., Pinder, L.W., 2004. Fireside Corrosion of Boiler Materials – Effect of Co-firing Biomass with Coal. UK Department of Trade and Industry. Report No. COAL R267 DTI/Pub URN 04/1795.

Doran, M., 2009. Feedstocks for thermal conversion. In: Bridgwater, A.V., Hofbauer, H., van Loo, S. (Eds.), Thermal Biomass Conversion. CPL Press, pp. 129–156.

EC, 2015. LIFE and Climate Change Mitigation. ISSN: 2314-9329. European Union, ISBN 978-92-79-43945-2. http://dx.doi.org/10.2779/59738.

EC Project SMT4-CT95-2001, 2000. Draft Code of Practice for Discontinuous Corrosion Testing in High Temperature Gaseous Atmospheres, TESTCORR. ERA Technology, UK.

EU Waste Framework Directive, 2008, 2008/98/EC article 4.

EUBIONET2, 2007. Biomass Fuel Supply Chains for Solid Biofuels – from Small to Large Scale. VTT Jyväskylä, Finland.

Farley, M., 2010. Overview of Capture Technologies for Pulverised Coal-Oxyfuel and Post Combustion Capture. Doosan Power Systems. Available from: http://www.specialmetalsforum.com/uploads/docs/12754049682.DoosanMFNamtecHarrogate2010.pdf (accessed 04.11.10.).

Finnie, I., 1995. Some reflections on the past and future of erosion. Wear 186–187, 1–10.

Foster, D.J., Livingston, W.R., Wells, J., Williamson, J., Gibb, W.H., Bailey, D., 2004. Particle Impact Erosion and Abrasion Wear –Predictive Methods and Remedial Measures. DTI, UK. Report No. COAL R241 DTI/Pub URN 04/701.

Francis, W., Peters, M.C., 1980. Fuel and Fuel Technology. Pergamon Press.

French, D.N., 1993. Metallurgical Failures in Fossil Fired Boilers. Wiley.

Hack, K., Jantzen, T., 2008. Development of toolboxes for the modelling of hot corrosion of heat exchanger components. In: Schütze, M., Quadakkers, W.J. (Eds.), Novel Approaches to Improving High Temperature Corrosion Resistance. European Federation of Corrosion Publications Number 47, Woodhead Publishing, pp. 550–567.

Heikinheimo, L., Baxter, D., Hack, K., Spiegel, M., Hämäläinen, M., Krupp, V., Arponen, M., 2008. Optimization of in-service performance of boiler steels by modelling high temperature corrosion (OPTICORR). In: Schütge, M., Quadakkers, W.J. (Eds.), Novel approaches to improving High Temperature Corrosion Resistance, European Federation of Corrosion Publications Number Quadakkers. 47, Woodhead Publishing, pp. 517−532.

Henderson, P.J., Karlsson, A., Davis, C., Rademakers, P., Cizner, J., Formanek, B., Gorannsson, K., Oakey, J., 2002. In-situ corrosion testing of advanced boiler materials with diverse fuels. In: Lecomte-Beckers, J., Carton, M., Schubert, F., Ennis, P.J. (Eds.), Materials for Advanced Power Engineering 2002. Forschungszentrum Jülich GmbH, pp. 785−800.

IEA, 2009. World Energy Outlook 2009.

Jacobsen, S., Brock, J.R., 1965. Journal of Colloid Science 20, 544−554.

James, P.J., Pinder, L.W., 1997. Materials at High Temperature 14, 117.

Jones, A., 2005. 54th BCURA Robens Coal Science Lecture. http://www.bcura.org/csl05.pdf (accessed 04.11.10.).

Kofstad, P., 1988. High Temperature Corrosion. Elsevier Applied Science.

Koppejan, J., van Loo, S., 2009. Biomass conversion overview. In: Bridgwater, A.V., Hofbauer, H., van Loo, S. (Eds.), Thermal Biomass Conversion. CPL Press, pp. 1−11.

van Krevelen, D.W., 1950. Fuel 29, 269−284.

Lant, T., Keefe, C., Davies, C.J., McGhee, B., Simms, N.J., Fry, A.T., 2011. Modeling fireside corrosion of heat exchanger materials in advanced energy systems. In: Gandy, D., et al. (Eds.), Proc. EPRI 6th Int. Conf. on 'Advances in materials technology for fossil power plants', vol. 255−267. ASM International, Materials Park, OH.

Larson, O.H., Montgomery, M., 2006. Materials problems and solutions in biomass fired plants. In: Lecomte-Beckers, J., Carton, M., Schubert, F., Ennis, P.J. (Eds.), Materials for Advanced Power Engineering 2006. Forschungszentrum Jülich GmbH, pp. 245−260.

Lindberg, D., Backman, R., Chartrand, P., 2006. Thermodynamic evaluation and optimization of the ($NaCl + Na_2SO_4 + K_2SO_4 + Na_2S_2O_7 + K_2S_2O_7$) system. J. Chem. Thermodynamics 38, 1568−1583.

Lindberg, D., Backman, R., Chartrand, P., 2007. Thermodynamic evaluation and optimization of the ($NaCl + Na_2SO_4 + Na_2CO_3 + KCl + K_2SO_4 + K_2CO_3$) system. J. Chem. Thermodynamics 39, 1001−1021.

Linjewile, T.M., Valentine, J., Davis, K.A., Harding, N.S., Cox, W.M., 2003. Prediction and real-time monitoring techniques for corrosion characterisation in furnaces. In: Norton, J.F., Simms, N.J., Bakker, W.T., Wright, I.G. (Eds.), Life Cycle Issues in Advanced Energy Systems, Science Reviews, pp. 175−184.

Livingston, W.R., 2009. Fouling corrosion and erosion. In: Bridgwater, A.V., Hofbauer, H., van Loo, S. (Eds.), Thermal Biomass Conversion. CPL Press, pp. 157−176.

Livingston, W.R., 2010. Advanced Biomass Co-firing Technologies for Coal-fired Boilers (publication details to be confirmed).

van Loo, S., Koppeian, J., 2007. The Handbook of Biomass Combustion and Co-firing, Earthscan.

Maciejewska, A., Veringa, H., Sanders, J., Peteves, S.D., 2006. Co-firing of Biomass with Coal: Constraints and Role of Biomass Pre-treatment. European Commission. EUR 22461 EN.

McGhee, B., 2009. US-UK advanced materials for low emission power plant: boiler corrosion/monitoring/Co-firing. In: US DoE 23rd Conf. Fossil Energy Mat.

Meadowcroft, D.B., Manning, M.I., 1983. Corrosion Resistant Materials for Coal Conversion Systems. Applied Science Publishers.

Nalbandian, H., 2004. Air Pollution Control Technologies and Their Interactions. IEA Clean Coal Centre. CCC/92.

Natesan, K., Purohit, A., Rink, D.L., 2003. Coal-ash corrosion of alloys for combustion power plants. In: US Department of Energy Fossil Energy Conference 2003.

Nelson, W., Cain, C., 1960. Trans ASME. Journal of Engineering for Power 82 (Series A), 194.

Nicholls, J.R., Simms, N.J., 2010. Gas turbine oxidation and corrosion. In: Richardson, T.J.A. (Ed.), Shreir's Corrosion. Elsevier, pp. 518–540.

Overgaard, P., Sander, B., Junker, H., Friborg, K., Larsen, O.H., 2004. Two years' operational experience and further development of full-scale co-firing of straw. In: 2nd World Conf. on Biomass for Energy, Industry and Climate Protection, Rome, May 2004.

Pipatti, R., Sharma, C., Yamada, M., 2006. IPCC guidelines for national greenhouse gas inventories. In: Waste, Chapter 2: Waste Generation, Composition and Management Data, vol. 5. IPCC, UN.

Prewin, 2011. Performance, reliability and emissions reduction in waste incinerators. Available from: http://www.prewin.eu/ [accessed 16 May 2011].

Raask, E., 1985. Mineral Impurities in Coal Combustion. Hemisphere Publishing Corporation.

Saunders, S.R.J., Simms, N.J., Osgerby, S., Oakey, J.E., 2002. Degradation of boiler and heat exchanger materials: data generation, databases and predictive modelling. In: Lecomte-Beckers, J., Carton, M., Schubert, F., Ennis, P.J. (Eds.), Materials for Advanced Power Engineering 2002. Forschungszentrum Jülich GmbH, pp. 801–813.

Shingledecker, J.P., Wright, I.G., 2006. Evaluation of the materials technology required for a 760 °C power steam boiler. In: Lecomte-Beckers, J., Carton, M., Schubert, F., Ennis, P.J. (Eds.), Materials for Advanced Power Engineering 2006. Forschungszentrum Jülich GmbH, pp. 107–119.

Simms, N.J., Kilgallon, P.J., Oakey, J.E., 2007a. Fireside issues in advanced power generation systems. Energy Materials: Materials Science and Engineering for Energy Systems 2, 154–160.

Simms, N.J., Kilgallon, P.J., Oakey, J.E., 2007b. Degradation of heat exchanger materials under biomass co-firing conditions. Materials at High Temperatures 24, 333–342.

Simms, N.J., Fry, A.T., 2010. Modelling fireside corrosion of heat exchangers in co-fired pulverised fuel power systems. In: Lecomte-Beckers, J., Carton, M. (Eds.), Materials for Advanced Power Engineering 2010. Forschungszentrum Jülich GmbH.

Simms, N.J., 2011a. Solid fuel composition and power plant fuel-flexibility. In: Oakey, J.E. (Ed.), Power Plant Life Management and Performance Improvement. Woodhead Publishing Ltd, pp. 3–37.

Simms, N.J., 2011b. Environmental degradation of boiler components. In: Oakey, J.E. (Ed.), Power Plant Life Management and Performance Improvement. Woodhead Publishing Ltd, pp. 145–179.

Sloss, L., 1992. Halogen Emissions from Coal Combustion. IEA Coal Research. IEACR/45.

Speight, J.G., 1994. The Chemistry and Technology of Coal. M Dekker Inc.

Speight, J.G., 2005. Handbook of Coal Analyses. John Wiley.

Spiegel, M., 2010. Corrosion in molten salts. In: Richardson, T.J.A. (Ed.), Shreir's Corrosion. Elsevier, pp. 316–330.

Stack, M.M., Lekatos, S., Stott, F.H., 1995. Erosion-corrosion regimes: number, nomenclature and justification ? Tribology International 28, 445–451.

Stam, A., 2013. Nextgenpower–demonstration and component fabrication of nickel alloys and protective coatings for steam temperatures of 750 °C. In: EPRI 7th International Conference on Advances in Materials Technology for Fossil Power Plants (22–25 October 2013, USA).

Stam, A., Livingston, W.R., Cremers, M.F.G., Brem, G., 2010. Review of models and tools for slagging and fouling prediction in biomass co-combustion. Review Article for IEA. Task 2010 32, 1–18.

Stringer, J., Wright, I.G., 1995. Current limitations of high temperature alloys in practical applications. Oxidation of Metals 44, 265–308.

Stringer, J., 1995. Practical experience with wastage at elevated temperatures in coal combustion systems. Wear 186–187, 11–27.

Syrett, B.C., 1987. 'Corrosion in Fossil fuel power plant', metals handbook In: Corrosion, ninth ed., vol. 13. ASM International.

Tomeczek, J., Palugniok, H., 2002. Kinetics of mineral matter transformation during coal combustion. Fuel 81, 1251–1258.

Tomeczek, J., Palugniok, H., Ochman, J., 2004. Modelling of deposits formation on heating tubes in pulverized coal boilers. Fuel 83, 213–221.

Tomeczek, J., Wacławiak, K., 2009. Two-dimensional modelling of deposits formation on platen superheaters in pulverized coal boilers. Fuel 88, 1466–1471.

Vassilev, S.V., Baxter, D., Vassileva, C.G., 2013. An overview of the behaviour of biomass during combustion: Part I. Phase-mineral transformations of organic and inorganic matter. Fuel 112, 391–449.

Vattenfall Research and Development, 2005. ChlorOut. Available from: http://www.vattenfall.com/en/file/ChlorOut_8459980.pdf (accessed 04.11.10.).

Waldron, D., 2010. Options for biomass firing in utility boilers. In: Proceedings of Bioten Conference (To be published).

White, L.P., Plaskett, L.G., 1981. Biomass as Fuel. Academic Press.

Wood, N.B., 1981. Journal of the Institute of Energy 76, 76–90.

Wright, I.G., Shingledecker, J.P., 2015. Rates of fireside corrosion of superheater and reheater tubes: making sense of available data. Materials at High Temperatures 32 (4), 426–437.

Young, D., 2008. High Temperature Oxidation and Corrosion of Metals. Elsevier.

Zhou, H., Jensen, P.A., Frandsen, F.J., 2007. Dynamic mechanistic model of superheater growth and shedding in a biomass fired grate boiler. Fuel 86, 1519–1533.

Fuel flexible gas production: biomass, coal and bio-solid wastes

9

Shusheng Pang
University of Canterbury, Christchurch, New Zealand

9.1 Introduction

The total world energy use in 2010 was reported to be 524 quadrillion Btu (5.526×10^{20} J or 9.5×10^{10} barrels of oil equivalent) and this is projected to increase to 630 quadrillion Btu in 2020 and to 820 quadrillion Btu in 2040 (EIA, 2013). Eighty percent of this energy consumed in 2010 was from fossil fuels including oil, gas and coal which, based on the data of EIA (2013), released 31 billion metric tons of carbon dioxide to the atmosphere. The heavily reliance of fossil fuels by humankind has already caused serious consequences such as climate change due to emissions of greenhouse gas (GHG) and looming energy shortage as a result of reserve depletion. Therefore, alternative energy resources which have low or zero GHG emissions have been actively sought as substitute of fossil fuels.

Biomass, one of the most abundant and renewable energy resources, has promising potential for future fuels and energy. The biomass is originated from plants such as trees, bagasse, grass and agricultural crops, which absorb carbon dioxide needed for photosynthesis in their growth. In this way, the whole cycle — from feedstock growing through energy processing to energy consumption — is largely carbon neutral.

However, biomass has low density and is generally distributed in wide areas; therefore, collection, storage and transportation involve high cost and high energy consumption. At present, biomass contributes about 10% to the overall energy demand (Tustin, 2012), and is mainly used for electricity and heat through direct combustion. Converting biomass to gaseous fuel through gasification and pyrolysis has advantages such as flexibility in product application, high energy efficiency and low negative impacts on the environment. However, commercialization of biomass gasification and pyrolysis has been facing challenges due to high costs of processing the low energy density biomass feedstock.

One of the solutions for increasing biomass utilization for gaseous fuels is to use blended biomass and coals or blended biomass and bio-solid wastes, considering that coal has a much higher density, and utilization of bio-solid waste has great benefits for the environment. Coal is an important conventional fossil fuel with abundant reserves estimated around 1000 billion tons in the world, which could last 180 years based on the current consumption rate (Lee et al., 2007). The common usage of coal is through direct combustion for electrical generation or as a feedstock for the production of coke and coal gas. In recent years, due to the concerns about the depletion of crude oil reserves in the world, coal has regained strong interest for economic advantages, especially for low-rank coal. However, the utilization of coal causes emission of pollutant substances, such as SO_x and NO_x in addition to CO_2.

Co-utilization of biomass with coal for gaseous fuels can be achieved by co-gasification and co-pyrolysis. Combined use of coal and biomass has great potential and benefits as it can overcome their mutual disadvantages (Kumabe et al., 2007). First, for the economic advantages, utilization of blended coal and biomass can achieve more flexible and reliable operation for a large-scale energy plant. For instance, the biomass energy plant which is built near a forestry or wood-processing industrial area can use the supplementary low-cost coal when the biomass feedstock is in shortage, avoiding the high delivery cost of biomass from long distances.

Second, for energy efficiencies biomass has a lower density, hence lower energy content; so adding coal to the biomass can increase the specific energy content of the product or reduce the auxiliary energy inputs. In addition, co-gasification process can be enhanced by blending biomass and coal as the feedstock because biomass contains a high content of metal elements which have a catalytic effect on the reactivity of coal gasification. Therefore, the carbon conversion rate to a gaseous product is increased, and the yield of tar and residual char are decreased (Brown et al., 2000). Third, Pinto et al. (2003) and Xu et al. (2015) have reported that the composition of gasification-producer gas could be altered by adjusting the blending ratio of the biomass with coal.

Last, but important for environmental benefits, biomass is sustainable and largely carbon neutral; thus, using biomass for energy reduces CO_2 emissions (Collot et al., 1999). In addition, biomass has low contents of sulphur and nitrogen as well as ash; therefore, the combined usage of biomass with coal in an energy plant can reduce the emissions of NO_x and SO_x as well.

Co-gasification of biomass with bio-solid wastes (dried sewage sludge and municipal solid wastes) has been reported by Saw et al. (2012) and Nipattummakul et al. (2010). Sewage sludge is generated from wastewater treatment and contains organic matters. Presently, it is commonly incinerated for heat and power generation and disposed at landfill. Utilization of sewage sludge has benefits for both the environment and energy generation. Municipal solid waste (MSW) may contain waste plastics, food packaging and wood. The characteristics and performance in energy-conversion processes differ significantly; therefore, this chapter focuses on dried sewage sludge and bio-solid wastes. Readers who are interested in other solid waste streams can refer to Arena et al. (2010) for gasification and to Sharypov et al. (2002, 2003, 2006) and Marin et al. (2002) for co-pyrolysis of waste polymers with biomass.

This chapter will describe and discuss gasification, co-gasification, pyrolysis and co-pyrolysis of the three types of solid fuels (biomass, coal and bio-solid wastes) for gaseous fuel products. These technologies are illustrated in Figure 9.1 in which combustion is also included for comparison. To discuss and understand the performance of solid fuels in the conversion process and product composition, the characteristics of each solid fuel are described in the following section.

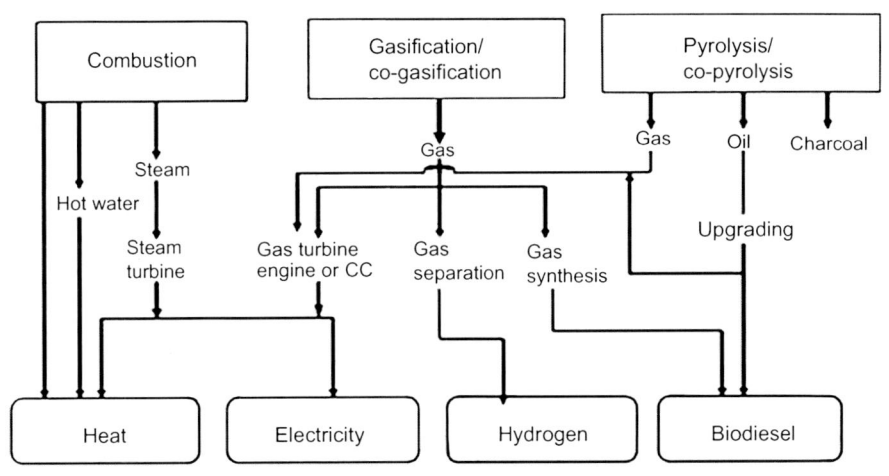

Figure 9.1 Illustration of gasification, co-gasification, pyrolysis and co-pyrolysis of the three types of solid fuels (biomass, coal and bio-solid wastes) for various energy products.

9.2 Characteristics of biomass, coal and bio-solid wastes

Biomass is defined as matter originating from living plants, including tree stems, branches, leaves as well as residues from agricultural harvesting and processing of seeds or fruits. Due to the diversity and complexity of biomass resources, this chapter will focus on woody biomass including stems and branches of trees and residues from wood processing. Biomass properties as related to thermal conversion processes are summarized in Table 9.1 for proximate analysis and Table 9.2 for ultimate analysis (Xu et al., 2015; Saw and Pang, 2013; Saw et al., 2012; Franco et al., 2003; Emami-Taba and Irfan, 2013; McLendon et al., 2004).

Coal is a well-known solid fossil fuel and its properties vary substantially with type, location and age. In this chapter, only selected coal types are discussed based on literature review. The physical and chemical properties of coals can be found in textbooks; thus, only results of proximate and ultimate analyses are described here for comparison with other solid fuels (Xu et al., 2015; Yan et al., 1998; Emami-Taba and Irfan, 2013).

Solid wastes may include bio-solid wastes (such as dried sewage sludge, food residues and waste wood) and inorganic solid wastes such as waste plastics. The performance of these two types of solid wastes is significantly different in processing to gaseous fuels, and this chapter will focus on bio-solid wastes only (Saw et al., 2012; Nipattummakul et al., 2010; Adegoroye et al., 2004; Manyà et al., 2006; Comos, 2012).

Physical properties of coal and bio-solid wastes are relatively constant; however, the properties, dimensions and initial moisture content of biomass vary significantly between different sources (Robertson and Manley, 2006; Li et al., 2006; Pang and Mujumdar, 2010). In log harvesting from forests, the biomass collected is in the

Table 9.1 Proximate analysis results for selected solid fuels: biomass, coal and bio-solid wastes

Solid fuels	Volatile matter (%)	Fixed carbon (%)	Ash (%)	Moisture (%)	HHV (MJ/kg)	References
Pine (softwood)	71.9–77.9	13.8–16.0	0.34–0.50	7.8–12.0	16.3–20.2	Xu et al. (2015), Franco et al. (2003), and Emami-Taba and Irfan (2013)
Eucalyptus (hardwood)	74.8–81.5	12.7–13.9	0.38–0.7	5.4–10.6	19.4–21.3	Xu et al. (2015), Franco et al. (2003), and Emami-Taba and Irfan (2013)
Lignite	32.9–35.3	28.3–34.1	4.2–4.9	19.1–34.6	17.3	Xu et al. (2015) and Saw and Pang (2013)
Sub-bituminous	38.6	42.4–49.0	5.4–8.9	3.5–13.6	26.1	Xu et al. (2015) and McLendon et al. (2004)
Dried sewage sludge	43.5–44.3	16.5–21.8	32–33.9	1.7–8	14.1–16.3	Saw et al. (2012) and Nipattummakul et al. (2010)

Table 9.2 **Ultimate analysis data for selected solid fuels: woody biomass, coal and bio-solid wastes (dry ash-free basis)**

Solid fuels	C (%)	H (%)	O (%)	N (%)	S (%)	References
Pine (softwood)	51.4–51.6	4.9–5.9	42.4–42.6	0.27–0.9	0.01	Xu et al. (2015), Franco et al. (2003) and Emami-Taba and Irfan (2013)
Eucalyptus (hardwood)	50.4–52.8	5.9–6.4	40.0–43.5	0.15–0.4	0.01	Xu et al. (2015), Franco et al. (2003) and Emami-Taba and Irfan (2013)
Lignite	66.6–68.4	4.8–4.9	25.2–27.1	0.7–0.72	0.8	Xu et al. (2015) and Saw and Pang (2013)
Sub-bituminous	64.3–73.3	4.5–5.1	16–17.9	1.0–1.27	1.9–2.4	Xu et al. (2015) and McLendon et al. (2004)
Dried sewage sludge	43.5–45.8	3.0–3.5	14.7–16.2	1.5–5.1	1.1–1.2	Saw et al. (2012), Nipattummakul et al. (2010)

Table 9.3 **Basic characteristics of wet woody biomass (Robertson and Manley, 2006; Li et al., 2006; Pang and Mujumdar, 2010)**

	Forest residue	Bark	Sawdust	Cut-offs
Size (mm)	Chips, <50	<500	≤3	Chips, <50
MC (%)	50–120	50–120	50–150	50–150
ρ^a (kg/m^3)	250–300	250	100–120	250–300

[a]Bulk oven dry density.

form of branches, roots and small top ends of logs, and these are normally cut into chips which are 2–3 mm in thickness, 20–30 mm in width and 30–50 mm in length. In wood processing, the biomass is generated from various operation steps, and its characteristics vary depending on the wood products processed. In sawmills, sawdust and cutoffs are generated from timber sawing and barks from debarking. The processing of laminated veneer lumber (LVL) produces biomass in forms of bark during debarking, cutoffs and core poles in veneer peeling. Large-sized cut-offs from wood processing are also cut into chips. The characteristics of woody biomass are described in Table 9.3.

9.3 Co-gasification of biomass and coal, and co-gasification of biomass and bio-solid wastes

Gasification is a thermochemical process that converts carbonaceous solid fuels such as coal and biomass into CO- and H_2-based gas mixtures through a series of reactions of the feedstock material with a controlled amount of gasification agent (O_2, air, steam). This gas mixture is termed as producer gas which can be used as a fuel gas or be further processed into more valuable energy products.

9.3.1 Gasification theories and technologies

There are two stages in gasification: initial devolatilization (or flash pyrolysis) and subsequent gasification reactions. In the initial devolatilization stage, the solid fuel is decomposed into solid carbon (char), H_2, CO, CO_2, CH_4, tars and other complicated hydrocarbons. In the subsequent gasification process, reactions occur between solid char and gases, and amongst gases including the gasification agent. A summary of the chemical reactions occurring in the gasification process is given in the following.

- Devolatilization (Flash pyrolysis — endothermic):

Biomass → C + H_2O, H_2, CO, CO_2, CH_4, tars, complicated hydrocarbon compounds.

Gasification reactions amongst gases:

Oxidation reactions of combustible gas species:

$$H_2 + 0.5O_2 \rightarrow H_2O \qquad \Delta H_r = -241.8 \text{ kJ/mol} \qquad (9.1)$$

$$CO + 0.5O_2 \rightarrow CO_2 \qquad \Delta H_r = -282 \text{ kJ/mol} \qquad (9.2)$$

$$CH_4 + 2O_2 \rightarrow CO_2 + H_2O \quad \Delta H_r = -802.3 \text{ kJ/mol} \qquad (9.3)$$

Water−gas-shift reaction (one of the key reactions in steam gasification):

$$CO + H_2O \rightarrow H_2 + CO_2 \quad \Delta H_R = -42 \text{ kJ/mol} \qquad (9.4)$$

Steam−methane-reforming reaction (slow reaction):

$$CH_4 + H_2O \rightarrow 3H_2 + CO \quad \Delta H_R = 232.8 \text{ kJ/mol} \qquad (9.5)$$

Gasification reactions between char and gases:

Combustion of char includes the total and partial oxidation of carbon:

$$C + O_2 \rightarrow CO_2 \quad \Delta H_R = -393.77 \text{ kJ/mol carbon} \qquad (9.6)$$

$$C + 1/2O_2 \rightarrow CO \quad \Delta H_R = -110.4 \text{ kJ/mol carbon} \qquad (9.7)$$

Steam gasification reaction (one of the key reactions):

$$C + H_2O \rightarrow H_2 + CO \quad \Delta H_R = 138.3 \text{ kJ/mol carbon} \qquad (9.8)$$

Bouduard reaction:

$$C + CO_2 \rightarrow 2 CO \quad \Delta H_R = 170.45 \text{ kJ/mol carbon} \qquad (9.9)$$

Methanation reaction:

$$C + 2H_2 \rightarrow CH_4 \quad \Delta H_R = -93.8 \text{ kJ/mol carbon} \qquad (9.10)$$

The above reactions are for a general situation, but some reactions may not occur in a specific type of gasifier or with the use of a specific gasification agent.

Different gasifiers may be categorized in different ways. However, most accepted classification is based on physical structure of the gasifier, namely, fixed-bed gasifier (updraft and downdraft), fluidized-bed gasifier (bubbling and circulating) and entrained-flow gasifier. For a given type of gasifier, the gasification agent can be air, oxygen or steam or a mixture of two. Using steam as the gasification agent has attracted great interest as it has advantages in that the producer gas has high content of hydrogen and thus high calorific value. However, the reactions of steam gasification overall are endothermic, which means external heat needs to be supplied for steam gasification. This can be achieved by installing a heat exchanger inside the gasifier or by circulation of bed material which acts as a heat carrier.

When air or oxygen or their mixture is used as the gasification agent, partial oxidation reactions occur which are exothermic, thus providing heat for other endothermic reactions. Air gasification produces a producer gas with calorific value of 4—7 MJ/Nm^3 suitable for boiler, gas engine and gas turbine applications (Zainal et al., 2001). If pure O_2 is used as the gasification agent, the calorific value of producer gas will increase (12—18 MJ/Nm^3) which is suitable as a synthesis gas for conversion to methanol and liquid biofuels, but the gasification operating cost will also be increased due to the pure O_2 production. The producer gas generated from steam gasification has a heating value of 10 to 18 MJ/Nm^3 with H_2 content of up to 60% (Saw and Pang, 2012a; Holfbauer and Knoef, 2005). Table 9.4 gives a summary of compositions of producer gases generated from gasification of biomass and coals from different types of gasifiers and using different gasification agents.

Table 9.4 Typical gas composition of producer gases from biomass gasification using different types of gasifiers and different gasification agents

Gasifier type	Gasification agent (references)	\multicolumn{5}{c}{Gas composition (mol/mol, %)}
		H_2	CO	CO_2	CH_4	N_2
Circulating fluidized bed	Air (Holfbauer and Knoef, 2005)	10—12	16—19	14—18	6—8	48—52
	Steam (Saw and Pang, 2012a)	35—55	15—30	15—25	8—12	0
	Oxygen (Meng et al., 2011)	24—28	20—22	40—44	—	—
Updraft fixed bed	Air (Holfbauer and Knoef, 2005)	19	23	12	5	41
Downdraft fixed bed	Air (Meng et al., 2011; Galindo et al., 2014)	15—18	15—21	13—15	1—2	44—56[a]
Entrained-flow gasifier[b]	Oxygen (Leijenhorst et al., 2015)	28—33	20—25	46	2	—

[a]Estimated by difference.
[b]Feedstock is bio-oil from pyrolysis of woody biomass and straw.

9.3.1.1 Downdraft fixed-bed gasifier

In fixed-bed gasifiers, the gasification reactions occur above a stationary grate. The fixed-bed gasifiers can be further divided into downdraft and updraft gasifiers depending on the flows of the gasification agent and the producer gas. In both types of gasifiers, solid fuel is fed at the top of the gasifier. In the downdraft gasifier, as shown in Figure 9.2, the gasification agent (air or O_2) is fed into the middle of the bed (combustion zone) above the stationary grate and the producer gas flows out of the gasifier from the bottom of the gasifier beneath the stationary grate. In this type of gasifier, the fed solid fuel moves downwards together with the gases through a drying zone, a pyrolysis zone, an oxidization (combustion) zone and a reduction zone. In the drying zone, moisture is vapourized and the solid fuel is dried. With downwards motion, the dry solid fuel is further heated and the dried solid fuel is decomposed to char and gases (pyrolysis). With continuous downwards motion, gasification agent is injected, thus partial combustion of char and some combustible gases occurs, providing needed heat to maintain the target gasification temperature. Then the gases and the char move to the reduction zone in which the gasification reactions occur and the producer gas is formed.

The temperature in each zone is different. In the drying zone, the temperature is normally at 200 °C or lower before the solid fuel is degraded. Temperature in the pyrolysis zone is up to 500–600 °C depending on the equivalence ratio (ER) (the ratio of oxygen provided to the stoichiometric oxygen demand). The oxidation zone has the highest temperature of up to 1500 °C at which tars and other heavy hydrocarbons are thermally cracked into lighter hydrocarbon gas species. Below the oxidation zone, the remaining char, ash, the producer gas and water vapour flow through the reduction zone in which the vapour can react with char (steam gasification reaction), CO (water–gas-shift reaction) and with CH_4 (steam–methane-reforming reaction) to form hydrogen which is desired.

Advantage of the downdraft fixed-bed gasifier is that the tars are cracked down in the oxidation zone, thus the producer gas has lower tar content compared to other types

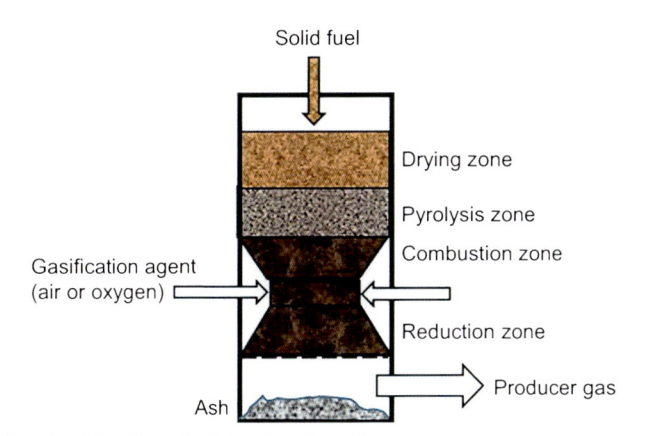

Figure 9.2 Sketch of the downdraft fixed-bed gasifier.

of gasifiers. However, the producer is easily contaminated by ash and other fine particles, and a separation device (e.g., two-stage cyclone and ceramic filter) is needed to clean the producer gas. Another setback with this type of gasifier is relatively high temperature of the exit producer gas, resulting in lower gasification efficiency. Due to the large variation of temperature profile within the gasifier, this type of gasifier is used at small to medium scale (100 kWth—5 MWth).

9.3.1.2 Updraft fixed-bed gasifier

Similar to the downdraft fixed-bed gasifier, the updraft fixed-bed gasifier also has a stationary grate and the solid fuel is fed from the gasifier top. However, the gasification agent (air or O_2) is introduced from the bottom of the gasifier, and the producer flows out of the gasifier from the upper part of the gasifier as shown in Figure 9.3.

The gasification process in the updraft fixed-bed gasifier also has four zones, namely, the drying zone, the pyrolysis zone, the oxidation zone and the reduction zone; however, the oxidation zone is at the bottom and the reduction zone is above it. At the top layer of the gasifier, the solid fuel is dried by the pyrolysis gases and upwards-moving gases from the lower reduction zone and the oxidization zone. At the same time, char from the pyrolysis zone moves downwards to the reduction zone and the oxidization zone in which gasification reactions occur.

The updraft fixed-bed gasifier is simple in structure and operation. The producer gas exiting the gasifier has low temperature, and thus this type of gasifier has high gasification efficiency. In addition, this type of gasifier can handle solid fuel with relatively high moisture content of up to 50%. However, the apparent disadvantage of this type of gasifier is the high tar content in the producer gas. The variation of temperature profile within the updraft gasifier is also significant, and the updraft fixed-bed gasifier is used at small to medium scale (1—10 MWth).

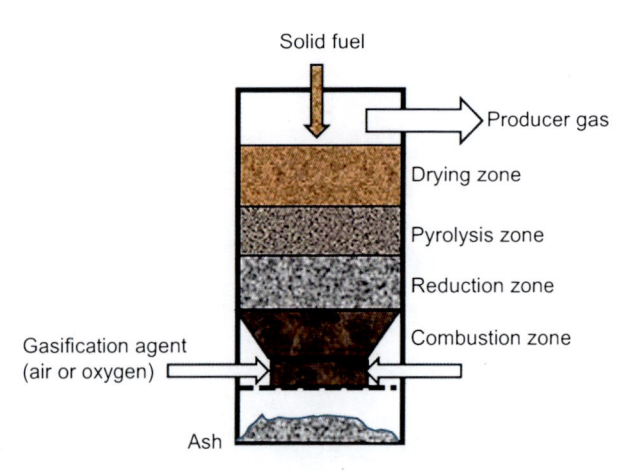

Figure 9.3 Sketch of the updraft fixed-bed gasifier.

9.3.1.3 Bubbling fluidized-bed gasifier

The bubbling fluidized-bed (BFB) gasifier is shown in Figure 9.4 in which the gasification agent is injected from the gasifier bottom and the producer gas exits from the gasifier top while the solid fuel is introduced to the bed. This type of gasifier is characterized by the bubbling of bed material and solid fuel by flowing gas through it when the gas velocity is sufficiently high. The bed material can be an inert medium such as sand or catalytic material such as CaO. The fluidization of the solid fuel and the bed material enhances heat and mass transfer between the solids and the gases, thus promoting the gasification reactions and maintaining target operation temperature.

The gas agent velocity which is able to cause the solids to bubble is called minimum fluidization velocity. However, in practical operation, the gas velocity is above this minimum fluidization velocity but below the threshold when the solid material is carried out of the gasifier, and this threshold velocity is termed terminate velocity. When the solid fuel enters the bed, it is rapidly heated by the bed material and pyrolyzed, generating char, tars, complex hydrocarbon compounds and non-condensable gases as the initial products. This process is followed by the gasification reactions in the upper layers of the bed. The reactions may also occur in the freeboard space above the bed if that space is high enough (Saw and Pang, 2012b).

The BFB gasifier can handle solid fuel with variable properties with good temperature control. Due to the uniform mixing between solids and gases, the temperature profile in the bubbling gasifier is uniform. Both the reaction rate and the carbon conversion efficiency are high. The tar content in the producer gas is moderate but the producer gas is prone to contamination of ash and fine particles.

The BFB gasifier can use air, or oxygen, or a mixture with steam as the gasification agent. The residence time is shorter and the gasification temperature

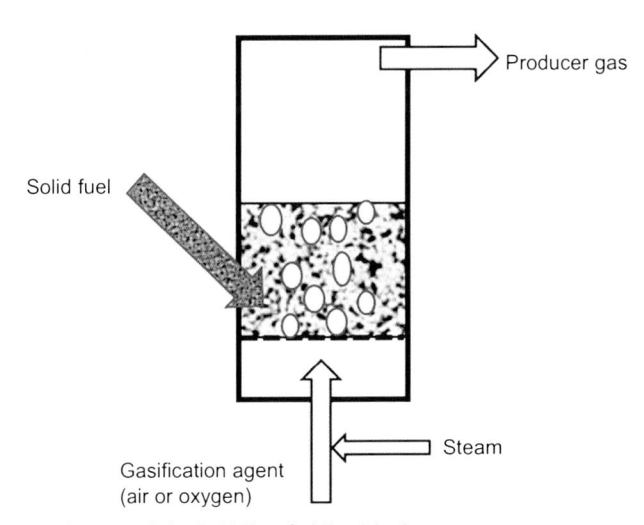

Figure 9.4 Flow diagram of the bubbling fluidized-bed system.

(800–900 °C) is lower than that of the fixed-bed gasifier; therefore, tar content in the gasification producer gas is generally higher than the downdraft fixed-bed gasifier. The BFB gasifier is suitable for medium- to large-scale plants up to 25 MWth. The capacity is limited by the gasifier diameter to control the gas velocity which should be lower than the terminate velocity to prevent solid particles from being carried out.

9.3.1.4 Circulating fluidized-bed gasifier

In a fluidized bed, when gas velocity is higher than the terminate velocity, the bed of solid particles expands to the full space of the reactor and solid particles are then carried out of the gasifier from the top. Gasifiers operating at this condition are called circulating fluidized-bed (CFB) gasifiers (Figure 9.5). To separate the solid particles from the gas, the producer gas and the solids are then directed into a cyclone in which the solid particles are discharged from the bottom and the producer gas flows out from the top. The solid bed material is recycled back to the gasifier while the ash is separated and collected for disposal. The producer gas leaves from the cyclone top, then cooled down and cleaned before further applications.

Characteristics of the CFB gasifier are similar to those of the BFB, but the gas velocity is much higher in the CFB gasifier so the gasifier diameter is much smaller. However, the construction and operation are more complicated than other gasifiers; thus, it is more suitable for large-scale plants.

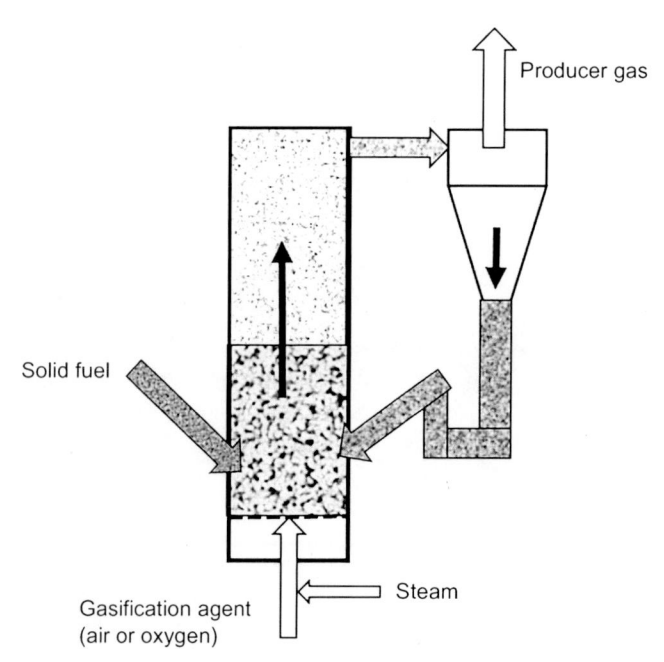

Figure 9.5 Sketch of a circulating fluidized-bed gasifier.

9.3.1.5 Dual fluidized-bed gasifier

A recent development of the fluidized-bed gasifiers is the integration of a BFB gasifier with a CFB combustor for heat supply; thus, pure steam can be used as the gasification agent in the BFB gasifier. This system is called dual fluidized-bed (DFB) gasifier as shown in Figure 9.6 and has attracted great interest in recent years (Saw and Pang, 2012a,b, 2013; Saw et al., 2012; Holfbauer and Knoef, 2005; Pfeifer et al., 2004; Rauch et al., 2013). In this type of gasifier, the solid fuel is fed to the bed of the BFB gasifier and steam is injected from the gasifier bottom as gasification agent. The char generated from the steam gasification together with the bed material flows to the CFB combustor through an inclined chute. In the CFB combustor, air is introduced and the char is combusted for heating up the bed material which is carried up and then out from the top by the flue gas to a cyclone. In the cyclone, the hot bed

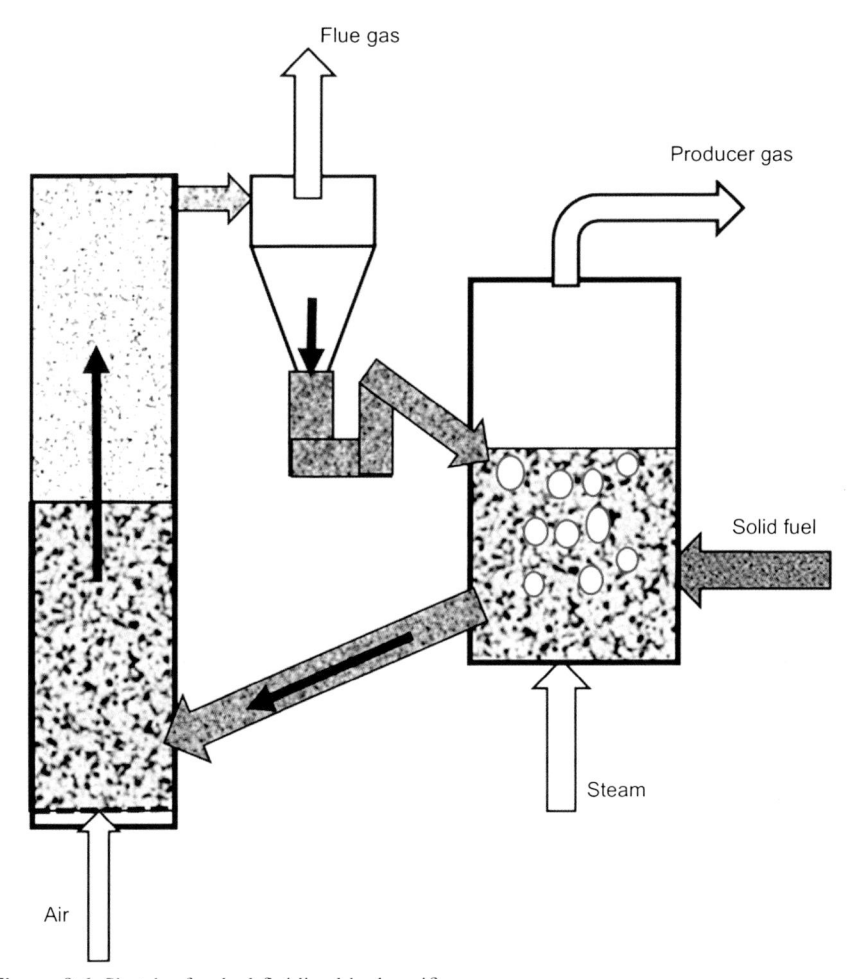

Figure 9.6 Sketch of a dual fluidized-bed gasifier.

material is separated from the flue gas and recycled back to the BFB gasifier through a siphon seal to provide needed heat for the steam gasification. The producer gas generated in the BFB gasifier flows from the gasifier top to another cyclone for removal of entrained ash and fine particles of char and bed material.

The advantage of the gasifier is that this gasifier can produce hydrogen-rich producer gas and achieve high overall energy efficiency in which the heat of the clean flue gas can be recovered. Another advantage is that catalytic bed materials can be applied for further increasing hydrogen content or reducing tar concentration in the producer gas (Saw and Pang, 2012a; Pfeifer et al., 2009). However, the system and operation are complicated, thus is suitable for large-scale plants when high-quality producer gas is targeted.

9.3.1.6 Entrained-flow gasifier

The entrained-flow gasifier is shown in Figure 9.7 in which liquid fuel, fine particles of solid fuels or slurry of solid and liquid fuels are first distributed uniformly at the gasifier top space and then gasified with O_2 as gasification agent. The gasification temperatures are between 1000 and 1500 °C, and the operation can be at atmospheric pressure or pressurized. The residence time of the gas is very short, usually a few seconds. With high operational temperature, the tars are cracked down to light hydrocarbons and thus producer gas is clean. The carbon conversion efficiency in the entrained-flow gasifier is also high. However, this high-temperature operation creates difficulties for material selection of the gasifier and the problem of ash melting. Due to the pre-treatment requirement for the fuels (liquid, fine particles or slurry state), the application of the entrained-flow gasifier is limited to large-scale plants.

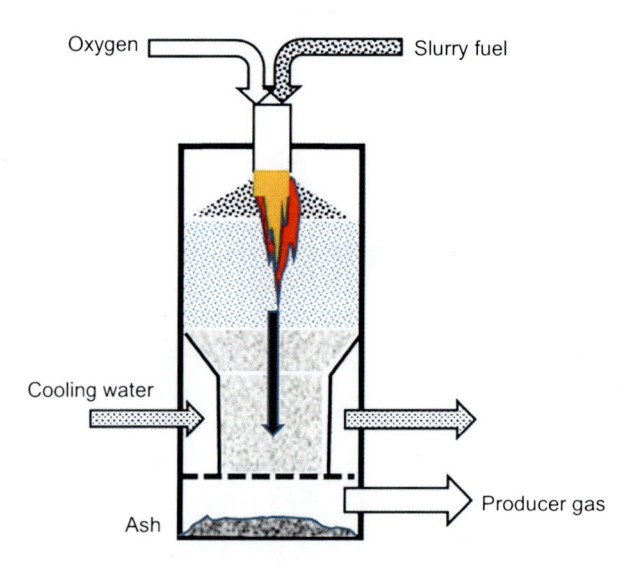

Figure 9.7 Sketch of an entrained-flow gasifier.

9.3.2 Co-gasification of biomass and coal

Gas composition and calorific values of the producer gas are key factors for co-gasification of biomass and coal. Effects of blending proportions and operation conditions have been investigated and recently reported; however, the results are not conclusive, possibly due to the different types of gasifiers and different gasification agents used by different researchers. Most of the reported work was conducted on fluidized-bed gasifiers as the objective for co-gasification of biomass and coal is to feed a large-scale gasification plant (Collot et al., 1999; Cormos, 2012; Saw and Pang, 2013; Xu et al., 2015; Xu, 2013).

Xu (2013) and Xu et al. (2015) have conducted both experimental and theoretical studies on co-gasification of blended biomass and coal using various feedstocks of coal (lignite and sub-bituminous) and biomass (pine and *Eucalyptus niten*). The selected feedstocks were mixed and pelletized at pre-set proportions and tested on a 100 kWth BFB gasifier (Xu, 2013; Xu et al., 2015) and on a 100 kWth DFB gasifier (Xu, 2013), respectively. Effects of fuel-blending ratio and operational temperature were investigated. For understanding of the co-gasification process and to predict the gasification performance at different operation conditions, a comprehensive mathematical model has been developed and validated. Selected results of producer gas composition from the experiments and the model simulation are shown in Figures 9.8 and 9.9 for co-gasification on the bubbling fluidized-bed gasifier and in Figure 9.10 on the DFB gasifier.

From the above results, it is observed that the blending ratio has significant impact on the gas composition both in the BFB gasifier and in the DFB gasifier. The two types of gasifiers show similar trends for the hydrogen content in which

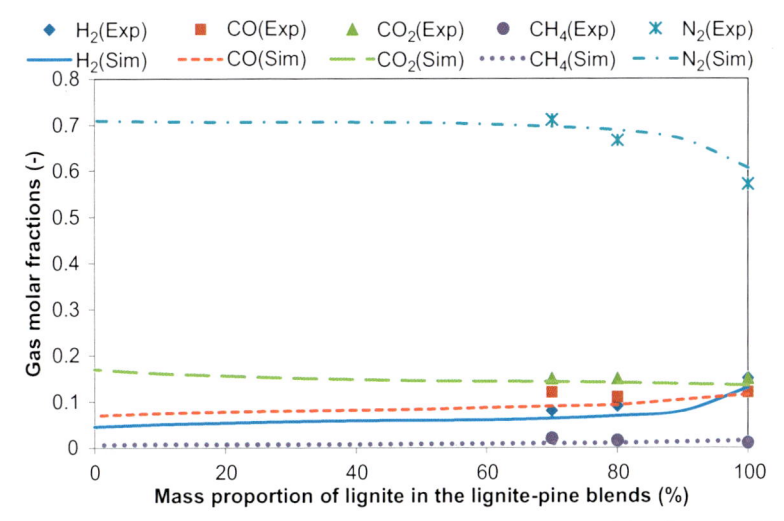

Figure 9.8 Model predicted and experimentally measured gas composition for co-gasification of lignite—pine pellets on a BFB gasifier at 900 °C (Xu, 2013; Xu et al., 2015).

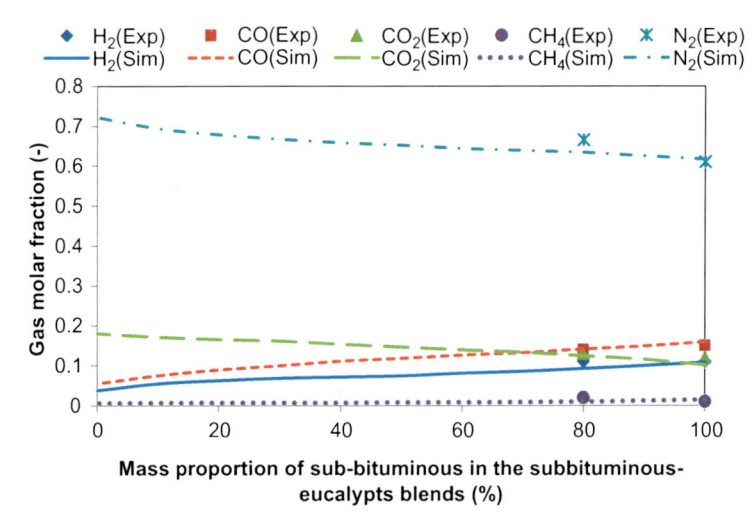

Figure 9.9 Model predicted and experimentally measured gas composition at steady state for gasification of subbituminous-*Eucalyptus* pellets on a BFB gasifier at 900 °C (Xu, 2013; Xu et al., 2015).

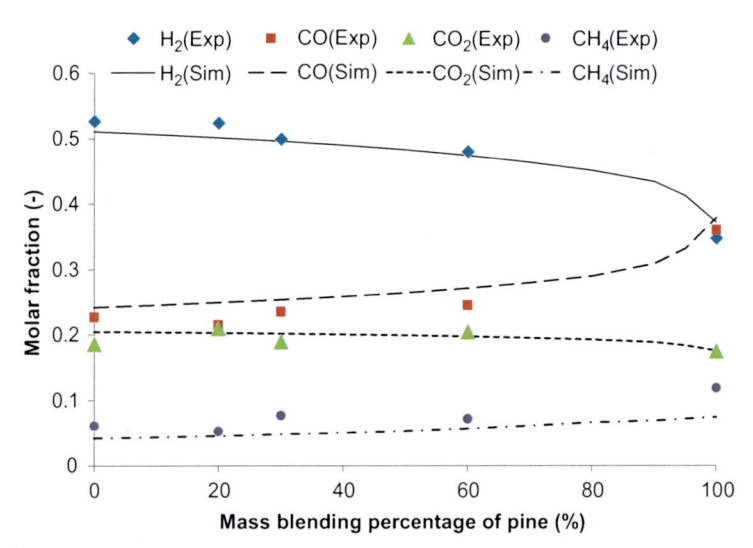

Figure 9.10 Model predicted and experimentally measured producer gas composition in steam co-gasification of blended lignite and pine pellets on a DFB gasifier at 800 °C (Xu, 2013).

the H$_2$ content decreased with increasing biomass-blending ratio (or decreasing coal-blending ratio). However, the hydrogen content from the BFB gasifier is much lower than that from the DFB gasifier as the former uses mixed air and steam as the gasification agent whereas the latter uses pure steam.

In the co-gasification in the BFB gasifier, the CO content was decreased and the CO$_2$ content was increased with the blending ratio of biomass, whereas in DFB gasification the CO content was increased but the CO$_2$ was maintained unchanged. The CH$_4$ content was also increased with the biomass proportion in the DFB gasification, but the blending ratio had insignificant impact on the CH$_4$ content in the BFB gasification.

It is also noticed that the blending method has some impact on the co-gasification as well. Xu et al. (2015) and Xu (2013) have found synergetic effect in co-gasification of blended coal and biomass which were pelletized for blending. It has been found that the char reactivity of the pelletized blends has non-linear relationship with the blending ratio, indicating the synergetic effect; however, the influence of biomass char becomes significant only at high biomass-blending ratio.

When the coal and biomass are mixed without pelletizing, the synergetic effect is not observed. Figure 9.11 shows the results reported by Aigner et al. (2011), who fed coal and biomass separately and blended them at the entrance of a DFB gasifier. From Figure 9.11, linear correlations are clearly observed between the gas composition and coal—biomass blending ratio. This behaviour was also simulated by the mathematical model developed by Xu (2013), confirming the non-synergetic effect with non-pelletizing of the blended fuels.

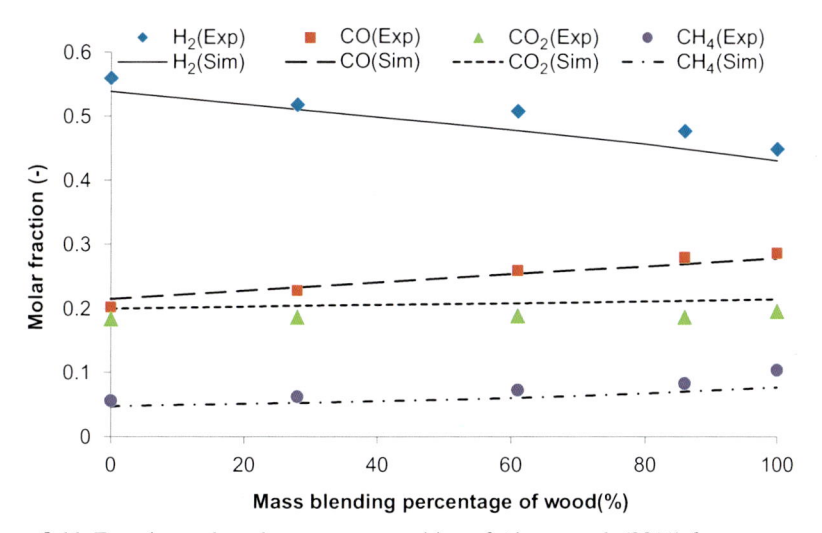

Figure 9.11 Experimental producer gas composition of Aigner et al. (2011) for steam gasification of non-pelletized lignite and pine pellets at various blending ratios. Model predicted results are also included to confirm the non-synergetic effect (Xu, 2013).

9.3.3 Co-gasification of biomass and dried sewage sludge

Dried sewage sludge is a mixture of carbonaceous, phosphorus and nitrogenous compounds and, in most cases, heavy metals and microbial organisms are present. Contamination of heavy metals, toxins, dioxins and microbial organisms in the bio-solid wastes means that these bio-solid wastes have risk of contaminants entering the food chain if used in farmland as fertilizers (Elled et al., 2007; Groß et al., 2008). However, the useful fixed carbon and hydrogen in the bio-solid wastes can be utilized as a solid fuel. At present, the bio-solid wastes are disposed in landfills or used for heat and power by combustion. Landfill disposal apparently occupies land and combustion causes environmental concerns due to the emissions of volatile organic compounds, NO_x and SO_x. Therefore, new processing technologies such as gasification and pyrolysis have recently been investigated.

Most bio-solids have high ash contents and low calorific values, thus the conversion technologies need to be capable of handling the ash to produce valuable gaseous fuels. In addition, environmental impact also needs to be taken into account due to the high contents of nitrogen and sulphur in the bio-solids. Co-gasification of bio-solid wastes with biomass or with coal has been investigated and the results are encouraging.

Saw et al. (2012) have examined the influence of bio-solids loading in blended woody biomass and bio-solids on steam gasification performance in a DFB gasifier. It is found that with the bio-solid proportion increasing from 0% to 100%, the H_2 content in the producer gas increased from 23% (pure wood) to 28% (pure bio-solid) (see Figure 9.12). However, due to the high ash content in the bio-solid wastes, the syngas yield and the cold gas efficiency in the co-gasification decreased dramatically at high loading of bio-solid waste. They concluded that adding 10–20% bio-solid wastes in the woody biomass will produce similar gas composition and gas yield compared to gasification of pure woody biomass. It is interesting to find that the gas calorific value

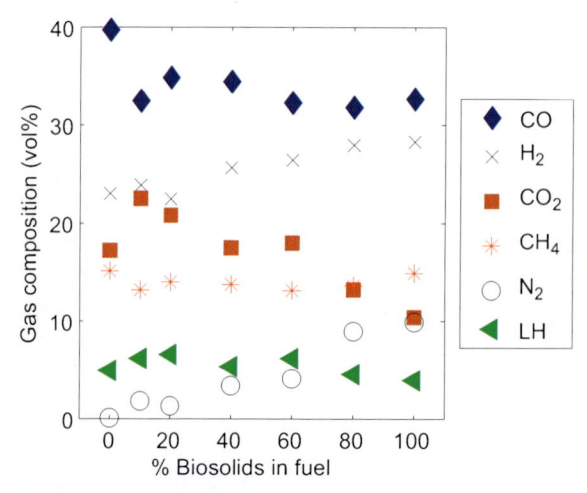

Figure 9.12 Effect of bio-solid proportion in the solid fuel on the composition of producer gas from steam gasification in a dual fluidized-bed gasifier (Saw et al., 2012).

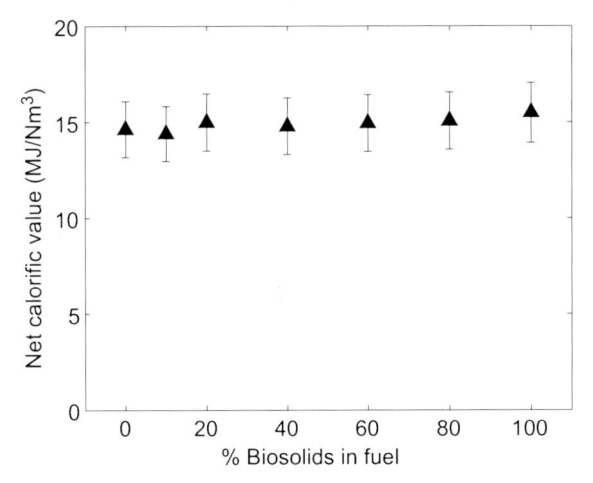

Figure 9.13 Effect of bio-solid proportion in the solid fuel on the calorific value of producer gas from steam gasification in a dual fluidized-bed gasifier (Saw et al., 2012).

increased slightly with increase in the bio-solid proportion in the blended fuel as shown in Figure 9.13.

The gas produced from steam gasification in a DFB gasifier has much higher H_2 and CO contents, thus it has higher calorific values in comparison with those reported by Seggiani et al. (2013), who conducted both experimental and theoretical studies on air gasification of sewage sludge in an updraft gasifier.

9.3.4 Issues in the co-gasification of blended solid fuels

Producer gas from gasification of biomass, coal and bio-solid wastes or their blends mainly consists of hydrogen (H_2), carbon monoxide (CO), carbon dioxide (CO_2), methane (CH_4) and other hydrocarbon gases, as well as tars and a trace amount of other impurities. The producer gas can be utilized in various ways and for different products: (1) it is used in gas turbines or internal combustion (IC) engines for power generation; (2) it is further purified for production of hydrogen gas; (3) it is reformed for synthetic natural gas; (4) it is used for synthesis of transportation fuels (such as Fischer–Tropsch liquid fuel). For different applications, gas quality specifications are different; however, all applications need the tars removed from the gas to different levels depending on the target application. Table 9.5 gives a summary of typical concentrations of impurities in the biomass gasification producer gas using different types of gasifiers (Hongrapipat, 2014; Cheah et al., 2009; Torres et al., 2007; van der Drift et al., 2001). However, the required levels of these contaminants in most of the applications are much lower than the levels presented in the raw producer gas. The specifications of feed gas for gas turbine and IC gas engine are given in Table 9.6 (Mitchell, 1998; Woolcock and Brown, 2013), and those for Fischer–Tropsch synthesis of liquid fuel are given in Table 9.7 (Boerrigter et al., 2003, 2004). Therefore, significant efforts have been spent on gas cleaning to meet these specifications.

Table 9.5 **Typical concentrations of impurities in biomass gasification producer gas using different types of gasifiers (Hongrapipat, 2014; Cheah et al., 2009; Torres et al., 2007; van der Drift et al., 2001)**

Gasifier type	Feedstocks	Tars (g/Nm^3)	NH$_3$ (ppmv)	H$_2$S (ppmv, dry basis)	HCl (ppmv, dry basis)
CFB	Wood, verge grass, sewage sludge	20–660	1000–13,000	50–230	1–200
Fluidized-bed	Wood	10	<1000	<50	<10
Updraft	Wood	50	120–160	20–50	
Downdraft	Wood, herbaceous feedstock	1	200–800	40–120 (Wood) 300–600 (Others)	

Table 9.6 **The specifications for IC engine and gas turbine (Mitchell, 1998; Woolcock and Brown, 2013)**

	Specifications	
Contaminant	Gas turbine	IC gas engine
Particulates	2 ppmw	<50 mg/Nm^3
Sulphur-containing compounds	20 ppmv	–
Nitrogen-containing compounds	50 mmpv	–
Hydrogen halides	1.0 ppmv	–
Alkali metals	0.024 ppmw	–
Alkali earth metals	1 ppmw	–
Trace heavy metals	<1 ppmw	–
Tar compounds	–	<100 mg/Nm^3

To remove the contaminants (tars, N-based and S-based gaseous compounds) from the producer gas, two types of measures have been proposed: (1) primary measures which are employed in the gasifier system with operational condition optimization and application of catalytic bed materials and; (2) secondary measures or downstream measures which are performed on the producer gas following the gasifier. However, the most effective gas cleaning will be the combination and optimization of the two

Table 9.7 Fischer–Tropsch feed gas specifications (Boerrigter et al., 2003, 2004)

Impurity	Removal level
Organic compounds[a] (tars)	Below dew point
N-compounds (NH$_3$, HCN)	<1 ppmV
S-compounds (H$_2$S, COS, CS$_2$)	<1 ppmV
Halogen (HCl, HBr, HF)	<10 ppbV
Alkaline metals	<10 ppbV
Solids (soot, dust, ash)	Essentially completely
Class 2[b] (hetero atoms)	<1 ppmV
CO$_2$, N$_2$, CH$_4$ and larger hydrocarbons	<15 vol%

[a]Organic compounds also include benzene, toluene and xylene (BTX).
[b]Class 2 tars comprise phenol, pyridine and thiophene.

measures. Presently, the gas cleaning operation still represents a substantial fraction of capital and operational costs in the gasification process.

Devi et al. (2003) performed an extensive review on primary measures to clean the producer gas from biomass gasification and found that the most important operational parameters affecting the tar content are gasification temperature, pressure, gasification agent, application of catalytic bed materials and/or bed material additives, ER and residence time. The extent of the influence of each parameter is also dependent on the type of gasifier used. In general, tar content of the producer gas tends to decrease with increasing gasification temperature and pressure. Potential catalytic bed materials and active bed-material additives which have noticeable influences include dolomite, olivine, char and Ni-based catalysts. In addition, the Ni-based catalytic materials are also reported very effective for decreasing the amount of nitrogenous compounds such as ammonia. However, the influence of residence time is inconclusive. Devi et al. (2003) noticed that the residence time had insignificant influence on the tar content, but Saw and Pang (2012b) found the tar content was decreased by 24% when the residence time was increased from 0.16 to 0.21 s. This may be due to the increased bed-material inventory to increase the gas residence time in the bed.

The secondary measures for gas cleaning include cold gas-cleaning systems, hot gas-cleaning systems and catalytic gas-cleaning systems. The cold gas-cleaning covers scrubbers in which liquid solvents are used to absorb the impurities, and these solvents may be water (Stevens, 2001), bio-diesel (Mwandila et al., 2014) or other organic liquids (Boerrigter et al., 2005). The advantages for cold gas cleaning are that the solvents can be recovered or reused and the operation is simple. However, it requires the producer gas from the gasifier to be cooled down to a temperature below 60 °C which will need a heat recovery system to increase the overall energy efficiency. In addition, if water is used as the solvent, disposal of wastewater causes environmental concerns.

The hot gas-cleaning technologies operate at temperatures above 1000 °C to crack down the tars and other gaseous impurities into light hydrocarbons and other less harmful simple gases (Torres et al., 2007; Woolcock and Brown, 2013). This technology has the advantages that no additional materials are needed and the resultant light hydrocarbons in the producer gas add to its calorific values. However, it operates at high temperatures and consumes heat for the heat-up.

The third type of gas cleaning is the application of a catalyst that operates at temperatures from 500 to 900 °C (Devi et al., 2003; Torres et al., 2007; Hongrapipat et al., 2014). This technology can improve the problems involved in the above two technologies and the costs may be high due to consumption of catalysts. Regeneration of used catalysts should be considered for commercial plants.

9.4 Co-pyrolysis of blended solid fuels

Pyrolysis is another thermochemical conversion process of carbonaceous substances under heating in the absence of oxygen in which the solid fuel decomposes into solid char, condensable vapours and non-condensable gases. The condensable vapours become liquid (commonly termed as bio-oil) at room temperature. The product distribution of gas, liquid and char is dependent on operating conditions (temperature, pressure, heating rate and residence time) and the type of solid fuel. Based on the heating rate, the pyrolysis process can be classified into conventional, slow, fast and flash pyrolysis with the heating rate from 1 to 1000 °C/s (Table 9.8). In general, with elevated temperature and higher heating rate, gaseous product yield is enhanced and the mean molecular weights decrease. Majority of the studies and commercial operations for pyrolysis have focused on liquid product (bio-oil) as the target product and in this case subsequent upgrading is needed if the bio-oil is used as a substitute for liquid fuel (Bridgwater, 2002, 2012; Mohan et al., 2006; Xu et al., 2013).

When the gaseous product is the target product, rapid heating and high-temperature operation should be used. In addition, catalysts may be applied (Chen et al., 2003; Demirbas, 2002; Garcia et al., 2002).

Table 9.8 Pyrolysis technologies and characteristics

Technology	Residence time	Heating rate	Temperature °C	Target products
Carbonation (torrefaction)	Days	Very low	<300 °C	Charcoal
Conventional pyrolysis	5–30 min	Low	400–500 °C	Bio-oil, gas and char
Fast pyrolysis	0.5–5 s	High	650 °C	Bio-oil
Flash pyrolysis	<1 s	Very high	650–100 °C	Gas, chemicals

When bio-oil or/and gases are the target products, the most commonly used reactors are bubbling fluidized-bed pyrolyzer, circulating fluidized-bed pyrolyzer and augur or screw-type reactor. The pyrolysis liquid products are a very complicated mixture of various organic compounds and thus upgrading processes can be complicated and costly for liquid fuel production. The non-condensable gas product after cooling can directly be used as gaseous fuel or processed for chemicals.

Co-pyrolysis of biomass and coal has been reported and will be discussed in the following section. However, no report has been found on co-pyrolysis of biomass and bio-solid wastes, possibly due to the increased complexity of the operation and products. Nevertheless, pyrolysis of dried sewage sludge and MSW has been found in the literature and the results are discussed in Section 9.4.2.

9.4.1 Co-pyrolysis of biomass and coal

Soncini et al. (2013) have investigated products distribution from co-pyrolysis of woody biomass (WB) of southern yellow pine and two types of coals (Mississippi lignite or LIG and sub-bituminous Powder River Basin coal or PRB) in a lab-scale BFB reactor. The operational temperatures were controlled at 600, 800 and 975 °C, respectively. The results are shown in Figure 9.14. From this study, it is found that with increase of biomass proportion in the biomass—coal blends, the gaseous product yield increases and char yield decreases, whereas liquid product yield remains unchanged. However, the gaseous product yield at higher temperature is higher and the other product (char and liquid) yields are lower (Figure 9.15). The gaseous products consist of hydrogen (H_2), methane (CH_4), methene (C_2H_2) and water vapour (H_2O). From the results, synergetic effect is also observed which is indicated by the non-linear relationship between product yields and the solid-fuel blending ratio.

Weilan et al. (2012) performed experiments on co-pyrolysis of Illinois #6 bituminous coal and switchgrass in a lab-scale BFB reactor, and found similar trends of gaseous product yield increasing with biomass proportion in the feedstock. The gases in this study consist of H_2, CO, CO_2 and CH_4.

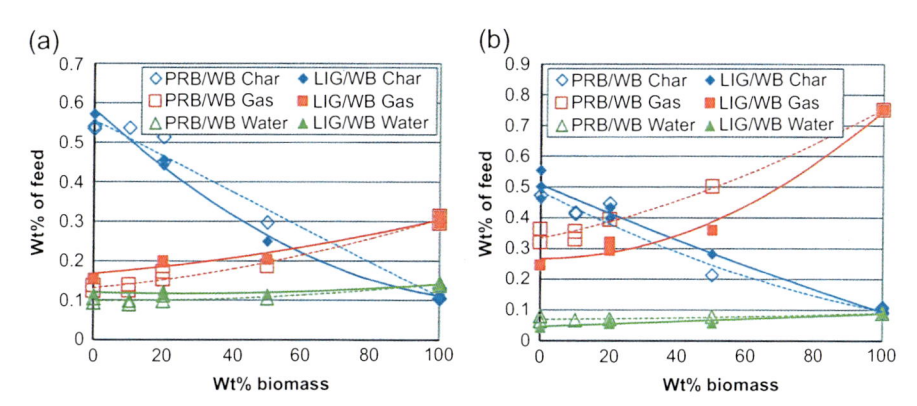

Figure 9.14 Product distribution of gases, liquid and char from co-pyrolysis of woody biomass and coal at (a) 600 °C and (b) 975 °C (Soncini et al., 2013).

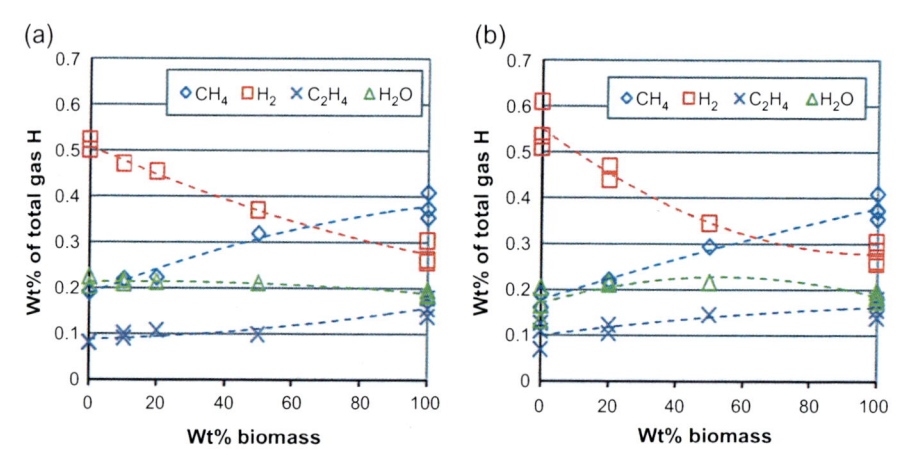

Figure 9.15 Composition of gaseous products from co-pyrolysis of woody biomass and coal at (a) 600 °C and (b) 975 °C (Soncini et al., 2013).

There are some challenges in the co-pyrolysis of biomass and coal. The first one is the rapid heating of the feedstock. This may be achieved by application of inert or catalytic bed material in a fluidized-bed reactor and in this case the carrier gas can be preheated. The second challenge is the target application of chars and liquid product to make the process economically viable. The solid residues may contain char as well as significant content of ash when the coal proportion is high in the blends. Liquid product may be upgraded to liquid fuel; however, the present upgrading technologies need further improvements to reduce the cost.

9.4.2 Pyrolysis of dried sewage sludge and municipal solid wastes (MSW)

Presently, sewage sludge and MSW are mostly incinerated and disposed in landfills. However, pyrolysis can be a promising alternative technology to recover the energy content in a cleaner way which produces fewer nitrogen oxides (NO_x) and sulphur oxides (SO_x) (Chen et al., 2014). The solid residues (char and ash) from the pyrolysis impose less hazard to the environment and soil when being disposed either in landfills or used as fertilizer.

Zhang et al. (2014) conducted flash pyrolysis of sewage sludge in a free-fall reactor at operation temperature of 1000–1400 °C. It is found that the gas product yield increases with operational temperature and, at 1300 °C, almost all of the volatile matters in the sludge are released as gaseous product (Figure 9.16). The gaseous product consists of H_2, CO, CO_2 and CH_4. With increase in pyrolysis temperature, the contents of H_2, CO and CO_2 tend to increase, but CH_4 content is decreased resulting in decrease in gas calorific value (Figure 9.17).

In a comprehensive review by Chen et al. (2014), it is recommended that the industrial pyrolysis facilities should be coupled with gasification or combustion; both processes should be equipped with gas scrubbing devices to reduce emissions. In addition,

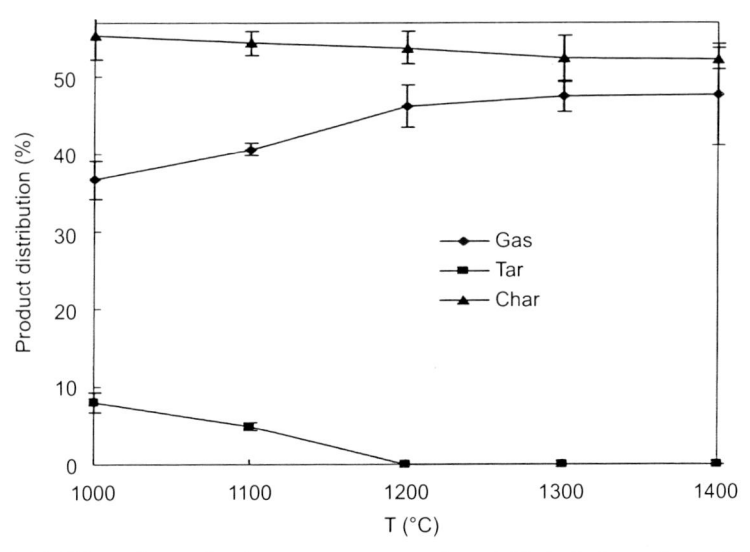

Fuel flexible gas production: biomass, coal and bio-solid wastes 265

Figure 9.16 Effect of operational temperature on product yields in flash pyrolysis of dried sewage sludge (Zhang et al., 2014).

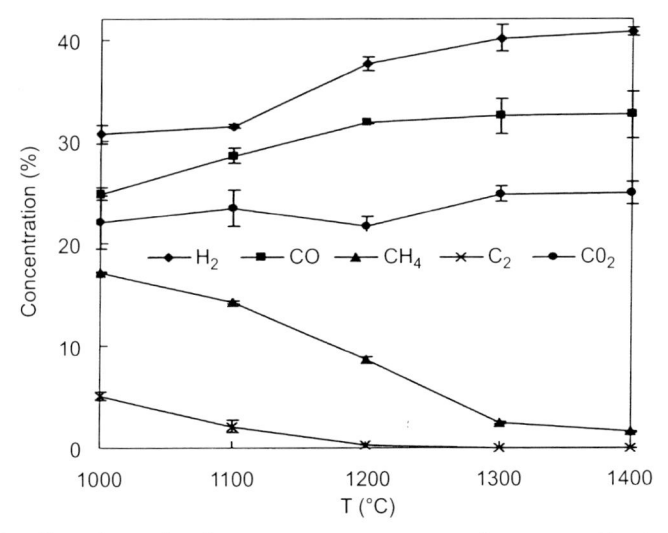

Figure 9.17 Effect of operational temperature on gaseous product composition in flash pyrolysis of dried sewage sludge (Zhang et al., 2014).

the char from the pyrolysis of MSW is of high calorific value and thus can be a promising solid-fuel resource. The challenge is to treat the char to remove heavy metals and organic pollutants to prevent potential contamination.

For some applications of the gaseous product, contamination of HCl, H_2S, SO_2 and NH_3 in the gaseous product should be controlled in a similar way to cleaning of the gasification producer gas.

9.5 Concluding remarks

Renewable solid fuels (biomass) and bio-solid wastes are potential feedstock sources for production of gaseous fuels. Thermal gasification and pyrolysis are the most promising conversion technologies for large-scale commercial plants. However, due to the low density of biomass, co-gasification and co-pyrolysis with coal will achieve the potential benefits of biomass and bio-solid wastes and make the commercial production of gaseous fuel economically viable. The technical challenge for co-gasification is the gas cleaning to remove tars and other impurities (NH_3, H_2S, HCl etc.) for all of the solid fuels as the feedstock. When solid wastes such as dried sewage sludge are added to the solid fuel, ash separation and treatment need to be considered.

Products from pyrolysis and co-pyrolysis of the carbonaceous solids include solid char, bio-oil and gases. When the gases are identified as the target product, the pyrolysis should be operated at high temperatures with fast heating rate. To make the production economically feasible, application of solid char and bio-oil should be taken into account and low-cost bio-oil upgrading technologies need to be developed. Application of mixed char and ash may be a challenge which should receive more attention in future research and development. Co-pyrolysis of bio-solid wastes with biomass or with coal has not been found in the literature, and this can be an area for future research.

References

Adegoroye, A., Paterson, N., Li, X., Morgan, T., Herod, A.A., Dugwell, D.R., 2004. The characterisation of tars produced during the gasification of sewage sludge in a spouted bed reactor. Fuel 83, 1949–1960.

Aigner, I., Pfeifer, C., Hofbauer, H., 2011. Co-gasification of coal and wood in a dual fluidized bed gasifier. Fuel 2404–2412.

Arena, U., Zaccariello, L., Mastellone, M.L., 2010. Fluidized bed gasification of waste derived fuels. Waste Management 30, 1212–1219.

Boerrigter, H., den Uil, H., Calis, H.P., 2003. Green diesel from biomass via Fischer-Tropsch synthesis: new insights in gas cleaning and process design. In: Bridgewater, A.V. (Ed.), Pyrolysis and Gasification of Biomass and Waste. CPL Press, Newbury, UK, pp. 371–383.

Boerrigter, P.C.H., Slort, D.J., Bodenstaff, H., Kaandorp, A.J., den Uil, H., Rabou, L.P.L.M., 2004. Gas Cleaning for Integrated Biomass Gasification (Bg) and Fischer-Tropsch (FT) Systems. Energy Research Centre of The Netherlands. ECN Report ECN-C-04–056.

Boerrigter, H., van Paasen, S.V.B., Bergman, P.C.A., Könemann, J.W., Emmen, R., Wijnands, A., 2005. 'OLGA' Tar Removal Technology Proof-of-Concept (PoC) for Application in Integrated Biomass Gasification Combined Heat and Power (CHP) Systems. Energy Research Centre of The Netherlands. ECN Report ECN-C-05–009.

Bridgwater, A.V. (Ed.), 2002. Fast Pyrolysis of Biomass: A Handbook, Volume 2. CPL Press, Newbury, UK.

Bridgwater, A.V., 2012. Review of fast pyrolysis of biomass and product upgrading. Biomass and Bioenergy 38, 68–94.

Brown, R.C., Liu, Q., Norton, G., 2000. Catalytic effects observed during the co-gasification of coal and switchgrass. Biomass and Bioenergy 18, 499–506.

Cheah, S., Carpenter, D.L., Magrini-Bair, K.A., 2009. Review of mid-to high-temperature sulfur sorbents for desulfurization of biomass-and coal-derived syngas. Energy and Fuels 23, 5291—5307.

Chen, G., Andries, J., Luo, Z., Spliethoff, H., 2003. Biomass pyrolysis/gasification for product gas production: the overall investigation of parametric effects. Energy Conversion and Management 44, 1875—1884.

Chen, D., Yin, L., Wang, H., He, P., 2014. Pyrolysis technologies for municipal solid waste, a review. Waste Management 34, 2466—2486.

Collot, A.G., Zhuo, Y., Dugwell, D.R., Kandiyoti, R., 1999. Co-pyrolysis and co-gasification of coal and biomass in bench-scale fixed-bed and fluidized bed reactors. Fuel 78, 667—679.

Cormos, C.C., 2012. Hydrogen and power co-generation based on coal and biomass/solid wastes co-gasification with carbon capture and storage. International Journal of Hydrogen Energy 37, 5637—5648.

van der Drift, A., van Doorn, J., Vermeulen, J.W., 2001. Ten residual biomass fuels for circulating fluidized-bed gasification. Biomass Bioenergy 20 (1), 45—56.

Demirbas, A., 2002. Gaseous products from biomass by pyrolysis and gasification, effects of catalyst on hydrogen yield. Energy Conversion and Management 43, 897—909.

Devi, L., Ptasinski, K.J., Janssen, F.J.J., 2003. A review of the primary measures for tar elimination in biomass gasification processes. Biomass and Bioenergy 24, 125—140.

EIA (US Energy Information Administration), 2013. International Energy Outlook 2013. Available from: http://www.eia.gov/forecasts/archive/ieo13 (accessed on 07.04.15.).

Elled, A.L., Amand, L.E., Leckner, B., Andersson, B.A., 2007. The fate of trace elements in fluidised bed combustion of sewage sludge and wood. Fuel 86, 843—852.

Emami-Taba, E., Irfan, M.F., 2013. Fuel blending effects on the co-gasification of coal and biomass, a review. Biomass and Bioenergy 57, 249—263.

Franco, C., Pinto, F., Gulyurtlu, I., Cabrita, I., 2003. The study of reactions influencing the biomass steam gasification process. Fuel 82, 835—842.

Galindo, A.L., Lora, E.S., Andrade, R.V., Giraldo, S.Y., Jaén, R.L., Cobas, V.M., 2014. Biomass gasification in a downdraft gasifier with a two-stage air supply, effect of operating conditions on gas quality. Biomass and Bioenergy 61, 236—244.

Garcia, L., Salvador, M.L., Bibao, R., Arauzo, J., 2002. Gas production from catalytic pyrolysis of biomass. In: Bridgwater, A.V. (Ed.), Fast Pyrolysis of Biomass, A Handbook Volume 2. CPL Press, Newbury, UK, pp. 393—406.

Groß, B., Eder, C., Grziwa, P., Horst, J., Kimmerle, K., 2008. Energy recovery from sewage sludge by means of fluidised bed gasification. Waste Management 28, 1819—1826.

Holfbauer, H., Knoef, H.A.M., 2005. Success stories on biomass agsification. In: Knoef, H.A.M. (Ed.), Handbook of Biomass Gasification. BTG Biomass Technology Group BV, The Netherlands, pp. 115—161.

Hongrapipat, J., 2014. Removal of NH_3 and H_2S from biomass gasification producer gas (Ph.D. thesis). University of Canterbury, Christchurch, New Zealand.

Hongrapipat, J., Yip, A.C.K., Marshall, A.T., Saw, W.L., Pang, S., 2014. Investigation of simultaneous removal of ammonia and hydrogen sulphide from producer gas in biomass gasification by titanomagnetite. Fuel 135, 235—242.

Kumabe, K., Hanaoka, T., Fujimoto, S., Minowa, T., Sakanishi, K., 2007. Co-gasification of woody biomass and coal with air and steam. Fuel 86, 684—689.

Lee, S., Speight, J.G., Loyalka, S.K., 2007. Handbook of Alternative Fuel Technologies. CRC Press, Boca Raton, Florida.

Leijenhorst, E.J., Assink, D., van de Beld, L., Weiland, F., Wiinikka, H., Carlsson, P., Ohrman, O.G.W., 2015. Entrained flow gasification of straw- and woodderived pyrolysis oil in a pressurized oxygen blown gasifier. Biomass and Bioenergy. http://dx.doi.org/10.1016/j.biombioe.2014.11.020 published online.

Li, J., McCurdy, M., Pang, S., 2006. Energy demand in wood processing plants. New Zealand Journal of Forestry 51 (2), 13—18.

Manyà, J.J., Sánchez, J.L., Ábrego, J., Gonzalo, A., Arauzo, J., 2006. Influence of gas residence time and air ratio on the air gasification of dried sewage sludge in a bubbling fluidised bed. Fuel 85, 2027—2033.

Marin, N., Collura, S., Sharypov, V.I., Beregovtsova, N.G., Kuznetsov, B.N., Baryshnikov, S.V., Cebolla, V.L., Weber, J.V., 2002. Copyrolysis of wood biomass and synthetic polymers mixtures. Part II, characterisation of the liquid phases. Journal of Analytical and Applied Pyrolysis 65, 41—55.

McLendon, T.R., Lui, A.P., Pineault, R.L., Beer, S.K., Richardson, S.W., 2004. High-pressure co-gasification of coal and biomass in a fluidized bed. Biomass and Bioenergy 26, 377—388.

Meng, X., de Jong, W., Fu, N., Verkooijen, A.H.M., 2011. Biomass gasification in a 100 kWth steam-oxygen blown circulating fluidized bed gasifier, effects of operational conditions on product gas distribution and tar formation. Biomass and Bioenergy 35, 2910—2924.

Mitchell, S.C., 1998. Hot Gas Cleanup of Sulphur, Nitrogen, Minor and Trace Elements. IEA Coal Research, Gemini House, London.

Mohan, D., Pittman, C.U., Steele, P.H., 2006. Pyrolysis of wood/biomass for bio-oil, a critical review. Energy and Fuels 20, 848—889.

Mwandila, G., Pang, S., Saw, W.L., 2014. Semi-empirical determination of overall volumetric molar transfer coefficient for the scrubbing of gaseous tars in biodiesel. Industrial and Engineering Chemistry Research 53, 4424—4428.

Nipattummakul, N., Ahmed, I., Kerdsuwan, S., Gupta, A.K., 2010. Hydrogen and syngas production from sewage sludge via steam gasification. International Journal of Hydrogen Energy 35, 11738—11745.

Pang, S., Mujumdar, A.S., 2010. Drying of woody biomass for bioenergy, drying technologies and optimization for an integrated bioenergy plant. Drying Technology 28, 690—701.

Pfeifer, C., Rauch, R., Hofbauer, H., 2004. In-bed catalytic tar reduction in a dual fluidized bed biomass steam gasifier. Industrial and Engineering Chemistry Research 43 (7), 1634—1640.

Pfeifer, C., Puchner, B., Hofbauer, H., 2009. Comparison of dual fluidized bed steam gasification of biomass with and without selective transport of CO_2. Chemical Engineering Science 64, 5073—5083.

Pinto, F., Franco, C., Andre, R.N., Tavares, C., Dias, M., Gulyurtlu, I., Cabrita, I., 2003. Effect of experimental conditions on co-gasification of coal, biomass and plastics wastes with air/steam mixtures in a fluidized bed system. Fuel 82 (15—17), 1967—1976.

Rauch, R., Hrbek, J., Hofbauer, H., 2013. Biomass gasification for synthesis gas production and applications of the syngas. WIREs Energy and Environment. http://dx.doi.org/10.1002/wene.97.

Robertson, K., Manley, B., 2006. Estimation of the availability and cost of supplying biomass for bioenergy in Canterbury. New Zealand Journal of Forestry 51 (2), 3—6.

Saw, W.L., Pang, S., 2012a. The influence of calcite loading on producer gas composition and tar concentration of radiata pine pellets in a dual fluidised bed steam gasifier. Fuel 102, 445—452.

Saw, W.L., Pang, S., 2012b. Influence of mean gas residence time in the bubbling fluidised bed on the performance of a 100-kW dual fluidised bed steam gasifier. Biomass Conversion and Biorefinery 2, 197—205.

Saw, W.L., McKinnon, H., Gilmour, I., Pang, S., 2012. Production of hydrogen-rich syngas from steam gasification of blend of biosolids and wood using a dual fluidised bed gasifier. Fuel 93, 473–478.

Saw, W.L., Pang, S., 2013. Co-gasification of blended lignite and wood pellets in a 100 kW dual fluidised bed steam gasifier, the influence of lignite ratio on producer gas composition and tar content. Fuel 112, 117–124.

Seggiani, M., Puccini, M., Vitolo, S., 2013. Gasification of sewage sludge, mathematical modelling of an updraft gasifier. Chemical Engineering Transactions 32, 895–900.

Sharypov, V.I., Marin, N., Beregovtsova, N.G., Baryshnikov, S.V., Kuznetsov, B.N., Cebolla, V.L., Weber, L.V., 2002. Co-pyrolysis of wood biomass and synthetic polymer mixtures. Part I, influence of experimental conditions on the evolution of solids, liquids and gases. Journal of Analytical and Applied Pyrolysis 64, 15–28.

Sharypov, V.I., Marin, N., Beregovtsova, N.G., Baryshnikov, S.V., Kuznetsov, B.N., Cebolla, V.L., Weber, L.V., 2003. Co-pyrolysis of wood biomass and synthetic polymers mixtures. Part III, characterisation of heavy products. Journal of Analytical and Applied Pyrolysis 67, 325–340.

Sharypov, V.I., Beregovtsova, N.G., Kuznetsov, B.N., Baryshnikov, S.V., Cebolla, V.L., Weber, J.V., Collura, S., Finqueneisel, G., Zimny, T., 2006. Co-pyrolysis of wood biomass and synthetic polymers mixtures Part IV, catalytic pyrolysis of pine wood and polyolefinic polymers mixtures in hydrogen atmosphere. Journal of Analytic and Applied Pyrolysis 76, 265–270.

Soncini, R.M., Means, N.C., Weiland, N.T., 2013. Co-pyrolysis of low rank coals and biomass, product distributions. Fuel 112, 74–82.

Stevens, D.J., 2001. Hot Gas Conditioning, Recent Progress with Larger-scale Biomass Gasification Systems. National Renewable Energy Laboratory, Colorado, USA. NREL/SR-510-29952.

Torres, W., Pansare, S.S., Goodwin, J.G., 2007. Hot gas removal of tars, ammonia, and hydrogen sulphide from biomass gasification gas. Catalysis Reviews, Science and Engineering 49, 407–456.

Tustin, J. (Ed.), 2012. IEA Bioenergy Annual Report 2012. IEA Bioenergy Secretary, Rotorua, New Zealand.

Weiland, N.T., Means, N.C., Morreale, B.D., 2012. Product distributions from isothermal co-pyrolysis of coal and biomass. Fuel 94, 563–570.

Woolcock, P.J., Brown, R.C., 2013. A review of cleaning technologies for biomass-derived syngas. Biomass and Bioenergy 52, 54–84.

Xu, Q., 2013. Investigation of Co-gasification Characteristics of Biomass and Coal in Fluidized Bed Gasifiers (Ph.D. thesis). University of Canterbury, Christchurch, New Zealand.

Xu, Q., Pang, S., Levi, T., 2015. Co-gasification of blended coal-biomass in an air/steam BFB gasifier, experimental investigation and model validation. AIChE J. http://dx.doi.org/10.1002/aic.14773 published online.

Xu, X., Zhang, C., Liu, Y., Zhai, Y., Zhang, R., 2013. Two-step catalytic hydrodeoxygenation of fast pyrolysis oil to hydrocarbon liquid fuels. Chemosphere 93, 652–660.

Yan, H., Heidenreich, C., Zhang, D.K., 1998. Mathematical modelling of a bubbling fluidized-bed coal gasifier and the significance of "net flow". Fuel 77 (9–10), 1067–1079.

Zainal, Z.A., Ali, R., Lean, C.H., Seetharamu, K.N., 2001. Prediction of performance of a downdraft gasifier using equilibrium modeling for different biomass materials. Energy Conversion and Management 42, 1499–1515.

Zhang, L., Xiao, B., Hu, Z., Liu, S., Cheng, G., He, P., Sun, L., 2014. Tar-free fuel gas production from high temperature pyrolysis of sewage sludge. Waste Management 34, 180–184.

Technology options and plant design issues for fuel-flexible gas turbines

10

Jenny Larfeldt
Siemens Industrial Turbomachinery AB, Finspong, Sweden

10.1 Introduction

Gas turbines offer an efficient conversion of natural gas into electricity approaching 40% electric efficiency, and in a plant configuration in which exhaust heat is used for generation of steam to a turbine the efficiency is nowadays almost 61%. The application areas of gas turbines are widening from natural gas-fired base-load operation to either fuel-flexible base load or a plant for covering daily variations. The introduction of renewables in the power grid, wind and solar, increases the market for fast start and ramping production. Potential future legislation or economical incitement for carbon capture will be a driver for special types of gas-turbine plants. Gas turbines are not only required to operate reliable with high performance and low emissions but also to do so with increased operational flexibility, varying fuels and potentially optimized for CO_2 capture.

10.2 Gas turbines in plants

Gas turbines are available in various sizes, from 1 up to 360 MW, and are either so-called industrial gas turbines or aero-derivatives. The latter are aero-engines that have been adapted for power generation or mechanical drive. Typically, aero-engines have higher simple-cycle efficiency due to higher-pressure ratio, whereas the industrial gas turbines have a lower pressure ratio and higher exhaust temperature to promote combined-cycle performance. This chapter will focus mainly on industrial gas turbine integration into plants, industrial gas turbines that can be either single shaft (see example in Figure 10.1) or multi-shaft (Cohen et al.).

Gas turbines can be combined with heat-recovery steam generators and/or steam turbines in a plant solution with the purpose of generating power or combined heat and power and at the same time satisfy the specific customers' technical requirements. Starting with a simple cycle as in the schematic shown in Figure 10.2(a), the cycle consists of a gas turbine with a generator connected on the cold end and a stack for the gas-turbine hot-end exhaust. The simple cycle installations are today exceeding 40% electric efficiency. The heat in the gas-turbine exhaust gases can be utilized to make steam in a so-called co-generation plant, shown schematically in

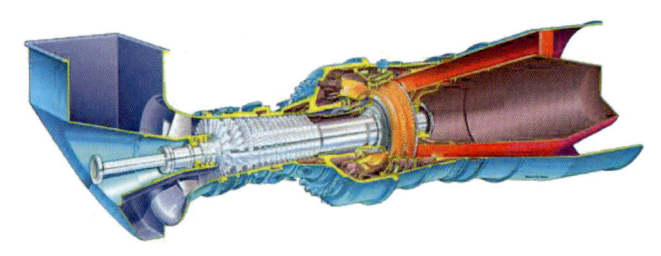

Figure 10.1 SGT-800 single-shaft 50 MW.
Courtesy of Siemens.

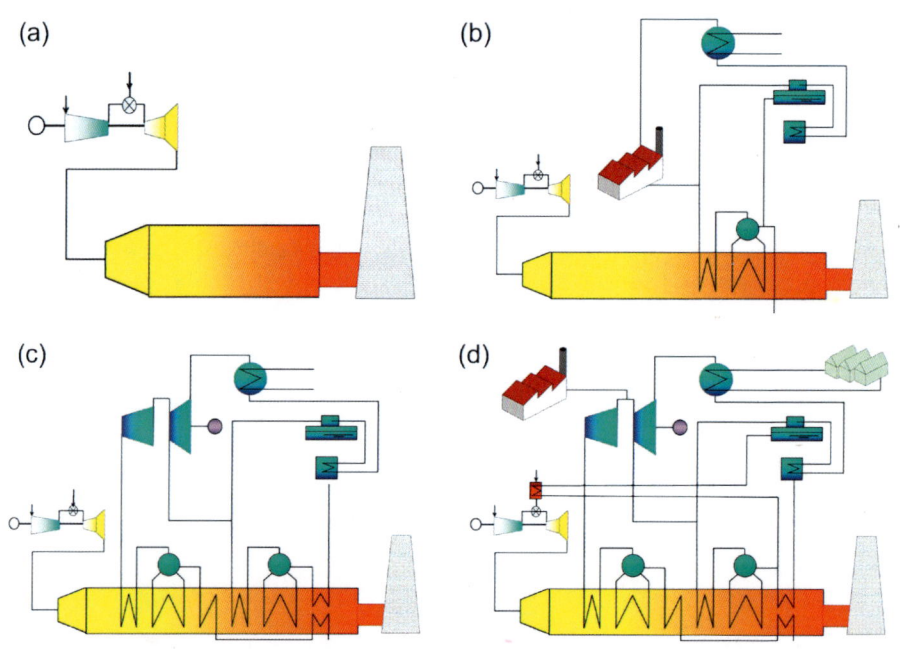

Figure 10.2 Illustration of cycles: (a) simple cycle, (b) simple-cycle cogeneration, (c) combined cycle and (d) combined-cycle cogeneration.

Figure 10.2(b). Cogeneration implies fuel savings and fuel efficiency increases from 40 up to about 90%.

If the exhaust heat is used to generate high-pressure steam which in turn drives a steam-turbine generator set (see Figure 10.2(c)), the electric power output increases. Extracting the steam at medium/low pressure for use in industries/district heating and cooling leads to a combined-cycle co-generation plant (Figure 10.2(d)), with as high as 94% fuel efficiency and up to 60% electrical efficiency. It should, however, be noted that the energy efficiencies in combined heat and power plants are very much dependent upon the specific situation and the local demand of electricity and heat.

10.3 Fuel-flexible gas turbines

Stationary gas turbines are continuously flowed through fixed-drive machines with high power densities, meaning that they deliver a large amount of energy in relation to their size and weight. The compact design involves the core components: compressor, combustor and turbine. The combustion process takes place at a pressure generated by the compressor and the airflow including the products of combustion (and excess air) is then delivered to the turbine which drives the compressor as well as generating power to the generator or other external equipment depending on application.

In order to offer a gas turbine product with increased fuel flexibility both the core engine and auxiliary systems must be taken in to account, however the primary issue is the stable operation of the combustor. The key performances of an industrial gas-turbine combustor are (Lefebvre and Ballal, 2010):

- High-combustion efficiency, meaning that the fuel should be completely burnt so that all its chemical energy is liberated as heat.
- An outlet temperature distribution (pattern factor) that is tailored to maximize the lifetime of the turbine guide vanes and blades.
- Low emissions of smoke and gaseous pollutant species.
- Freedom from pressure pulsations and other manifestations of combustion-induced instability.
- Wide stability limits, that is the flame should stay alight at pressures and air/fuel ratios corresponding to the whole gas-turbine operating range.
- Low pressure loss.
- Design for minimum cost and ease of manufacturing.
- Size and shape compatible with engine envelope.
- Maintainability.
- High availability and reliability.
- Reliable and smooth ignition at the ambient conditions of relevance for an industrial gas turbine.

So, how can this be realized? Experience from design of commercial combustion systems like power plant boilers is of little use when designing a gas-turbine combustion system. Unlike other applications, the exhaust gas stream temperature from the combustor has to be comparatively low to suit the highly stressed turbine materials. The exhaust gas stream has to be controlled so that the temperature and velocity distributions do not cause local overheating and for the turbine to deliver the desired power.

The temperature distribution in the combustor is not only of importance from a component lifetime perspective but also very important for emissions control. The prevailing technique in industrial gas turbines today is lean, premixed combustion systems which have replaced water or steam injection for emission control purposes. The emission of nitrous oxides, NO_x, is strongly related to the local temperature in the combustion zone. Premixing the fuel with more air than what is needed from a stoichiometric point of view leads to the desired flame with several hundred degrees lower temperature than the adiabatic flame temperature. Such lean flames are, however, prone to instabilities such as oscillating heat release which may in worst case interact with pressure fluctuations depending on the combustor design acoustic properties. The control of thermo-acoustic combustor oscillations is a key issue when developing low-emission gas turbines (Lefebvre and Ballal, 2010).

The global fuel-to-air ratio varies greatly with load because the fuel flow at part load is reduced at a higher rate than the airflow if no control measures are taken. The full-load lean flame thus becomes even leaner at part load, and, at some point, combustion in the main flame can no longer be sustained and the flame would be extinguished. The remedy for this unstable combustion situation is fuel staging. To maintain a stable combustion at all gas-turbine loads from idle to full load, pilot fuel is used. Pilot flames are more fuel rich and therefore more stable, although the downside is their contribution to the NO_x emissions. Typical expected emissions from a 50 MW_{el} single-shaft gas turbine is shown in Figure 10.3. The NO_x emission is about 12 ppm@15%O_2 at full load and down to 50% load corresponding to a lean premixed combustion generating also low emissions of CO and unburnt hydrocarbons. Below 50% load, it is necessary to support the main flames with the more fuel-rich pilot flames leading to an increase in NO_x emissions. At the same time, the emissions of CO and unburnt increase because overall combustion temperature gets colder and the residence time and/or mixing in the combustor is no longer sufficient to achieve a complete combustion of the fuel.

It can be mentioned that industrial gas turbines were historically designed for low emissions and high performance at full-load operation. When the grid receives more

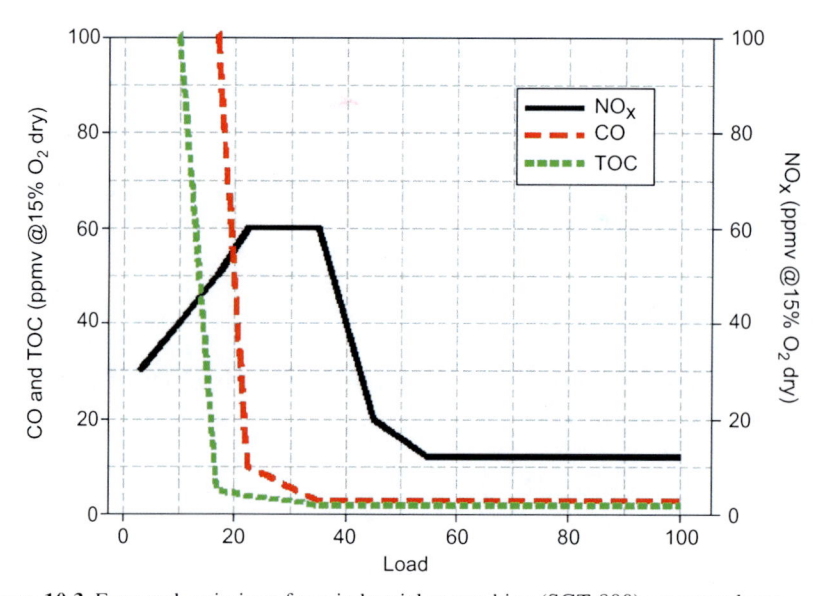

Figure 10.3 Expected emissions from industrial gas turbine (SGT-800) on natural gas.

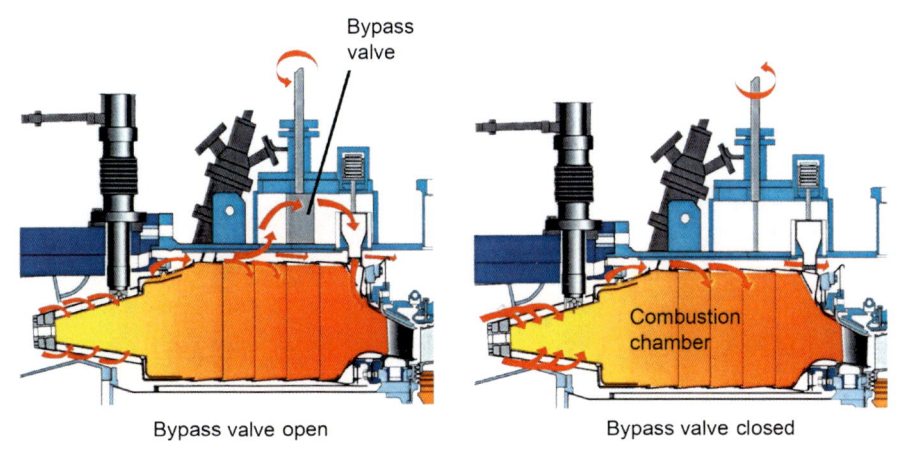

Figure 10.4 Gas-turbine combustor bypass.

electricity from renewable energy sources, the demand for power plants with fast response and low part-load emissions to balance the grid increases (International Energy Agency; IEA, 2014). As described previously, as load decreases the reduction in fuel flow is not matched by the reduction in airflow from the compressor. A bypass system (Figure 10.4) can be used at part load to reduce the amount of air in the combustion zone and thereby making the flame region more fuel rich, or rather, less lean. The airflow to the combustor can also be reduced by so-called "bleeding", when a portion of air is extracted from the compressor either back to the gas-turbine inlet, resulting in an increase in compressor outlet temperature, or to the gas-turbine outlet. A drawback of bleeding air compared to the combustor bypass is the negative impact on gas turbine efficiency since compression work is lost.

10.4 Gaseous fuels for gas turbine operation

Because the combustion in a gas-turbine combustor occurs at a high pressure, the fuel has to be supplied at a pressure high enough to overcome the pressure in the combustor as well as the pressure losses in the fuel feeding system.

The interchangeability of fuels can be characterized by the Wobbe index, accounting for not only the heating value but also the specific gravity of the fuel. The fuel gas Wobbe index, WI^0, is defined as the lower heating value (LHV) (volumetric) divided by the square root of the relative density:

$$WI^0 = \frac{LHV}{\sqrt{\rho_{rel}}} \left(\rho_{rel} = \frac{\rho_{gas}}{\rho_{air}} \right)$$

In Figure 10.5, a large number of fuels of relevance for potential gas-turbine customers are shown as Wobbe index versus LHV by weight. Natural gas has been the

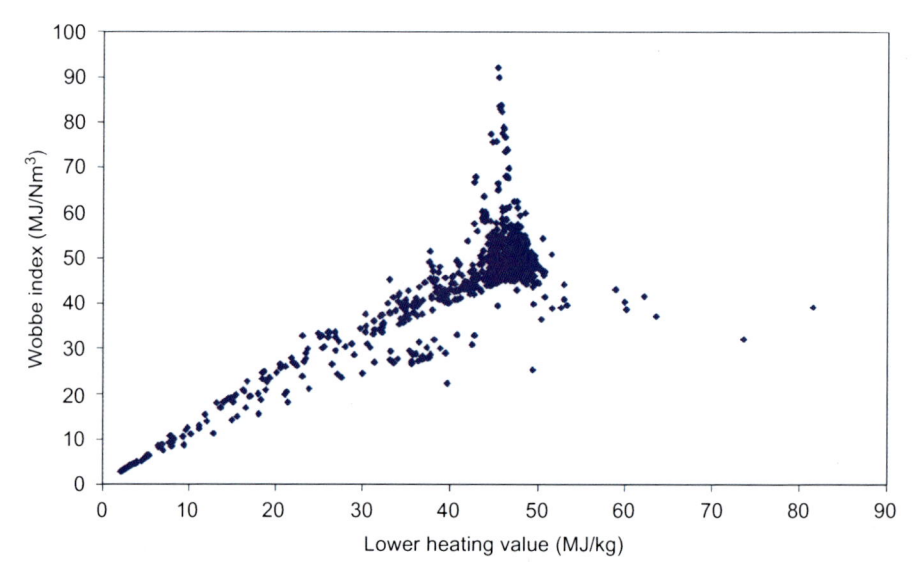

Figure 10.5 Wobbe index calculated using normal cubic metre at 101.325 kPa and 273 K (0 °C) versus lower heating value (LHV) for typical fuels.

typical fuel of choice for gas turbines, and its composition can vary from pipeline-quality gas consisting mostly of methane to raw natural gas and associated gas available close to gas wells and oil wells, respectively. This corresponds to the dense, central area of Figure 10.5 with Wobbe index about 42–53 MJ/nm^3 and lower heating value of about 45 to slightly above 50 MJ/kg fuel. Pipeline natural gas is often produced by removing heavier hydrocarbons, inert gases (carbon dioxide, water or nitrogen) and contaminants from the raw gas. For gases with higher heavy hydrocarbon content, such as associated gases, or even LPGs consisting of propane and butane, the Wobbe index increases but the LHV by weight is the same. These dots are seen in Figure 10.5 above the centre cluster of natural gas-like fuel dots.

Refineries and industrial chemical processes generate significant quantities of gas by-products with high hydrogen content and, in the case of refinery gases, also includes heavy hydrocarbons, both saturated and more-reactive unsaturated ones (olefins). Mixing natural gas with hydrogen results in a slightly lower Wobbe index but a higher LHV (right-hand side of Figure 10.5).

For natural gases with high inert content, and some process-waste gas, for instance boil-off gas from liquefied natural gas (LNG) production, both Wobbe index and LHV will be reduced. In Figure 10.5, these fuels are represented by the dots towards the lower left corner, the uppermost trend of these dots. The lower streak represents syngases typically derived from solid-fuel gasification or process gases in the steel industry, such as coke-oven gas. The syngases typically consist of carbon monoxide and hydrogen which keep the heating value per weight comparatively high but has lower Wobbe index than methane-containing fuels.

10.5 Gas turbine combustion-related challenges for gaseous fuel flexibility

The market demand for fuel-flexible gas turbines is increasing at the same time as environmental impact is desired to be kept low, that is low emissions in a wide operating range. Gas turbines are challenged to handle reactive gaseous fuels containing heavy hydrocarbons and/or hydrogen as well as fuels with high inert content. The combustion properties of relevance for gas-turbine combustors will vary with the fuel composition and need to be considered to avoid flashback, lean blow out and combustion dynamics. Flashback may be the result if a burner designed for natural gas is operated on fuels with shorter ignition delay time and/or higher flame speed. Lean blow out may occur in the same burner if fuel has lower heating value and/or low adiabatic flame temperature. These combustion properties will be describe in more detail in the following.

10.5.1 Auto-ignition delay time

For fuels with high content of heavy hydrocarbons in a premixed combustion system, ignition of the combustible mixture prior to entering the combustor must be prevented. All fuels can be characterized by their auto-ignition temperature, the lowest temperature at which the fuel can spontaneously ignite. In gas-turbine applications, the auto-ignition temperature of the fuel is usually lower than the combustion air temperature, that is the temperature that the air achieves from compression before entering the combustor. Thus, most fuels will ignite when they are premixed with this warm air, and this is a matter of time. This time is referred to as the auto-ignition delay time.

Auto-ignition time delay for methane, the main constituent of natural gas, at typical gas-turbine pressures and temperature is more than 1 s, which is a very long time compared to typical residence times in a combustor. For heavier hydrocarbons, such as butane and pentane, the auto-ignition delay time at the same combustion conditions is in the range of 10 ms. Clearly, the fraction of heavier hydrocarbons in the natural gas supports the ignition process (Spadaccini and Colket, 1994). The auto-ignition delay time is faster at typical turbine pressures than at atmospheric conditions (about 100 times at 30 bars compared to atmosphere). There is also a slightly faster ignition for fuel-rich mixtures than for lean.

The residence time for the mixed fuel and air in the burner, upstream from the combustor, has to be shorter than the auto-ignition delay time, and for a typical industrial turbine this is in the range of milliseconds.

10.5.2 High flame speed

The laminar flame speed for pure hydrogen is almost 10 times higher than for methane (Glassman and Yetter, 2008). At lean combustion conditions, such as in premixed gas-turbine combustors, the flame speeds are generally lower and the differences between the various fuels are lower. For mixtures of hydrogen and methane, the laminar flame speed will increase with increasing hydrogen content. The flame speed increases from

37 cm/s for pure methane up to 53 cm/s for a 40% H_2 mixture with methane, which is not a dramatic increase for a typical industrial gas-turbine combustion system. For mixtures with H_2 content above 40% the combustion situation will change and for instance increased NOx emissions can be expected.

In addition, the flammability limit is wider for hydrogen than for methane. In other words, hydrogen in air can be ignited and burn at a much wider range than methane. Although this fact increases the risks of handling hydrogen, in general it is in fact positive for a lean premixed combustion system because flames can be sustained at leaner conditions than for natural gas. In general, increased reactivity of fuels may lead to increased lean combustion stability particularly at part load as discussed previously (Jahnson, 2013).

10.5.3 Combustion dynamics and lean blow out

As already mentioned, premixed combustion systems work with the principle of burning lean. If a fuel with high inert content is utilized, the combustion becomes even more lean and closer to the limit to lean blow out. The flame stabilization in a typical industrial gas-turbine combustor involves competition between the rates of chemical reaction and rates of turbulent diffusion of species and energy. Any flow and mixture perturbation in the highly turbulent swirling flame will lead to oscillations in the heat release. The heat release oscillation will in turn disturb the pressure field, because gas temperature and pressure are thermodynamically interlinked. Disturbances in the pressure field lead to acoustic oscillation, and how much energy that is transferred from the combustion process to the acoustic pressure field is described by the so-called Rayleigh criterion. The combustor oscillation can be described as a generic feedback loop because these acoustic oscillations will affect the flow and mixture perturbations and the loop is closed. This feedback mechanism is called the "Richards—Lieuwen mechanism", and typical frequency range of interest for practical cases is 100—1000 Hz. Significant fundamental understanding of flame propagation and the stability characteristics of lean premixed systems has been gained in conventionally fuelled natural gas systems (Durbin and Ballal, 1996). Nevertheless, little is known about these issues for alternative fuels (Lieuwen and Zinn, 1998).

10.5.4 Flame temperature

Depending on the gas-turbine operating point and its efficiency, there is a limit to how low-grade fuel can be used. If the fuel has a very low heating value, the necessary flame temperature cannot be reached and, therefore, neither the design-point turbine inlet temperature. The gas turbine will in such case not be able to deliver full load.

This is illustrated by a comparison of four potential gas-turbine fuels surveyed in Table 10.1: two typical steel industry gases, blast furnace gas (BFG) and coke oven gas (COG), a syngas made from thermal gasification of solid fuel and natural gas (NG). As shown in the table, the LHV for these fuels ranges from BFG at 3 up to

Table 10.1 Fuels survey for adiabatic temperature in Figure 10.6

Fuel composition		BFG	COG	Syngas	NG
		mol%	mol%	mol%	mol%
Methane	CH_4		22.90	0.10	98.32
Ethane	C_2H_6				0.88
Propane	C_3H_8				0.28
Isobutane	iso-C_4H_{10}				0.05
n-Butane	n-C_4H_{10}				0.05
Isopentane	iso-C_5H_{12}				0.01
n-Pentane	n-C_5H_{12}				0.01
Hydrogen	H_2	11.50	59.20	37.10	0.00
Water	H_2O				
Hydrogen sulphide	H_2S		0.40	0.10	
Carbon monoxide	CO	21.40	5.30	43.60	
Nitrogen	N_2	43.90	9.90	0.40	0.41
Oxygen	O_2		1.10		
Carbon dioxide	CO_2	23.10	1.90	18.60	
Total		99.90	100.70	99.90	100.00
Calculated properties*					
LHV	MJ/kg	3	33	10	50
Wobbe index	MJ/m³	4	24	11	46

* Combustion and metering reference conditions 25 and 0 (DEgree symbol) C, respectively and 1013 mbar.

NG with 50 MJ/kg. The adiabatic flame temperatures of the fuels are shown in Figure 10.6 for two different combustor air-inlet temperatures; 773 and 693 K. The blast furnace gas has substantially lower adiabatic flame temperature; at typical lean premixed conditions with a fuel-to-air ratio of 0.5, the temperature is 1529 K for the case of 693 K inlet-air temperature. This temperature is normally too low to sustain a flame and not sufficient for the gas turbine to deliver full power. If the gas-turbine combustor operation point can be redesigned to stoichiometric conditions (fuel-to-air ratio 1), the adiabatic temperature is 1840 K, which is still on the lower limit.

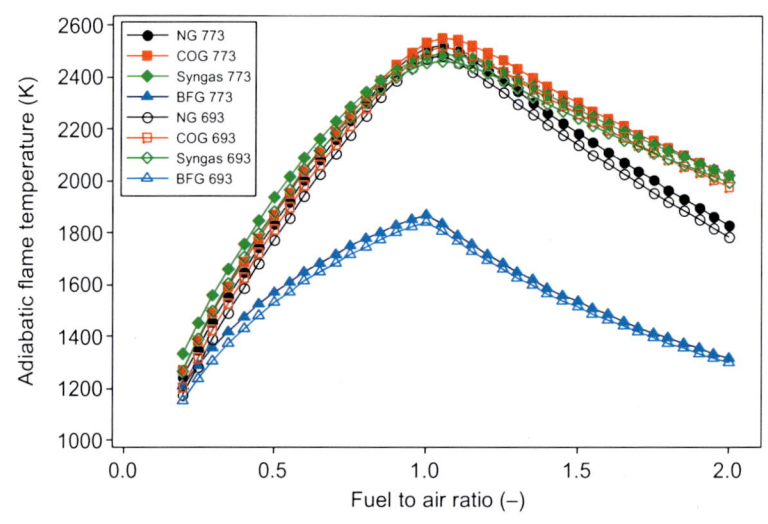

Figure 10.6 Adiabatic flame temperature versus combustion stoichiometry at 20 bars, 300 K fuel temperature and air temperature 773 and 693 K, respectively.

For syngas and coke-oven gas, the adiabatic temperature in Figure 10.6 indicates that flame temperatures in the same range as for natural gas can be reached, whereas for COG even slightly higher due to its high hydrogen content. These fuels can be used for gas-turbine operation with a design of the combustor fuel and airflow that considers the fuel-specific Wobbe index.

10.6 Other fuel flexibility impacts on the gas turbine

Using very low calorific fuels in gas turbines may move the operating point of the gas-turbine compressor closer to the stall limit. Due to the high inert-species content in the fuel gas, the mass flow of the fuel is substantially larger than for natural gas. The fuel flow is then no longer negligible compared to the airflow in the gas turbine and the operation point can be redesigned by either making the turbine wider and maintain the pressure ratio or changing the compressor design and thereby allowing a higher pressure ratio. For such cases, the effort of compressing the fuel up to the combustor pressure can be a substantial part of the plant economy.

For the gaseous fuel applications in gas turbines, it is important to avoid gas temperatures below the dew point, when the first droplets of hydrocarbons or water form in the fuel-feeding system. Such droplets may cause damage to the burner due to the high energy density of these liquid droplets compared to the gaseous fuels (an alternative for heavy hydrocarbons is, of course, to burn them as liquid fuel and then make sure the fuels stay in liquid form, i.e. to monitor the bubbling point). Dew point temperature depends on pressure and gas composition (Figure 10.7) (Näsvall and Larfeldt) and becomes more critical when larger fractions of heavy hydrocarbons are present in

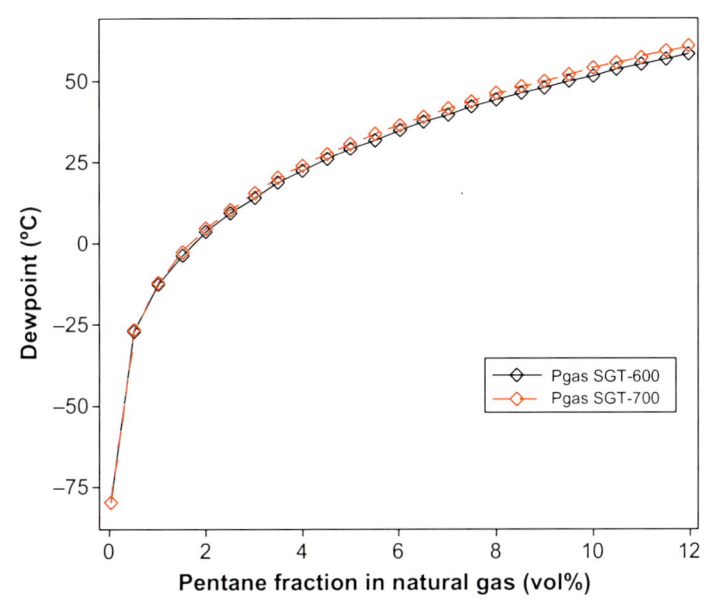

Figure 10.7 Dew point temperature for natural gas with varying pentane content versus pentane fraction for two typical gas turbine fuel feeding pressures.

the fuel. Based on the gas-turbine fuel feeding pressure, the design temperature is calculated as the highest dew point temperature plus a margin in the range of some 10 degrees. At no point in the fuel-feeding system must the fuel temperature be below this temperature.

Some fuels require adaptation of safety systems such as gas detection and fire extinguishing, such as for hydrogen with its high flammability or CO and H_2S due to their toxicity.

The choice of fuel also impacts the emissions in some cases, for instance if sulphur is present in the fuel, such as commonly in associated gases, there will be emissions of SO_x. In addition, durability of the gas-turbine components may be impacted if fuel or exhaust gases have contaminants such as corrosive components. This is very much depending on the gas-turbine operating point temperature and pressure, and the design material in hot parts of the gas turbine.

10.7 Fuel-flexible gas turbine installation

Gas turbines that can handle wider fuel variations can potentially be fed with several process gas streams in process plants such as steel industries, chemical industries or refineries. Here, the gas turbines are utilized either for power generation or as mechanical drives for the process and the more fuel flexible is the gas turbine, the higher efficiency the installation will generate, because waste gases in this case are not merely used for heat/steam production (or in worst case simply flared).

As an example, a liquefied natural gas (LNG) facility is typically located at a point at which natural gas can be extracted or collected in substantial amounts, and the manufactured LNG can be shipped (or transported by other means) to the end users. The manufacturing of LNG through cooling and condensation of the feed gas is a power-requiring process, and Figure 10.8 shows an example of a complete combined-cycle power plant for island-mode operation at an LNG facility. The plant in Figure 10.8 consists of six gas turbines, six heat-recovery steam generation (HRSG) boilers, three steam turbines and three air-cooled condensers as well as mechanical and electrical systems to form three autonomous power blocks.

The feed-gas composition, that is the natural gas, differs from the composition of the end product, that is the LNG, which gives rise to various other gas streams in the process depending on the type of LNG plant. Natural gas consists mainly of hydrocarbons dominated by methane, but also substantial amounts of ethane, propane, butane, pentane and heavier. As discussed previously, the heavier the hydrocarbon the more prone to form liquid (dew point). Typically, the feed gas also has inert gases, water vapour, nitrogen and carbon dioxide, some sulphur-containing gaseous components such as the very toxic components hydrogen sulphide and mercury. As LNG is produced, inert gases will be removed because nitrogen will simply stay in gaseous form (condensation point is lower than methane) and water and carbon dioxide will be removed in the "dehydration" and "acid gas removal" units, respectively. After cleaning the feed gas from toxic components, ethane, propane and butane, depending

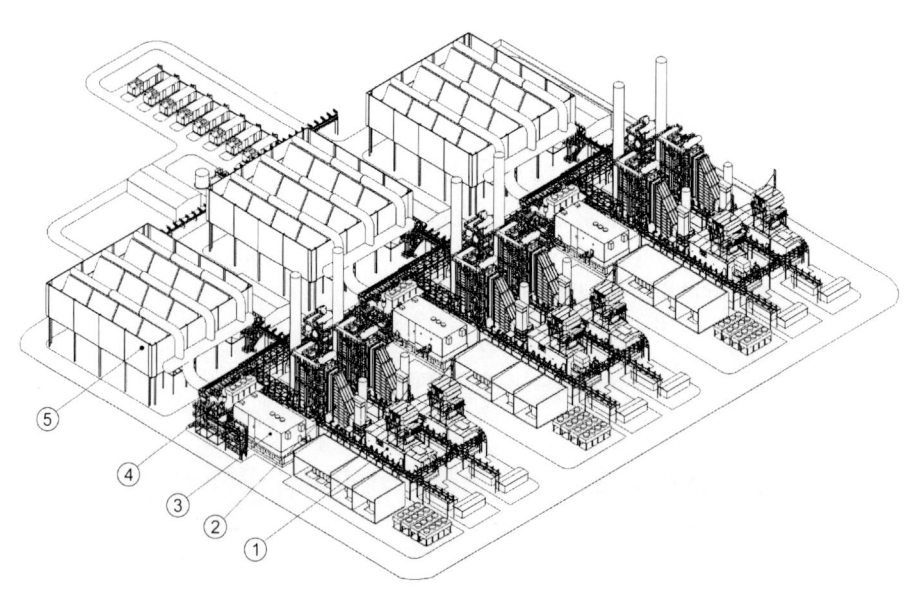

Figure 10.8 Combined-cycle power plant for island-mode operation of an LNG facility. Modules: 1 − gas turbine (GT), 2 − heat recovery steam generator (HRSG), 3 − steam turbine (ST), 4 − Feed water and steam tail, 5 − air-cooled condensers (ACC).

on the amounts, can either be in the LNG product or separated and sold as liquid petroleum gas (LPG). For heavier hydrocarbons, pentane and larger can be separated and used for local power and heat production.

The power plant in Figure 10.8 utilizes isopentane (iC5) from the LNG process, mixed with natural gas as fuel in the gas turbines. The iC5 is also used as fuel in the supplementary firing in the heat recovery steam generator when required. The power plant is tailored to minimize all sorts of gas flaring and to handle all sorts of load-shedding cases on this island-mode system. It is beneficial for the client to use advanced heavy-duty industrial gas turbines with high operational flexibility for this type of plant due to high efficiency. High efficiency implies low CO_2 emissions and fuel costs not only at full load but also at lower loads due to a controlled mass flow through the gas turbine (variable inlet compressor guide vane) with maintained high exhaust temperature. Part-load operation in an island-mode operation with the three gas turbines enables a quick and powerful load response from gas turbines in upset conditions.

10.8 Gas turbine with external heating integrated in plants

Using an external source of heat in a gas-turbine cycle opens up for a wide variety of fuel sources; however, the complexity of the plant increases. The compressed air can be extracted from the gas-turbine central casing to an external source of heat. After heating, the hot gas returns to the core gas turbine and generates power in the turbine. Such a heat source might, for instance, be the sun or external combustors for solid fuels such as biomass, coal and various solid wastes.

10.8.1 Concentrated solar plant

Installed solar thermal power today is dominated by parabolic trough mirrors and electrical generation by steam turbines. Integrating a gas turbine would increase efficiency, reduce water consumption and maximize the flexibility, but the key barrier is the relatively high cost of such a plant (Kehlhofer et al., 2009). The principle of a concentrated solar plant utilizing gas turbines instead of steam turbines for power generation is shown in Figure 10.9. Solar power is collected by a heliostat field of mirrors that directs the sun light onto a solar receiver. The ambient air (state 0 in Figure 10.9) is filtered (state 1) and then compressed (state 2) and extracted from the gas-turbine central casing to the solar receiver (state 3) in which it is heated (state 4). A part of the compressor airflow that is used for cooling of gas-turbine hot parts is extracted as in the conventional application. Depending on design of the concentrated solar power, the air is heated to temperatures not more than 1200 °C (Avila-Marin, 2011) (state 5) which makes complementary firing necessary to reach a combustor outlet temperature (state 6) corresponding to full power production. The hot gas is then entering the turbine (state 7) and expanded through the turbine (state 8) and the exhaust vented to the

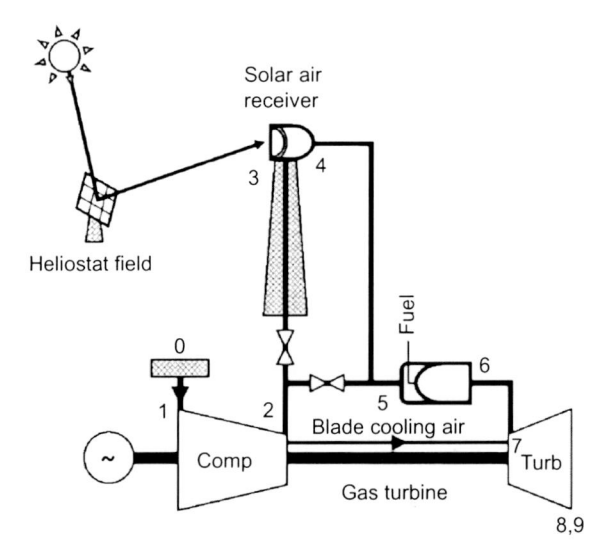

Figure 10.9 Layout of a simple-cycle hybrid solar gas turbine.
Ref: Spelling, J. D., 2013. Hybrid Solar Gas-Turbine Power Plants (Ph.D. thesis), ISBN:978-91-7501-704-4.

atmosphere (state 9) or to an HRSG. Due to the intermittency of solar power, it is desired to integrate a storage capacity in the plant concept.

10.8.2 Pressurized fluidized-bed combustion

In the pressurized fluidized-bed combustion (PFBC) technology, solid fuel is burnt in an external combustor utilizing the air from the gas turbine. Figure 10.10 surveys the flow scheme of the PFBC including the integrated GT-35P gas turbine. Compressed air of about 12–16 bars is fed from the gas turbine to the pressure vessel in which the combustor, a fluidized bed, is situated. The air fluidizes and entrains the bed materials, consisting of coal ash and limestone/dolomite sorbent, and is partly consumed in the combustion reactions with the solid fuel and generation of calcium sulfate $CaSO_4$.

Submerged in the fluidized bed is a tube bank connected to a steam cycle, and, after leaving the bed, the hot gases and entrained-ash particles pass into two-cyclone trains removing about 98% of the entrained ash. The hot gases leave the pressure vessel via a coaxial pipe and are expanded through the turbines. Needless to say, the turbines are exposed to harsh conditions due to the entrained ash. The exhaust gas passes through a heat recovery system pre-heating the feed water in the steam cycle and after additional cleaning is released to the atmosphere.

In the GT-35P, compression is generated from two compressors, a low-pressure compressor and an intercooler followed by a high-pressure compressor. The low-pressure compressor runs at variable speed and is mechanically connected to the low-pressure turbine. GT-35P is a two-shaft gas turbine and the high-pressure shaft

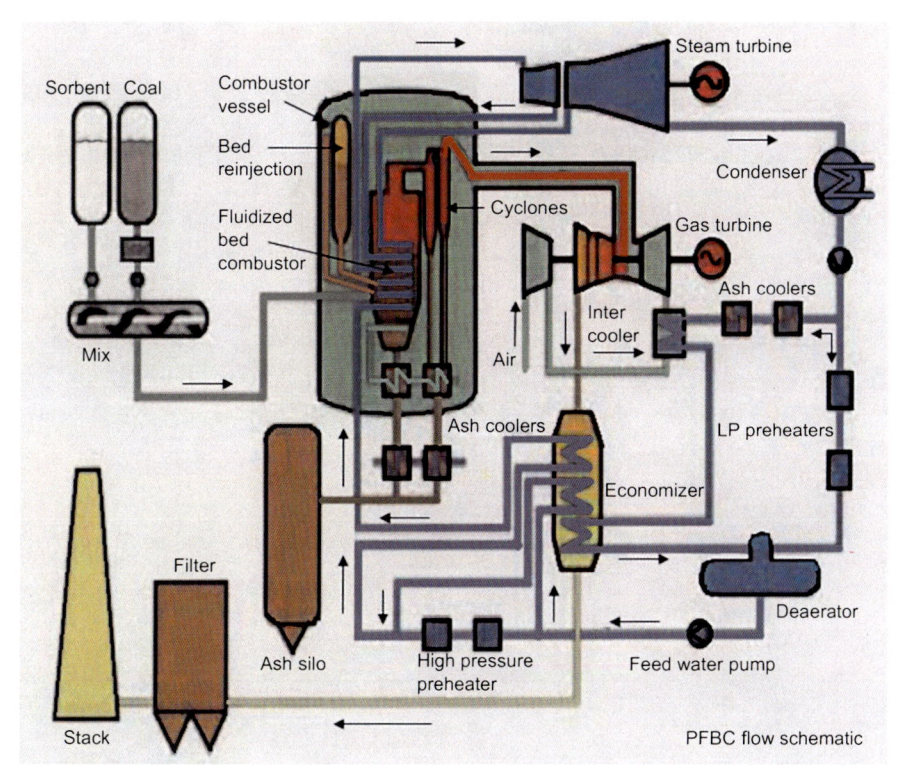

Figure 10.10 Schematic of gas-turbine integration in a pressurized fluidized-bed combustion (PFBC) plant.

runs at constant speed connecting the high-pressure turbine to the high-pressure compressor and the generator. A special feature of the GT-35P is that the last, low-pressure, turbine has a variable guide vane utilized for mass-flow control of the entire cycle. If the fuel flow is increased, the turbine guide vane will close to increase the pressure ratio over the turbine. More work is now allocated to the low-pressure shaft increasing the speed and thereby the mass flow of air in the entire gas-turbine cycle.

The first PFBC plant was commissioned in 1989 in Stockholm and is still in operation. The PFBC is a so-called "clean coal technology" because it is capable of burning high-sulphur coal to generate electricity with high efficiency and low emissions of SO_x and NO_x.

10.8.3 Integrated gasification combined cycle

Another technology targeting clean coal utilization is the integrated gasification combined cycle (IGCC) (Rao, 2012). Here, the solid fuel and oxygen is fed to a gasifier that produces a syngas. Gas turbines in IGCC plants that burn syngas can be divided into cycles with air-side integration or the non-integrated. In the air-side integrated plants, the oxygen to the gasifier is generated from an air separation unit (ASU) that extracts pressurized air from the gas-turbine compressor. In the air-side non-integrated plants, the ASU has its own compressor feeding the ASU with pressurized air taken from ambient conditions.

For an air-integrated IGCC plant, air for the ASU is completely extracted downstream of the gas-turbine compressor (see Figure 10.11). After the ASU, the oxygen is fed to the gasifier. The waste nitrogen is partly reintroduced to the gas-turbine fuel by compressing and mixing with the undiluted syngas. With an air-side IGCC, the air extraction compensates the higher fuel mass flow due to lower heating values of the syngas in comparison to standard fuels. Consequently, the turbine mass flow is about the same as for natural gas or fuel oil operation and the same compressor as for conventional these fuels can be used without any modification for this syngas application. This plant concept with 100% air and nitrogen integration has also an efficiency bonus in comparison with a plant with partial or without air-side integration.

In the air-side non-integrated plants, the ASU has its own compressor feeding the ASU with pressurized air taken from ambient conditions. This solution is often used in

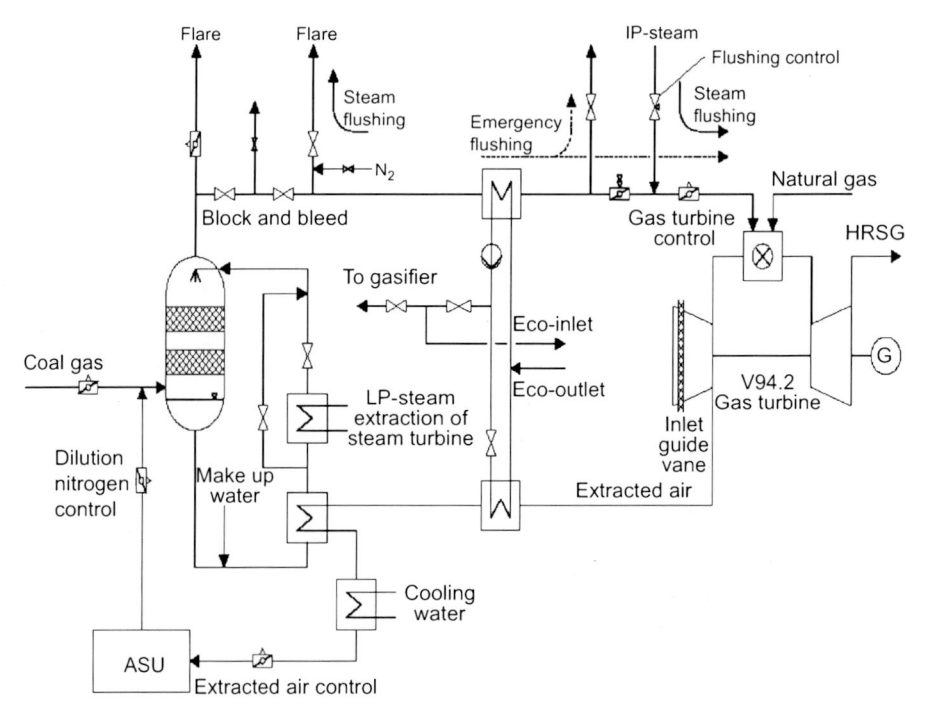

Figure 10.11 Schematic of an integrated gasification combined cycle with an air-side integrated gas turbine.
Ref: Huth, M., Helios, A., Gaio, G., Karg, J., 2000. Operation experience of Siemens IGCC gas turbine using gasification products from coal refinery residues. In: Asme turboexpo, 2000-GT-26, Munich.

refinery applications with medium calorific heating value syngas and is from a plant integration viewpoint a simpler solution, because the interaction between gas-turbine extraction and ASU operation is separated. In this case, there is no air extraction from the compressor and to compensate the higher fuel mass flow a rebuild of the compressor or increasing the turbine flow coefficient is often necessary to handle surge issues (Reiss et al., 2002).

The efficiency of an IGCC is comparable to a conventional pulverized coal-fired boiler plant with supercritical steam data which is about 47—49% for double reheat with steam pressure 300 bar and maximum steam temperature of 600 °C. Therefore, the more complex IGCC plants struggle with being cost-effective. The advantage of an IGCC plant compared to conventional coal plants occurs when pre-combustion CO_2 capture is included. A so-called water—gas-shift reactor can be included after the gasifier to increase the hydrogen content in the syngas and reduce the CO concentration as CO is converted to CO_2. The CO_2 is captured in a CO_2-capture plant and a sulphur acid gas cleanup plant is also included generating a fuel for the combined-cycle plant that will be close to pure hydrogen. The fuel gas can in this case be diluted with the nitrogen from the ASU to facilitate the combustion process.

10.9 CO_2 capture in gas-turbine integrated plants

A natural gas-fired gas turbine in a combined cycle has less than half the CO_2 emissions in grams per produced kWh compared to a coal-fired conventional plant. With carbon capture and storage (CCS) technology the CO_2 footprint of a plant can be even lower. Today, CCS installations are primarily motivated by regulations following a growing awareness of climate change, and partly for using CO_2 with enhanced oil recovery.

Technology for CCS can be divided into either pre-combustion as exemplified by IGCC above, post-combustion capture of CO_2 or oxy-fuel combustion. As of today most of the research done by the power equipment suppliers involved in gas-turbine plant development is directed towards pre- and post-combustion concepts identified as the quicker route to reduce CO_2 emissions. However, by comparison and based on the best information available today, gas turbine-based oxy-combustion with CO_2 capture would hold competitive potentials regarding plant efficiency and cost of electricity.

Post-combustion carbon capture can be applied to conventional combustion plants by separating of CO_2 in the exhaust gas using an absorbent. Both efficiency and cost of the CO_2 separation process strongly depends on the concentration of CO_2 and the volumetric flow of gas to be treated. The high exhaust-gas flow and, therefore, the low CO_2 concentration of gas turbine combined-cycle plants compared to conventional coal-fired plants is a disadvantage. A remedy for this is to use exhaust-gas recirculation (EGR) in which exhaust gas is extracted after the heat recovery steam generator and returned at the compressor air intake after removal of water content. The literature indicates that about 40% of the exhaust gas can be recirculated without major changes in gas-turbine design. The CO_2 concentration in the exhaust gas can be increased from below 5% to almost double using EGR. At the same time, oxygen concentration to the combustor will be decreased to about 16 vol% compared to 21 vol% in air which may affect the combustion process.

If, instead, combustion is performed in pure oxygen in an oxy-fuel combined-cycle plant, the exhaust gas will consist of CO_2, water and small amounts of argon (Jericha et al., 2007). This means that the CO_2 can be removed by condensing the water. In fact, CO_2 is used as a working medium in the gas-turbine cycle, instead of air as illustrated by the schematic in Figure 10.12. In normal combustion, the temperature is limited by nitrogen, which adsorbs a portion of the energy released. To limit the flame temperature in oxy-fuel combustion, the main part of the CO_2 and/or steam is recirculated from the exhaust gas.

Two main concepts based on oxy-fuel combustion have emerged in combined-cycle power plants; the semi-closed oxy-fuel combustion combined cycle (SCOC-CC) and the Graz cycle. Both cycles include an ASU which removes nitrogen from the air. In the Graz cycle, the gas turbine is cooled with steam from the steam cycle. The SCOC-CC in Figure 10.12 has many similarities to normal combined cycles, but the gas-turbine cycle differs in two respects: (1) the working fluid is, as mentioned already, almost pure CO_2; and (2) the combustion chamber is operated under near-stoichiometric conditions with oxygen as the oxidizer and CO_2 as the coolant. The CO_2 is compressed in the gas-turbine compressor to about 40 bars, which is higher than in a normal combined-cycle gas turbine. This pressure is required to reach the desired temperature for steam generation with CO_2 as working medium similar to a conventional air-based cycle.

There are no existing oxy-fuel plants based on a gas-turbine cycle and technical challenges exist related to gas-turbine core-component design including cooling layout and material properties.

10.10 Other integrated cycles

It should be mentioned that in an attempt to reduce plant complexity, wet cycles have been studied in which gas and steam are expanded in the same turbine (Barlett, 2002). There are numerous variants of concepts for wet cycles; for instance, the steam-injected gas turbine shown in Figure 10.13. Steam generated in the HRSG is introduced in the turbine inlet together with the combustor exhaust gases. Water vapour is then condensed from the exhaust gas and returned into the steam cycle in which steam is generated from the exhaust gas heat. This puts high requirements on a condenser that can handle non-condensable gases as well as steam. In addition, water treatment becomes crucial because water is recirculated and might be contaminated by the gaseous exhaust gases depending on the air and fuel quality. Add to that, the high pressure and the steam-rich atmosphere in the gas-turbine core components will require design adaptations or even development efforts.

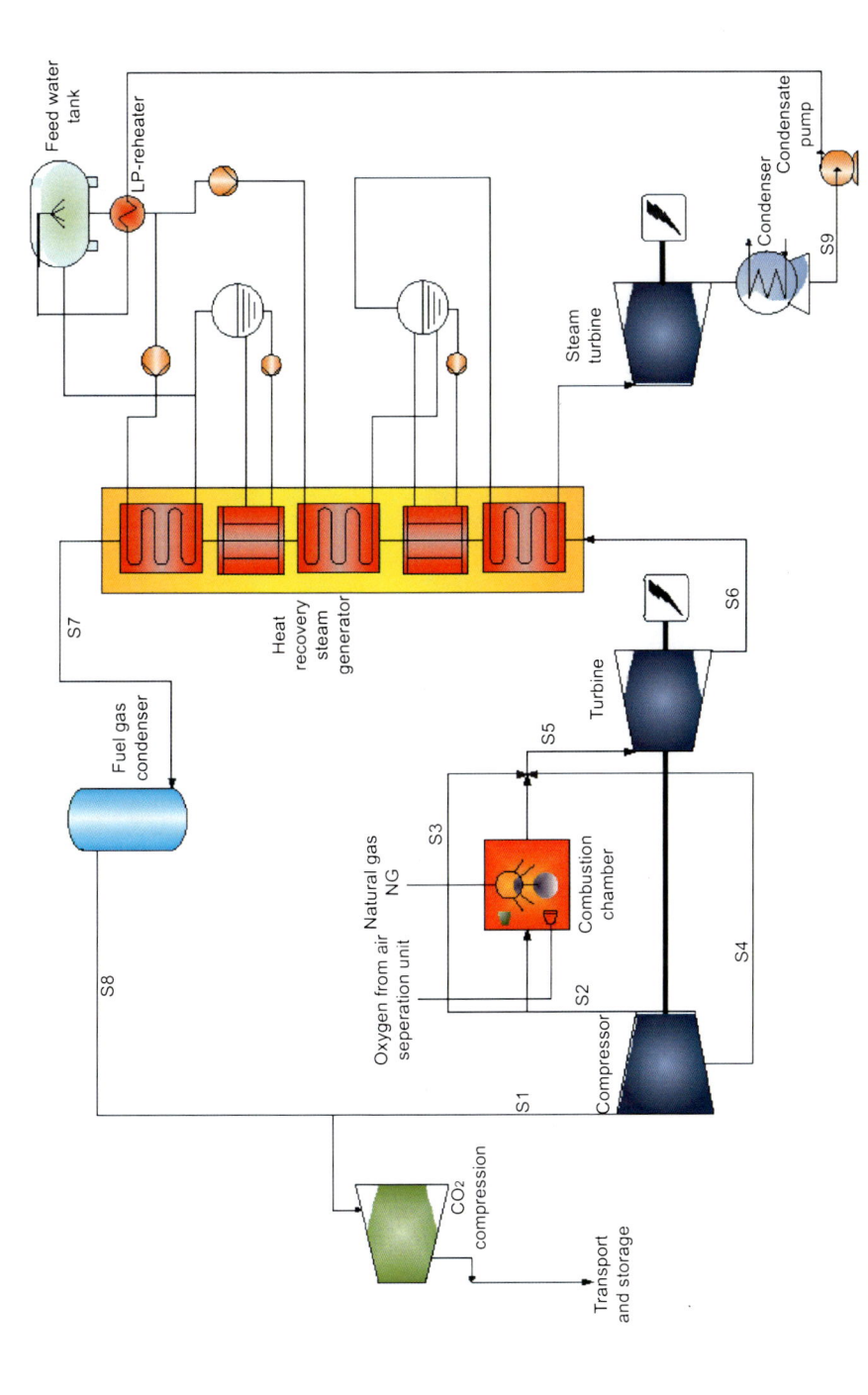

Figure 10.12 Principal flow scheme for a semi-closed oxy-fuel combustion combined cycle (SCOC-CC). Ref: Sammak, M., Jonshagen, K., Thern, M., Genrup, M., Thorbergsson, E., Grönstedt, T., Dahlquist, A., 2011. Conceptual design of a mid-sized, semi-closed oxy-fuel combustion combined cycle. In: Asme turboexpo, GT2011-46,299, Vancouver.

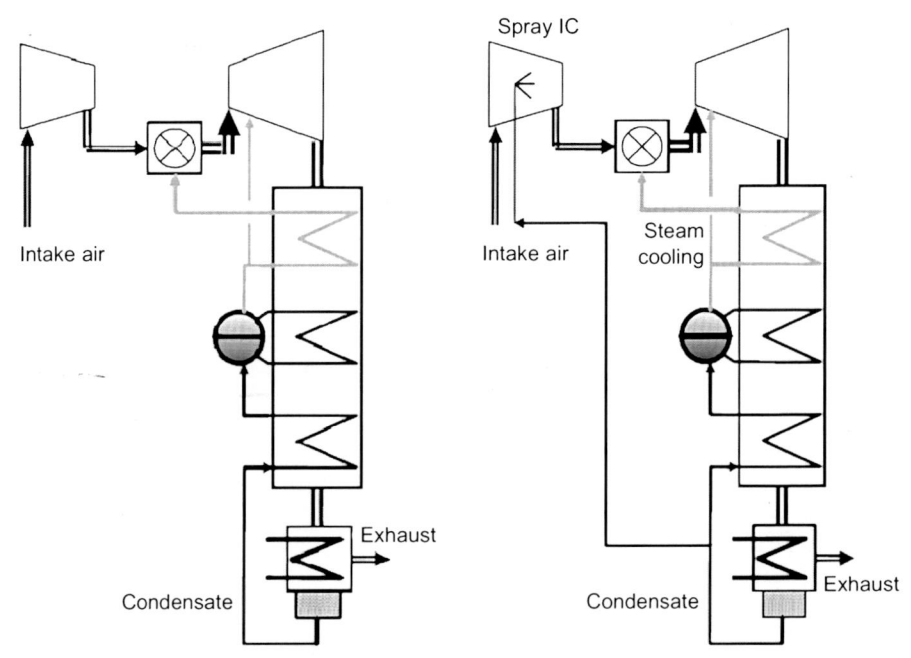

Figure 10.13 The steam-cooled gas-turbine cycle (STIG) with spray intercooling (right-hand side).
Barlett, 2002.

References

Avila-Marin, A., 2011. Volumetric receivers in solar thermal power plants with central receiver system technology: a review. Solar Energy 85, 891—910.
Barlett, M., 2002. Developing Humified Gas Turbine Cycles (Ph.D. thesis).
Cohen, H., Rogers, G.F.C., Saravanamuttoo, H.I.H., 2009. Gas Turbine Theory.
Durbin, M., Ballal, D., 1996. Studies of lean blowout in a step swirl combustor. Journal of Engineering for Gas Turbines and Power 118.
Glassman, I., Yetter, R., 2008. Combustion (Glassman, I., Combustion, Academic Press: New York, 1996).
Energy Technology Perspectives, 2014. IEA.
Jahnson, P. (Ed.), 2013. Modern Gas Turbine Systems. Woodhead Publishing Limited, ISBN 978-1-84569-728-0.
Jericha, H., Sanz, W., Bauer, B., Göttlich, E., 2007. Qualitative and quantitative comparison of two promising oxy-fuel power cycles for CO_2 capture. In: ASME Paper GT2007-27375, ASME Turbo Expo, Montreal, Canada.
Kehlhofer, R., Hannemann, F., Stirnimann, F., et al., 2009. Combined-Cycle Gas and Steam Turbine Power Plants, third ed. Pennwell Corporation, Tulsa (from James Spelling, Hybrid Solar Gas-turbine Power Plants, PhD Thesis, 2013).
Lefebvre, A.H., Ballal, D.R., 2010. Gas Turbine Combustion — Alternative Fuels and Emissions.

Lieuwen, T., Zinn, B.T., 1998. The role of equivalence ratio oscillation in driving combustion instabilities in low NO$_x$ gas turbiens. In: Proceedings of the Combustion Institute, Pittsburgh, PA, vol. 27, pp. 1809–1816.

Näsvall, H., Larfeldt, J., 2011. (ISSN 1653-1248) Power Generation Utilizing Process Gas to Avoid Flaring, SYS08–841, Värmeforsk.

Rao, A. (Ed.), 2012. Combined Cycle Systems for Near-Zero Emission Power Generation, ISBN 978-0-85709-013-3.

Reiss, F., Griffin, T., Reyser, K., 2002. The Alstom GT13E2 medium BTU gas turbine. In: GT-2002-30108, ASME Turboexpo, Amsterdam.

Spadaccini, L.J., Colket, M.B., 1994. Ignition delay characteristics of methane fuels. Prog. Energy Combust. Sci. vol 20, pp. 431–460.

Fuel flexibility with dual-fuel engines

11

Jacob Klimstra
Jacob Klimstra Consultancy, Broeksterwald, The Netherlands

11.1 Introduction

Stationary reciprocating engines currently find applications as prime movers for electricity generators, compressors and pumps. In the late part of the nineteenth century, following the invention of the four-stroke gas engine by Nicolaus August Otto in 1876, many medium-size companies used gas engines as a direct driver of mechanical tools. In the beginning of the twentieth century, electrical motors took over the task of these engines. However, large-bore reciprocating engines started to be applied to drive the electricity generators in many local power plants. With the advent of large integrated electricity transmission and distribution grids, the small local power plants were abolished and large steam turbine-driven generators in central power plants became common practice. The use of reciprocating engines for electricity generation became restricted to areas with small grids with relatively low demand and for applications in which cheap fuel from sewage plants and landfills was available. After the oil crises in the 1970s, concern about the availability and cost of fuels evoked the stimulation of the energy-efficient co-generation of power and heat. Many gas engine-driven generators were installed at locations with a substantial heat requirement. Diesel engines running on heavy fuel oil were installed in developing countries and on islands to drive generators for covering the need for electricity. It is expected that the role of reciprocating engines in electricity generation will increase, because the technique is very suitable to serve as a backup for the variable output of renewable electricity sources based on wind and solar radiation.

Fuel flexibility is a positive asset of reciprocating engines. An example is a site at which the production of biogas fluctuates whereas the demand for electricity is constant. Natural gas can then be used as a backup fuel. The maximum use of associated gas at oil production wells in a dual-fuel engine that can at the same time run on crude oil is another example. Such a solution avoids flaring of the gas and reduces oil consumption. In crucial applications, such as power supply to hospitals and military centres, the use of dual-fuel engines increases the reliability of electricity supply. In case the gas supply fails, oil can be used as a backup fuel. Running on natural gas gives substantially lower emissions of NO_x and of particulates than running on liquid fuel.

This chapter will further explain the fuel conversion process in the different engine types and the techniques needed for creating fuel flexibility.

11.2 The four-stroke spark-ignited gas engine

In a spark-ignited gas engine, a mixture of fuel gas and air is drawn into the engine cylinders via positive displacement of the pistons. This mixture is subsequently compressed by the pistons to the point at which the geometrical cylinder volume is close to its minimum and then ignited with a spark plug. The pressure in the cylinder rises due to the heat release by the combustion process. The cylinder contents are then expanded so that the pressure decreases again. As soon as the cylinder volume comes close to its maximum, the exhaust valves open so that the cylinder contents can escape. During the following stroke, the piston pushes the bulk of the rest gas out of the cylinder and a new cycle can follow.

The fuel efficiency η of such a process is in theory only determined by the compression ratio ε, which is the ratio of maximum volume and minimum volume of the cylinder and by the medium property k. Pure air at ambient conditions has a k value of 1.40. Combustion end products can have a k value of 1.32 (Figure 11.1).

$$\eta = \left\{ 1 - \left(\frac{1}{\varepsilon}\right)^{(k-1)} \right\} \cdot 100\%$$

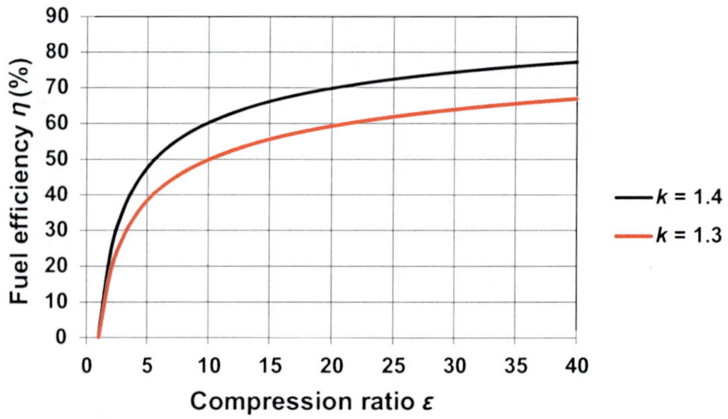

Figure 11.1 The theoretically attainable fuel efficiency of the idealised reciprocating engine process depends only on the compression ratio ε and the medium property k.

The compression ratio cannot be increased to very high values because of risk of mechanical overload of the engine and tendencies towards auto-ignition of the fuel–air mixture leading to combustion knock. Combustion knock, that is premature instantaneous ignition of a part of the cylinder contents, is destructive for engines. The compression ratio of stationary gas engines is generally limited to 12. In practice, heat losses to the cylinder wall and friction losses lower the theoretically attainable fuel efficiency by 10–15 percentage points. A fuel efficiency of 50% is currently considered as the maximum attainable value for a four-stroke engine. Such a high value can only be reached with engines with a relatively large bore that run on a fuel-lean mixture. A large bore and a lean mixture substantially reduce the relative heat loss to the cylinder walls.

The power capacity of four-stroke stationary gas engines ranges between 15 kW and 20 MW. The smaller engines can have running speeds up to 3600 revolutions per minute (rpm) and the largest ones have a minimum running speed of 450 rpm. The maximum running speed of an engine is determined by the mean value of the piston speed. A mean piston speed of 10 m/s is considered as the maximum for stationary reciprocating engines. Higher values would exponentially increase the friction loss and introduce lubrication problems.

11.3 The diesel engine

In diesel engines, fuel is injected into the cylinder close to the moment that the pistons have compressed the air in the cylinders to the minimum volume. The sprays of fuel from the injection nozzle subsequently evaporate and mix with the air in a diffusion process before the fuel can burn. The temperature of the air is sufficiently high that the fuel—air mixture exceeds its auto-ignition temperature so that combustion starts without an external ignition source. The prerequisite of this process is a relatively high compression ratio to reach a sufficiently high air temperature, whereas the fuel should have a high willingness for auto-ignition. The willingness to auto-ignite is expressed in the so-called cetane number. Petrol, for instance, has such a low cetane number that it is unsuitable as a diesel fuel. Natural gas has an even lower cetane number than petrol, so that a diesel engine based on using only natural gas is not feasible (Karim, 1983).

Dedicated diesel engines can run on heavy fuel oil, crude oil, light fuel oil and bio oils. Diesel engines can also be modified to use a large fraction of natural gas. This gas is then mixed with the intake air. The liquid-fuel injector serves as the igniter. There is a minimum in the fraction of liquid fuel for such dual-fuel engines. If the liquid flow through the injectors becomes too small, the penetration of the fuel oil jets into the air will be insufficient. Next to that, the injector tips can overheat due to the lack of cooling by the liquid flow. Clogging of the injector tips will then occur resulting in malfunctioning of the ignition process.

Special engines, the so-called gas—diesel (GD) engines, use high-pressure systems (350 bar) to inject the gaseous fuel into the cylinders. The gas and liquid injectors have been integrated for such engines. The liquid fuel serves again to start the combustion process. In such engines, the ratio of gas and liquid fuel can vary in a wide range. The combustion process of a GD engine is a diffusion process, as in an engine running only on liquid fuel.

Most stationary diesel engines are four-stroke engines, in which each cylinder requires two engine-shaft revolutions to complete a full cycle. Sporadically, large-bore two-stroke engines are used as prime movers for stationary applications. Such engines need only one revolution of the crankshaft to complete a full cycle. Recent developments have also resulted in a design that enables such engines to accept fuel flexibility, including the use of natural gas.

11.4 Fuel specifications

11.4.1 Properties of gaseous fuels

Fuel gas has to react with oxygen to release its energy. Ambient air is normally the source of this oxygen. Fuel gas is therefore premixed with air before it enters the cylinders of the engine. The ratio of air to gas has to be in a well-defined range, otherwise the mixture cannot be ignited or the combustion process becomes destructive or produces undesired emissions. The theoretically minimum amount of air required for complete combustion is called the stoichiometric air requirement. Table 11.1 gives the stoichiometric air requirement of a number of gaseous fuels. A stoichiometric mixture has by definition an air-to-fuel ratio λ of 1.0. A mixture with A % more air than stoichiometric has an air-to-fuel ratio λ of $1.0 + A/100$. Therefore, a mixture of methane and air with $\lambda = 2.0$ consists of 19.36 m^3 of air for 1 m^3 of methane.

Small-bore automotive-size stationary engines generally run on a stoichiometric mixture of fuel gas and air. A three-way catalyst in the exhaust ensures that excessive amounts of nitrogen oxides (NO$_x$) resulting from high temperatures, and CO and hydrocarbons resulting from incomplete combustion, are decreased to below legal limits. Most larger stationary gas engines run on a so-called lean mixture, with λ values between 1.6 and 2.1. Such lean mixtures are difficult to ignite with a simple spark plug as applied in automotive engines. Swirl-chamber spark plugs and special rich-running pre-chambers are generally used to ensure proper ignition. The advantages of a lean mixture are that the peak temperatures of the combustion process are relatively low, which substantially reduces the thermal stress on the engine and ensures that the NO$_x$ production is drastically limited. Next to that, a lean mixture has a much higher knock resistance than a stoichiometric mixture. With a lean mixture, the power output of an engine can be much higher than with a stoichiometric mixture. The output is increased by increasing the pressure of the mixture entering the cylinders with a so-called turbocharger.

Table 11.1 **Stoichiometric air requirement of a number of gaseous fuels (for 'standard' air with a relative humidity of 50% at 20 °C)**

Fuel type	Symbol	Stoichiometric air requirement, m^3/m^3
Methane	CH$_4$	9.68
Ethane	C$_2$H$_6$	17.06
Propane	C$_3$H$_8$	24.66
Butane	C$_4$H$_{10}$	32.67
Hydrogen	H$_2$	2.38
Carbon monoxide	CO	2.39

Fuel flexibility with dual-fuel engines

Engines are equipped with devices that ensure that the air-to-fuel ratio remains within a narrow window. Many engines use a gas—air mixer, a carburettor that is based on a Venturi and a zero-pressure regulator for linking the gas flow to the airflow. Engine management systems based on, for example, the relationship between intake pressure and engine-power output or based on the cylinder-head temperature, correct the λ value for small changes in air properties (temperature, humidity) and fuel gas properties.

The Wobbe index is a major property of a gaseous fuel. A constant Wobbe index ensures that the energy flow with the gas for a given pressure drop over a given restriction remains the same. The definition of the Wobbe index WI is:

$$\text{WI} = H \bigg/ \sqrt{\left(\frac{\rho_\text{gas}}{\rho_\text{air}}\right)}$$

in which H is the calorific value of the gas expressed in MJ/m^3, whereas ρ_gas and ρ_air are viz. the density of the gas and of air expressed in kg/m^3 at standard conditions. For H, the upper or the lower calorific value of the fuel gas can be used. In this chapter, the upper (superior) calorific value is used. The upper calorific value includes the heat that will be released by a combustion process by condensing the water vapour in the combustion end products. If the WI changes, a carburettor with fixed settings will induce a change in air-to-fuel ratio λ according to:

$$\lambda_\text{new} = \frac{\text{WI}_\text{old}}{\text{WI}_\text{new}} \cdot \lambda_\text{old}$$

Methane has a WI of 50.69 MJ/m^3 (reference conditions 15 °C) based on the upper calorific value, whereas biogas consisting of 60% methane and 40% CO$_2$ has a WI of only 23.36 MJ/m^3 (Table 11.2). Switching from a WI of 50.69 MJ/m^3 to 23.36 MJ/m^3 without correcting the carburettor setting while running at a λ

Table 11.2 **The Wobbe index of some gaseous fuels (conditions: calorific value density m^3 at 1013 mbar, temperature 15 °C, starting condition 15 °C)**

Gas type	Wobbe index, MJ/m^3
Methane	50.69
Ethane	65.12
Propane	76.90
Butane	87.83
Hydrogen	45.83
Carbon monoxide	12.18
Biogas (60% methane, 40% CO$_2$)	23.36

of 1.8 will increase the λ to 3.9. The result is an immediate stop of the engine because of complete misfiring. A mixture of biogas and air with a λ of 3.9 cannot be ignited leading to sustained combustion. Most engine management systems can correct for WI variations of $\pm$ 3%. Larger deviations from the mean value of the WI require adjustment of carburettor systems. Natural gas supplied via pipeline systems can have a WI ranging between 41 MJ/m^3 and 53 MJ/m^3, depending on the source. However, such a wide WI range is never supplied to customers because it would jeopardise safety, reliability, emission levels and fuel efficiency. It is almost common practice to limit the local WI variations to $\pm$ 3%.

The methane number (MN) is another important quality indicator of natural gas. The knock resistance of gaseous fuels for stationary engines is expressed in this methane number. Pure methane is a knock-resistant fuel and has by definition a methane number of 100. The much more knock-sensitive hydrogen has by definition a methane number of 0. Mixtures of methane and carbon dioxide have a methane number higher than 100 because the high heat capacity of CO_2 works as a knock inhibitor. Biogas, which can contain up to 40% of CO_2, has therefore a high knock resistance (Leiker et al., 1971). Gaseous fuels containing higher hydrocarbons than methane, such as ethane, propane and butane, have a lower knock resistance than methane. A computer programme with an algorithm based on experimental work can closely predict the methane number of a gaseous fuel which composition is known. Best performance of reciprocating engines is reached for methane numbers exceeding 80. For lower values of the methane number, the compression ratio has to be lowered or the power output has to be reduced. Both measures affect the fuel efficiency (Table 11.3).

Other quantities important for gaseous fuels are the dew point and the sulphur contents. Liquid fuels have much higher volumetric energy content than gaseous fuels. If such liquids enter a gas engine, destruction is imminent. Water flow along with the gas can cause freezing of a carburettor. Water entering the cylinders will deteriorate the lubrication of the cylinders. Some installations have been equipped with liquid separators to avoid any risk that liquids enter the engine. Gas suppliers try to keep the dew point well below normal ambient temperatures. However, temperature decrease by

Table 11.3 **The methane number of some gaseous fuels**

Gas type	Methane number
Methane	100
Ethane	44
Propane	32
Butane	8
Hydrogen	0
Carbon monoxide	75
Biogas (60% methane, 40% CO_2)	140

pressure reduction and fuel-supply lines outside buildings can still cause condensation when the gas contains higher hydrocarbons and water vapour.

11.4.2 Properties of liquid fuels

Liquid fuels are generally classified as distillate fuels and residual fuels. Important characteristics of a liquid fuel are its kinematic viscosity and its density. The ignitability of distillate fuels is expressed in the so-called cetane number. Cetane has by definition a cetane number of 100, whereas alpha-methyl naphthalene was given a cetane number of 0. The cetane number of automotive diesel fuels generally ranges between 46 and 60. The ignitability of residual fuels can be approached with the calculated carbon aromaticity index (CCAI) based on the fuel density ρ in kg/m^3 at 15 °C and the kinematic viscosity ν in mm^2/s. T is the absolute temperature in kelvin:

$$\text{CCAI} = \rho - 140.7 \log\{\log(\nu + 0.85)\} - 80.6 - 483 \log \frac{T}{323}$$

Fuels with a CCAI below 840 will easily ignite in a diesel engine. Diesel engines designed for heavy fuel oil can accept CCAI values of up to 870. Increasing the fuel temperature T helps to improve the ignitability. Heavy fuel oil has a lower calorific value that is some 5% lower than that of light fuel oil. Other liquid fuel-quality characterisers are ash content, water content, sulphur content and acidity. Diesel engines can only run on heavy fuel oil if the oil is filtered and its viscosity has been adapted in a fuel-treatment system.

11.5 Systems for creating fuel flexibility

11.5.1 Deviating gases

In case a gas engine should normally run on biogas, while natural gas is used as the backup fuel, a double supply line with each its own zero-pressure regulator and main adjustment bolt can be applied. Because the calorific value of biogas is lower than that of natural gas, the flow of biogas can be a factor two times higher than that of natural gas. The gas mixer has to be able to accept this higher flow. Figure 11.2 gives the basic setup. The time required from the moment of ignition to reach a progressive flame front is called the apparent heat-release delay. This heat-release delay depends on the composition of the fuel gas. Richer mixtures have a shorter heat-release delay than leaner mixtures. CO_2 in the fuel gas increases the heat-release delay. For biogas, the ignition timing has therefore to be advanced compared with that of running on natural gas to create the proper phasing of the combustion process in the cylinder cycle.

11.5.2 Gas and liquid fuel operation

A diesel engine can, in principle, use a large fraction of gas as long as no knocking occurs. Gases with a low methane number can easily cause knocking in a dual-fuel

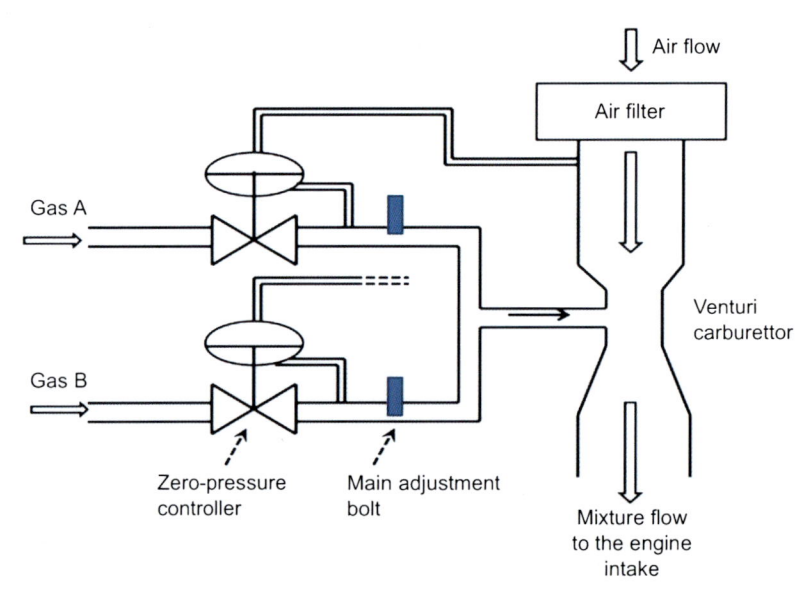

Figure 11.2 A system with an in-parallel gas supply to accommodate two gases with a different Wobbe index.

engine. With a low fraction of gas, knocking will not occur, but the tendency to knock will increase with a substantial fraction of gas. The compression ratio of diesel engines cannot be lowered at will; the temperature of the combustion air at the end of the compression stroke should be high enough to allow proper evaporation of the liquid fuel flowed by ignition. Liquid fuels have to turn into a gas before a chemical reaction with oxygen can occur. If the fraction of gaseous fuel in a dual-fuel diesel engine is small, pockets of very lean mixtures will occur resulting in incomplete combustion and consequently relatively high emissions of hydrocarbons. This also reduces the fuel efficiency of the engine. If the fraction of gas is relatively high, insufficient combustion air can be present in the area of the then-small diesel jets near the injector thus hampering the onset of ignition. The timing of the liquid fuel injection has to be adjusted depending on the effect of the gaseous fuel on the ignition delay.

A diesel engine can be made into a dedicated dual-fuel engine by using a special design for the liquid fuel injector. In this case, the injector has two nozzles, a small one and a large one, as shown in Figure 11.3. Every cylinder cycle, a small nozzle injects a fixed amount of less than 1% of the fuel energy at full load into the combustion chamber. This small nozzle is called the pilot injector. In the diesel-fuel mode, the large injector injects the remaining liquid fuel requirement into the combustion chamber. In the gas mode, natural gas is injected into the combustion air upstream of the intake valves to supply the bulk of the fuel needs. In that case, the pilot injector serves as the ignition source for starting the combustion process. The fuel for the pilot injector is light diesel oil; the fuel for the main diesel injector can range from light oil to heavy fuel oil. Such engines can instantaneously switch over from running for 99% on gas to

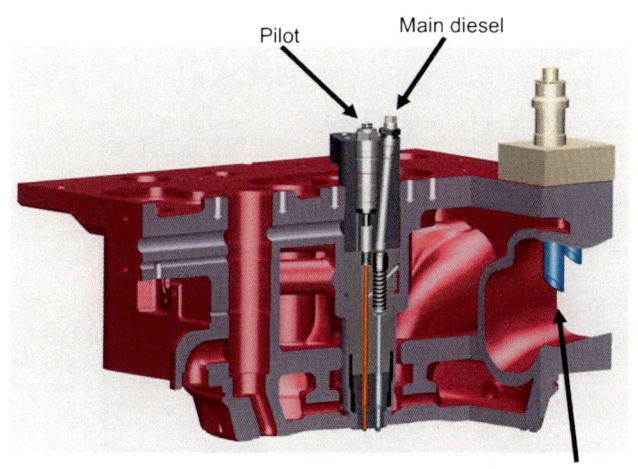

Fuel flexibility with dual-fuel engines

Figure 11.3 The cylinder head of a dedicated dual-fuel engine with a separate pilot injector to allow running on 99% gas.
Picture courtesy of Wärtsilä.

running on liquids only and vice versa. This is advantageous in case of, for example, calamities with the gas supply. Typical applications can be found in hospitals and military applications in which a guaranteed supply of electricity is crucial.

A gas—diesel engine can, in contrast with a dedicated dual-fuel engine, run on a wide range of fractions of the gaseous fuel. This engine type does not have a special pilot injector and therefore the minimum amount of liquid fuel is 5%. The special combined injector for diesel fuel and gas is shown in Figure 11.4. Running on only 5%

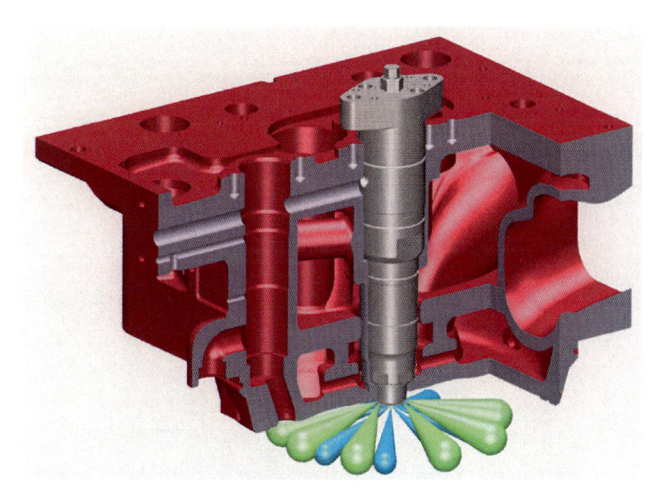

Figure 11.4 The cylinder head of a gas—diesel engine with a combined injector for liquid fuel and 350 bar gaseous fuel.
Picture courtesy of Wärtsilä.

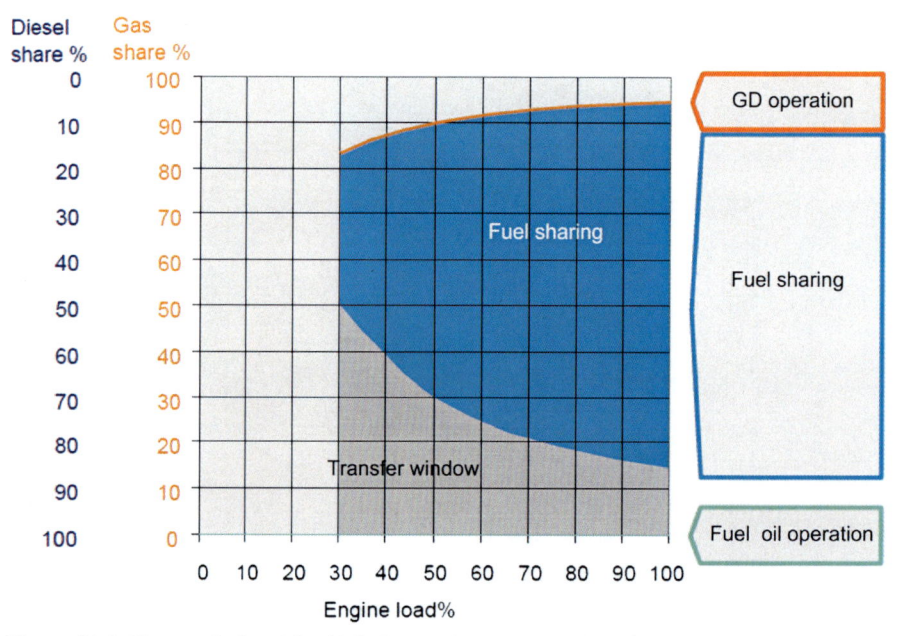

Figure 11.5 The gas-fuel and liquid-fuel operating window of a gas–diesel engine. Picture courtesy of Wärtsilä.

liquid fuel is called the gas mode. In fuel-sharing mode, the fraction of gas can vary in a wide range of the fuel energy demand, as illustrated in Figure 11.5. The diesel-type combustion process of this engine makes the engine insensitive to the knock resistance of the fuel. It means that associated gas of variable quality as present at oil production wells can be utilised thereby removing the need for flaring (Klimstra, 2009). Such GD engines are available in the power range between 6 and 18 MW.

11.6 Plant performance

Stationary reciprocating engines generally have to be able to adjust their output to the demand for power. Turbocharged reciprocating engines with a high work output per unit of cylinder volume and running at fuel-lean conditions have a high efficiency thanks to relatively low heat losses to the cylinder wall and low friction losses. If fuel flexibility is applied to such engines, the fuel efficiency might decrease slightly compared with engines dedicated to using a single fuel type. On the one hand, engines running on biogas only can have a higher compression ratio than an engine that has to be able to run also on a gas with a methane number of 70. On the other hand, the presence of CO_2 in the biogas can slow down the combustion process resulting in a lower efficiency. The compression ratio and power output of diesel engines running on liquid fuels only is not limited by the knock sensitivity of the fuel, as in the case of dual-fuel operation.

Fuel flexibility with dual-fuel engines 303

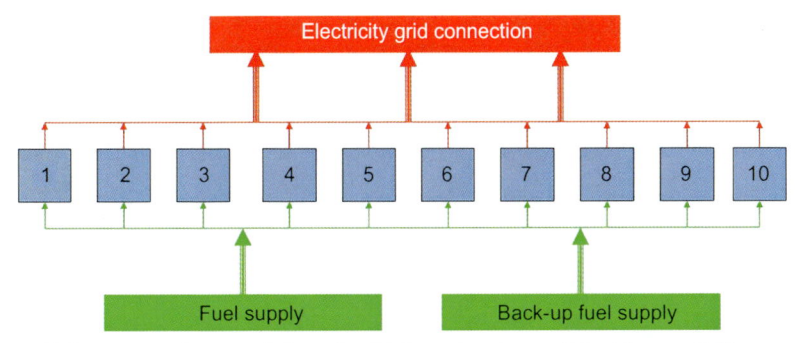

Figure 11.6 A power plant consisting of multiple engines in parallel with a possibility to run on different fuels.

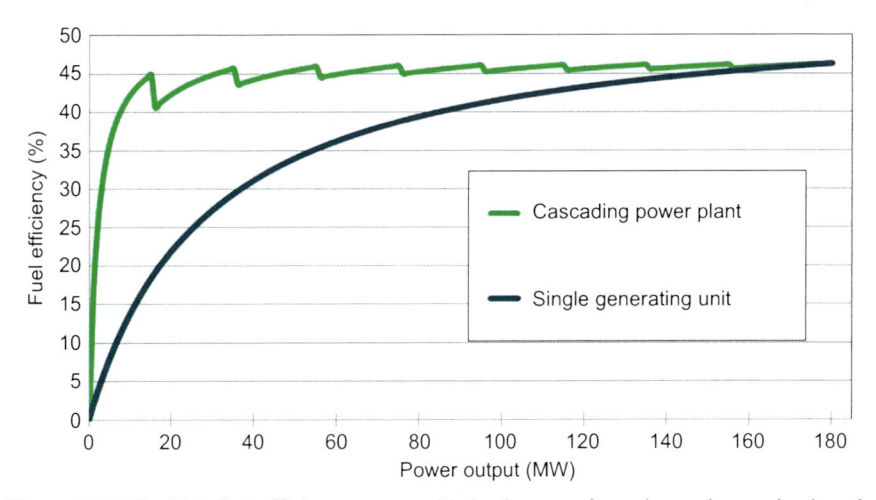

Figure 11.7 The high fuel efficiency over a wide load range of a reciprocating engine-based power plant.

The easy starting and stopping of reciprocating engines makes them very suitable for constructing a power plant based on multiple units in parallel. A power plant based on multiple units as schematically shown in Figure 11.6 can easily adapt its output to the desired value by operating only the number of engines required for that output. In that way, the running engines can operate close to their optimum load. The combined power plant has therefore high operational efficiency over a wide output range (Figure 11.7). Such a modular build-up of a power plant also offers extra possibilities for fuel flexibility. If for some reason the availability of one fuel type is changing, the number of engines running on that fuel can be adapted. In the case of 10 engine-driven generators in parallel, just one of them could in principle be operated on an alternative fuel. The reliability of such a fuel-flexible power plant is very high, because outage of one unit reduces the maximum available output by only 10%.

11.7 Conclusions

1. Reciprocating internal-combustion engines can be operated as dual-fuel engines that accept gaseous as well as liquid fuels. A dual-fuel engine always uses some liquid fuel for starting the combustion process.
2. Dedicated dual-fuel engines can use less than 1% of liquid fuel and instantaneously switch over from liquid-fuel operation to gaseous-fuel operation and vice-versa under load conditions.
3. Care should always be taken that the dual-fuel engine is properly tuned to the relevant fuel properties. The Wobbe index, the methane number, the cetane number and the carbon aromaticity index are important qualifiers in this respect.
4. A power station based on multiple dual-fuel engines in parallel offers high flexibility, both in fuel use and in output. This increases the reliability of power supply.

References

Karim, G.A., June 1983. The dual fuel engine of the compression ignition type − prospects, problems and solutions − a review. In: Presentation at the Conference on Compressed Natural Gas as a Motor Fuel, Pittsburgh.

Klimstra, J., June 10, 2009. Flare gas utilisation for mechanical drives and electricity production. In: Oil & Gas Emissions Seminar, Aberdeen UK.

Leiker, M., Christoph, K., Rankl, M., Cartellieri, W., Pfeifer, U., 1971. The evaluation of the anti-knocking property of gaseous fuels by means of the methane number and its practical application to gas engines. In: A2 Conference CIMAC, Stockholm, Sweden.

Index

'*Note*: Page numbers followed by "f" indicate figures, and "t" indicate tables.'

A
Aero-derivatives, 271
Agglomeration, 153
Agricultural byproducts and residues, 67
Air separation unit (ASU), 285–286
Air-to-fuel ratio, 296–297
AISI. *See* American Iron and Steel Institute (AISI)
Alkalis, 181
Alternative energy sources. *See* Nonconventional energy sources
Aluminium oxide (Al_2O_3), 43–44
American Iron and Steel Institute (AISI), 227–228, 233
American Society for Testing and Materials (ASTM), 34, 203
Anthracite, 9, 37–38, 38f
Apparent heat-release delay, 299
Ash, 40
  composition, 43–44
  fusibility test methods, 43
  fusion temperatures, 42–43
Aspen, 71–72
ASTM. *See* American Society for Testing and Materials (ASTM)
ASU. *See* Air separation unit (ASU)
Atmospheres Explosibles (ATEX), 117
Auto-ignition delay time, 277

B
Babcock & Wilcox Company (B & W), 38
Bagasse, 70
Barnett Shale of Texas, 10
Base-load power plants, 16
BD. *See* Bulk density (BD)
Benson-type boiler operating, 195
BFB. *See* Bubbling fluidised-bed (BFB)
BFG. *See* Blast furnace gas (BFG)
Big bale, 68

Bio-crude. *See* Bio-oil
Bio-oil, 13–14, 161
Biofuel, 3–4
Biogas, 158–160, 298
  constituents and variable composition, 159t
  fuel analysis, 160
  quality control, 160
  technologies, 159
Biological degradation, 126–127
Biomass, 13–14, 17–18, 59, 122, 241. *See also* Synthesis gas (Syngas)
  biological degradation, 126–127
  biomass properties and measurement of properties, 70–89
  coal and wood fuels comparison, 61f
  explosivity, 133–139
  future trends, 90–91
  handling and storage at coal-fired power plants, 121
    considerations, 121–122
    conveying, 123
    hardware modifications, 124–125
    silos/storage, 123
    unloading/discharge, 122
  industrial-scale experience with pre-treated biomass, 126
  influence of fuel characterization, 64f
  measurement of properties and applied standards, 82–89
    bulk density, 87–89
    calorific value, 84–86, 85f
    moisture content, 86–87
    particle-size determination, 87
    proximate and ultimate analysis, 83–84, 83f
  mechanical durability and storage, 132–133
  modern fluidised-bed boilers for, 185–186
  modern stokers for, 178–180, 179f

Biomass (*Continued*)
　pneumatic conveying, 127–132
　pre-treatment technologies, 125
　properties, 70–73
　pyrolysis, 263–264
　　European and global woody, 64–67
　　herbaceous and fruit biomass, 67–70
　　supply chain for straw bales, 69f
　sampling and sample reduction, 73–82, 78f
　solid biomass fuel supply chain options, 62t–63t
Biomass fuel transport and handling
　biomass sources and types, 108–112
　　biomass materials and classifications, 111–112
　　class identification, 111
　　handling equipment selection, 112
　causes of handling problems with biomass, 101–102
　challenges of biomass handling, 102–108
　choosing right solutions, 118
　considerations for fuel compatibility, 113–118
　future trends, 119–120
　importance to cost-effective biomass fuel valorization, 99–100
　need to 'know your enemy', 118–119
　solids-handling processes in biomass generation plant, 100f
　special features of biomass as fuel, 100–101
Biomass fuels, 3–4, 59, 61, 207–210, 209t
　chemical composition, 76t–77t
　properties, 74t–75t
Biomass gasification, 150–155. *See also* Coal gasification; Opportunity fuels
　fuel properties, 154–155
　technologies, 153–154
Biomass-fired systems, 225
Bitumen, 11–12
Bitumen-rocks oil, 12
Bituminous coal, 9, 37
Blast furnace gas (BFG), 45, 278–279
Bleeding, 274–275
Blended solid fuels
　co-gasification, 259–262
　co-pyrolysis, 262–265
Blending ratio, 255–257
Blue-water gas, 45

Bouduard reaction, 247
Bread and butter feedstock, 155
Briquettes, 67
British Standards Institution (BSI), 203
Brownian–eddy diffusion process, 218
BSI. *See* British Standards Institution (BSI)
Bubbling fluidised-bed (BFB), 182
　gasifier, 251–252, 251f, 255–257
　technologies, 182–184, 182f
Bulk density (BD), 87–89
Byproduct gas from gasification, 46

C
Calcium (Ca), 71
Calciumoxide (CaO), 43–44
Calorific value (CV), 84–86, 85f, 105, 203
Cane trash, 70
CaO. *See* Calciumoxide (CaO)
Carbon, hydrogen, nitrogen, sulfur (CHNS), 152
Carbon aromaticity index (CCAI), 299
Carbon capture and sequestration (CCS). *See* Carbon capture and storage (CCS)
Carbon capture and storage (CCS), 53–54, 287
Carbon dioxide (CO_2), 259
　capture in gas-turbine integrated plants, 287–288
Carbon monoxide (CO), 145, 147–148, 158–159, 259
Carbonaceous feedstocks, 145–146, 156–157, 167
Carbonaceous material, 145
Carbonization, 161
Carburetted water gas, 45
Carburettor, 297–298
Catalytic partial oxidation reaction (CPOX reaction), 166–167
CCAI. *See* Carbon aromaticity index (CCAI)
CCP. *See* Cereal co-products (CCP)
CCS. *See* Carbon capture and storage (CCS)
CEN. *See* European Committee for Standardisation (CEN)
Centrifuge, 51
Cereal co-products (CCP), 207
Cetane number, 295, 299
CFB. *See* Circulating fluidised-bed (CFB)

Index

CfD. *See* Contract for Difference (CfD)
Chipping, 65−67, 66f−67f
Chlorine (Cl), 71
ChlorOut process, 230−231
CHNS. *See* Carbon, hydrogen, nitrogen, sulfur (CHNS)
Chopped straw, 110f
CHP plant. *See* Combined heat and power plant (CHP plant)
Chromia (Cr_2O_3), 222
Circular economy, 119−120
Circulating fluidised-bed (CFB), 182
   gasifiers, 252, 252f
   technologies, 182−184, 184f
Clean coal technology, 285
CNG. *See* Compressed natural gas (CNG)
Co-firing, 219−220, 230
   of biomass, 124−125
   fuels, 231
   test, 126
Co-gasification. *See also* Gasification
   biomass, 255−259
   coal, 255−257
   dried sewage sludge, 258−259
   feedstocks, 168−169
   Fischer−Tropsch feed gas specifications, 261t
   issues in blended solid fuels, 259−262
   technology, 17−18, 155−156
Co-generation plant, 271−272
Co-milling, 191
Co-pyrolysis of blended solid fuels, 262
   biomass, 263−264
   coal, 263−264
   dried sewage sludge pyrolysis, 264−265
   MSW pyrolysis, 264−265
Coal beneficiation, *See* Coal—preparation
Coal gasification, 17, 146. *See also* Biomass gasification; Opportunity fuels
   products, 46
      properties, 148−150
   technologies, 146−148
Coal washing, *See* Coal—preparation
Coal-handling infrastructure, 124−125
Coal(s), 8, 29, 32−38, 202−207, 204t
   analysis, 38
   and biomass, 59
   characterization, 38−44
   classification, 32−38

fuels derived from, 44−48
gaseous fuels from, 44−46
grade, 9
and lignite reserves, 32
liquefaction, 160−161
mills, 124
minerals in, 206t
mining, 48−51
   supply chain, 49f
   surface mining, 48−51
   underground, 48, 50f
moisture, 39
preparation, 51
pyrolysis, 263−264
transportation, 51−53
ultimate analysis, 41
world availability, 29−32
Coatings protection, 233
COG. *See* Coke oven gas (COG)
Cohesive materials without extreme shape, 109
Coke, 44
Coke breeze, 44
Coke oven gas (COG), 44−45, 278−279
Combined cycle, 271, 282
   power plant, 282f
Combined heat and power plant (CHP plant), 193
Combined-cycle power plant, 16−17
Combustion
   dynamics, 278
   process, 273
   systems, 231
   zone, 165
Complexity of petroleum, 5
Component-and material-monitoring methods, 232−233
Compressed natural gas (CNG), 4
Concentrated solar plant (CSP), 283−284
Contract for Difference (CfD), 99
Conventional energy sources, 5. *See also* Unconventional energy sources
   coal, 8−9
   natural gas, 6−8
   oil from shale, 11
   overview of refinery, 6f
   petroleum, 5−6
   shale gas, 10−11
Conventional fuel sources, 3

Conveying, 123
Conveyor, 106
"Core flow" pattern, 113, 114f
Cost-effective biomass fuel valorization, 99–100
CPOX reaction. *See* Catalytic partial oxidation reaction (CPOX reaction)
Crude oil. *See* Petroleum
Crude petroleum, 7t
CSP. *See* Concentrated solar plant (CSP)
CV. *See* Calorific value (CV)

D

daf. *See* Dry ash-free (daf)
Dangerous Substances and Explosible Atmospheres Regulations (DSEAR), 117
db. *See* Dry basis (db)
DBFZ, 65
Dedicated biomass burners, 191
Degradation mechanisms and modelling
  deposition, 217–221
  erosion–abrasion–wear, 229
  fireside corrosion, 222–229
  nominal compositions of heat-exchanger tube materials, 223t
  oxidation, 221–222
Deposition, 217–218
  deposit compositions, 219–221
  on superheater tube in coal-fired boiler, 220f
  for vapours, 218
Deviating gases, 299
DFB gasifier. *See* Dual fluidized-bed gasifier (DFB gasifier)
Diesel engine, 295
Digester gas. *See* Biogas
Direct co-firing in PC boilers, 190–192
Direct inertial impaction, 217–218
Drax power plant, 99–100, 122
Dried sewage sludge
  co-gasification, 258–259
  pyrolysis, 264–265
Dry ash-free (daf), 38
Dry basis (db), 38
Dry small steam nuts (DSSN), 9
DSEAR. *See* Dangerous Substances and Explosible Atmospheres Regulations (DSEAR)

DSSN. *See* Dry small steam nuts (DSSN)
Dual fluidized-bed gasifier (DFB gasifier), 253–257, 253f
Dust explosion, 101, 116
Dust formation upon handling, 134

E

E85 vehicles. *See* Flexible-fuel vehicle (FFV)
ECN. *See* Electrochemical noise (ECN); Energy Research Centre of the Netherlands (ECN)
EDX analysis. *See* Energy dispersive X-ray analysis (EDX analysis)
EGR. *See* Exhaust-gas recirculation (EGR)
Electrical energy, 16
  other fuels, 17–19
  power plant operations, 16–17
Electrochemical noise (ECN), 232
Electrostatic charge buildup, 121
Electrostatic precipitation (ESP), 179–180
Energy dispersive X-ray analysis (EDX analysis), 232
Energy Research Centre of the Netherlands (ECN), 126
Energy wood bundling, 65
Engineering, procurement and construction (EPC), 118
Entrained-bed process, 147
EPC. *See* Engineering, procurement and construction (EPC)
Equivalence ratio (ER), 249
ER. *See* Equivalence ratio (ER)
Erosion, 229
Erosion–abrasion–wear, 229
ESP. *See* Electrostatic precipitation (ESP)
Ethanol, 4
EU. *See* European Union (EU)
EU COST. *See* European Cooperation in Science and Technology (EU COST)
European and global woody resources and supply chains, 64–67
European Committee for Standardisation (CEN), 203
European Cooperation in Science and Technology (EU COST), 227–228
European Union (EU), 31
Exhaust-gas recirculation (EGR), 287–288

Index

Explosion
　pressure, 137–138
　protection, 116
Explosivity, 133
　flame-front velocity, 137–139
　MEC, 136–137, 140f
　moisture content, 136
　native dust and dust formation upon handling, 134
　of raw biomass chips *vs.* torrefied biomass pellets, 134–135
Extraneous ash, 70
Extreme-shape particles, 109–110

F

Falling-stream samplers, 81
Feed-gas composition, 282–283
Feed-in tariff (FIT), 99
Feeder, 106
Feedstock, 166–167
Ferric hydroxide (Fe$_2$O$_3$), 43–44
FFV. *See* Flexible-fuel vehicle (FFV)
Fireside corrosion, 222
　melting points of deposit constituents, 226t
　superheater–reheater corrosion, 224–229
　waterwall corrosion, 222–224
Fischer–Tropsch process (FT process), 15, 46–48
　chemical reaction equation, 47
　feed gas specifications, 261t
　synthesis, 163–164, 168
　reaction, 161–162
FIT. *See* Feed-in tariff (FIT)
Fixed-bed process, 147
Flame temperature, 278–280
Flame-front velocity, 137–139
Flex cars flexi-fuel vehicles. *See* flexible-fuel vehicle (FFV)
Flex vehicles. *See* Flexible-fuel vehicle (FFV)
Flexible-fuel, 4
Flexible-fuel vehicle (FFV), 4
Fluid bed process, 147
Fluid temperature (FT), 43
Fluidised-bed boilers, 177–178
Fluidised-bed combustion. *See also* Grate combustion; Pulverised fuel combustion (PF combustion)
　basics, 181

BFB combustion, 182–184, 182f
CFB combustion, 182–184, 184f
modern fluidised-bed boilers for biomass, 185–186
Flywheel, 155
Former Soviet Union (FSU), 29
Fossil fuel feedstocks
　coal
　　characterization, 38–44
　　classification, 32–38
　　coal-producing and-importing countries, 31f
　　world availability of, 29–32
　　top coal-exporting countries, 30f
　　total world electricity generation, 30f
Fossil fuels, 29
Fouling, 220–221
Four-stroke spark-ignited gas engine, 294–295
Free water, 70–71
Free-flowing particles without extreme shape, 108–109
Free-swelling index (FSI), 42–43
Fruit biomass resources and supply chains, 67–70
FSI. *See* Free-swelling index (FSI)
FSU. *See* Former Soviet Union (FSU)
FT. *See* Fluid temperature (FT)
FT process. *See* Fischer–Tropsch process (FT process)
Fuel analysis, 160
Fuel compatibility considerations
　dust control, ATEX and DSEAR, 116–117
　explosion protection, 116
　limitation of storage time, 113–116
　need for stock rotation, 113–116
　special care in relation to large vessels, 116
　tests for checking compatibility, 117–118
Fuel conversion process, 293
Fuel efficiency, 294
Fuel flexibility, 177, 183, 194, 293
　diesel engine, 295
　environmental issues, 19
　four-stroke spark-ignited gas engine, 294–295
　fuel specifications, 296–299
　plant performance, 302–303
　systems for creation

Fuel flexibility (*Continued*)
    deviating gases, 299
    gas and liquid fuel operation, 299−302
    trends and technological challenges, 19−20
Fuel flexible energy
    conventional energy sources, 5−11
    electrical energy, 16−19
    Fischer-Tropsch process, 15
    unconventional energy sources, 11−14
Fuel flexible gas production
    bio-solid wastes
        characteristics, 243−246
        co-gasification, 246−262
    biomass, 241
    biomass
        characteristics, 243−246
        co-gasification, 246−262
        co-pyrolysis of blended solid fuels, 262−265
    coal, 241−242
    coal
        characteristics, 243−246
        co-gasification, 246−262
        for energy efficiencies, 242
        gasification, co-gasification, pyrolysis and co-pyrolysis, 243f
    GHG, 241
    solid fuels
        proximate analysis results for, 244t
        ultimate analysis, 245t
Fuel-flexible gas turbines, 273−275
    combined-cycle power plant, 282f
    gaseous fuels for gas turbine operation, 275−276
    installation, 281−283
Fuel-to-gas ratio, 150
Fuel(s), 3, 181. *See also* Gaseous fuels
    derived from coal, 44−48
    fuel-preparation systems, 231
    gas, 296
    preparation, 213
    special features of biomass as, 100−101
    specifications
        gaseous fuels properties, 296−299
        liquid fuels properties, 299
    substitution, 230−231
    suppliers, 67
    transport−handling−storage systems, 231
Future energy system, 53

G
Gas
    agent velocity, 251
    and liquid fuel operation, 299−300
        cylinder head of dedicated dual-fuel engine, 301f
        diesel engine, 300−301
        gas−diesel engine, 301f−302f
        mode, 301−302
    yield and oil yield, 13
Gas turbines, 271
    CO_2 capture in gas-turbine integrated plants, 287−288
    combustion-related challenges for gaseous fuel flexibility, 277
        auto-ignition delay time, 277
        combustion dynamics, 278
        flame temperature, 278−280
        high flame speed, 277−278
        lean blow out, 278
    with external heating integration in plants, 283−287
    fuel flexibility impacts on, 280−281
    integrated cycles, 288
    in plants, 271−272
    SGT-800 single-shaft 50 MW, 272f
Gas-to-liquids (GTL), 47−48
Gas-turbine integrated plants, CO_2 capture in, 287−288
Gas−diesel engine (GD engine), 295, 301−302
Gaseous fuels. *See also* Fuel(s)
    from coal
        blast furnace gas, 45
        byproduct gas from gasification, 46
        coke oven gas, 44−45
        producer gas, 45−46
        water gas, 45
    for gas turbine operation, 275−276
    properties, 296
        liquid fuels, 298−299
        MN, 298, 298t
        stoichiometric air requirement, 296, 296t
        WI, 297, 297t
Gasification, 9, 145−146, 192, 246. *See also* Co-gasification
    biomass, 150−155
    CFB, 192−193
    coal, 146−150

Index   311

gasification-based refinery, 168
opportunity fuels, 155–158
reactions, 247
route, 193
theories and technologies, 246
   BFB gasifier, 251–252, 251f
   CFB gasifiers, 252, 252f
   DFB gasifier, 253–254, 253f
   downdraft fixed-bed gasifier, 249–250, 249f
   entrained-flow gasifier, 254, 254f
   gas composition of producer gases, 248t
   gasifiers, 247–248
   updraft fixed-bed gasifier, 250, 250f
Gasifier, 15, 247–248
GCV. *See* Gross calorific value (GCV)
GD engine. *See* Gas–diesel engine (GD engine)
GHG. *See* Green house gas (GHG)
Grate combustion. *See also* Fluidised-bed combustion; Pulverised fuel combustion (PF combustion)
   basics, 177–178
   modern stokers for biomass, 178–180, 179f
Grate-fired boilers, 177–178
Green house gas (GHG), 53, 241
   effect, 14
Grindability, 42
Gross calorific value (GCV), 41–42
GT-35P gas turbines, 284–285
GTL. *See* Gas-to-liquids (GTL)

H

H/C ratios. *See* Hydrogen/carbon ratios (H/C ratios)
H_2/CO ratio. *See* Hydrogen/carbon monoxide ratio (H_2/CO ratio)
Hard and brown coal, 34
Hardgrove, 42
Hardware modifications, 124–125
Hartmann tube apparatus, 133
Harvesting, 65–67
   of delimbed stems, 65
HC. *See* Hydrocarbons (HC)
Heat-recovery steam generation boilers (HRSG boilers), 282
Heating value, 41–42
Hemispherical temperature (HT), 43

Herbaceous and fruit, 67–70
Herbaceous biomass resources and supply chains, 67–70
Herbaceous by-products and residues, 67
Heterogeneous vapour condensation, 218
HHV. *See* High heating value (HHV)
High flame speed, 277–278
High heating value (HHV), 41–42
High-temperature Fischer–Tropsch reaction (HTFT reaction), 47, 164–165
High-velocity oxygen fuel thermal spraying (HVOF thermal spraying), 233
Homogeneous vapour condensation, 218
HRSG boilers. *See* Heat-recovery steam generation boilers (HRSG boilers)
HT. *See* Hemispherical temperature (HT)
HTFT reaction. *See* High-temperature Fischer–Tropsch reaction (HTFT reaction)
HVOF thermal spraying. *See* High-velocity oxygen fuel thermal spraying (HVOF thermal spraying)
Hydrocarbons (HC), 296
Hydrogen (H_2), 14, 145, 147–148, 167, 259, 263
Hydrogen sulfide (H_2S), 6–7, 158–159
Hydrogen/carbon monoxide ratio (H_2/CO ratio), 151, 163, 163t
Hydrogen/carbon ratios (H/C ratios), 203

I

IC engines. *See* Internal combustion engines (IC engines)
IEA. *See* International Energy Agency (IEA)
IGCC. *See* Integrated gasifier combined cycle (IGCC)
Indigenous coal, 29
Indirect co-firing technologies, 192–194
Industrial gas turbines, 271
Industrial-scale experience with pre-treated biomass, 126
Initial deformation temperature (IT), 43
Integrated cycles, 288
Integrated gasifier combined cycle (IGCC), 54, 285–287
Internal combustion engines (IC engines), 259
International Energy Agency (IEA), 207

International Organization for Standardization (ISO), 34, 59, 232–233
International Panel on Climate Change (IPCC), 210
INTREX™, 184, 185f
IPCC. *See* International Panel on Climate Change (IPCC)
Iron-based catalysts, 163
ISO. *See* International Organization for Standardization (ISO)
Isopentane (iC5), 283
IT. *See* Initial deformation temperature (IT)

J
Jet, 9

K
Kerogen, 12
Knock resistance, 296, 298
Kymijärvi II unit, 194, 194f

L
Laminated veneer lumber (LVL), 243–246
Landfill gas (LFG). *See* Biogas
Large-haul discharging, 122
Large-scale solid-fuel combustion technologies, 177
  fluidised-bed combustion, 181–186
  grate combustion, 177–180
  PF combustion, 187–196
Lean blow out, 278
Lean mixture, 296
LHV. *See* Lower heating value (LHV)
Lignite, 8–9, 36–37, 36f
  pre-drying, 54–55
Linear polarisation resistance (LPR), 232
Liquefaction, 9, 160–161
Liquefied natural gas (LNG), 19, 276, 282
Liquefied petroleum gas (LPG), 4, 7–8
Liquid fuels, 298–299
Liquid petroleum gas (LPG), 282–283
LNG. *See* Liquefied natural gas (LNG)
Local dust formation, 122
Logging residues, 65
Long-distance conveying of biomass, 123
Longwall mining, 48
Low-temperature Fischer–Tropsch reaction (LTFT reaction), 47, 164–165

Lower heating value (LHV), 41–42, 275
LPG. *See* Liquefied petroleum gas (LPG); Liquid petroleum gas (LPG)
LPR. *See* Linear polarisation resistance (LPR)
LTFT reaction. *See* Low-temperature Fischer–Tropsch reaction (LTFT reaction)
LVL. *See* Laminated veneer lumber (LVL)

M
maf. *See* Mineral matter/ash-free (maf)
Magnesium (Mn), 71
Magnesium oxide (MgO), 43–44
Maize, 70
Marsh gas. *See* Biogas
Mass flow pattern, 114, 115f
MEC. *See* Minimum explosible concentration (MEC)
Mechanical durability and storage, 132–133
Mechanical dust collector, 179–180
Methanation reaction, 247
Methane (CH_4), 158–159, 259, 263
Methane number (MN), 298
  gaseous fuels, 298t
MIE. *See* Minimum ignition energy (MIE)
Milled palm nut kernels, 110f
Mineral matter, 149
Mineral matter/ash-free (maf), 38
Minimum explosible concentration (MEC), 136–137, 137f, 140f
Minimum fluidization velocity, 251
Minimum ignition energy (MIE), 133–134, 134f–135f
Mining, 31–32
Mining methods, 48
Miscanthus, 68
MN. *See* Methane number (MN)
Modern water-cooled cyclone, 186f
Modified Fischer assay test method, 13
Moisture
  content, 86–87, 136
    analysis, 78
  determination, 39–40
  effects, 106–108
Molten salt processes, 147–148
MSW. *See* Municipal solid waste (MSW)
Multi-fuel CFB boilers, 184
Multifuel vehicle, 4

Index

Municipal solid waste (MSW), 18, 155–158, 178, 210, 242, 264
  pyrolysis, 264–265

N
Native dust, 134
Natural gas (NG), 6–8, 278–279, 293, 295, 297–299
Natural gas liquids (NGLs), 10
NCV. *See* Net Calorific Value (NCV)
Net Calorific Value (NCV), 41–42
NG. *See* Natural gas (NG)
NGLs. *See* Natural gas liquids (NGLs)
Nitrogen oxides (NO$_x$), 264, 296
Noncatalytic partial oxidation reaction (TPOX reaction), 166–167
Nonconventional energy sources, 3

O
O/C ratios. *See* Oxygen/carbon ratios (O/C ratios)
OFA systems. *See* Over-fire air systems (OFA systems)
Oil
  sand, 11
    bitumen, 5
  from shale, 3, 11
  shale, 12–13
  and water, 13
Olive cake, 73
Opencast, 48
  mines large equipment, 51
Opportunity fuels, 155–158. *See also* Biomass gasification; Coal gasification
  product properties, 157–158
  technologies, 156–157
Organic sedimentary rock, 8
Over-fire air systems (OFA systems), 178
Oxidation, 221–222
Oxyfuel combustion, 54
Oxygen-blown entrained gasifiers, 170
Oxygen-containing substances, 167–168
Oxygen/carbon ratios (O/C ratios), 203

P
Parallel co-firing, 194–195
Particle-size determination, 87

PC boilers. *See* Pulverised-coal boilers (PC boilers)
PC combustion. *See* Pulverised coal combustion (PC combustion)
PCC. *See* Pulverized coal combustion (PCC)
PDI. *See* Pellet Durability Index (PDI)
Peak-load power plants, 16
Peat, 34–36, 36f
Pellet Durability Index (PDI), 132, 133f
Pellets, 59
Petroleum, 5
  coke, 42, 155
  products and fuels, 4
  refinery, 5
PF combustion. *See* Pulverised fuel combustion (PF combustion)
PFBC. *See* Pressurized fluidized-bed combustion (PFBC)
Pilot injector, 300–301
Plasma
  arc processing, 157
  gasification, 157
  technology, 156–157
Pneumatic conveying, 127–132
Post-combustion carbon capture, 287–288
Postcombustion capture, 54
Potassium oxide (K$_2$O), 43–44
Powder River Basin coal (PRB coal), 263
Power plant
  fuel combustion, 216f
  fuel preparation, 213
  pulverised-fuel power plant, 214f
  steam–water system, 214f
  superheaters–reheaters–waterwalls, 213–217
  types, component operating environments and fuel options, 212
  waste-fired grate unit, 216f
PRB coal. *See* Powder River Basin coal (PRB coal)
Precombustion capture of CO$_2$, 54
Pressure swing adsorption unit (PSA unit), 167
Pressurized fluidized-bed combustion (PFBC), 284–285
Pressurized operation, 153–154
Primary reformer, 165
Producer gas, 45–46, 248t
Proximate analysis, 38, 40, 83–84, 83f

Proximate coal analysis, 40
Proximity, 18
    principle, 155
PSA unit. *See* Pressure swing adsorption unit (PSA unit)
Pulverised coal combustion (PC combustion), 187
Pulverised fuel combustion (PF combustion), 181. *See also* Fluidised-bed combustion; Grate combustion
    general, 187−189
    technology options for co-firing, 196, 196f
        advantages and disadvantages, 197t
        direct co-firing in PC boilers, 190−192
        general, 189−190
        indirect co-firing technologies, 192−194
        parallel co-firing, 194−195
        PC boiler conversion into BFB boiler, 196
Pulverised fuel-fired boilers, 222
Pulverised torrefied biomass pellets, 134−135, 138
Pulverised-coal boilers (PC boilers), 188f, 189, 193f
    advantages and disadvantages, 197t
    co-firing options in, 196
    conversion into BFB boiler, 196
    direct co-firing in, 190−192
    down-shot PC furnace, 189f
    indirect co-firing technologies, 192−194
    parallel co-firing, 194−195
Pulverized coal combustion (PCC), 29
Pure methane, 298
Pyrolysis, 155−156, 262, 262t. *See also* Co-pyrolysis
Pyrolysis oil. *See* Bio-oil

Q
Quality control, 160
Quartz (SiO_2), 181

R
Rayleigh criterion, 278
RDF. *See* Refuse-derived fuel (RDF)
Re-burning, 191−192
Reactors, 146−147
Reclaimed wood, 109−110
Refinery, 5

Refuse-derived fuel (RDF), 18, 99, 155, 178, 211
Renewable Heat Incentive (RHI), 99
Renewable Obligation Certificates (ROCs), 99
Residues, 61, 67
Reversibility, 107−108
Revolutions per minute (rpm), 295
RHI. *See* Renewable Heat Incentive (RHI)
Richards−Lieuwen mechanism, 278
Roadside landing, 65−67
Rock asphalt, 11−12
ROCs. *See* Renewable Obligation Certificates (ROCs)
ROM. *See* Run of mine (ROM)
Room-and-pillar mining, 48
rpm. *See* Revolutions per minute (rpm)
Rule of thumb, 59, 110
Run of mine (ROM), 51

S
Saltation velocity, 127−128
Sasol's South African facility, 47−48
SCOC-CC. *See* Semi-closed oxy-fuel combustion combined cycle (SCOC-CC)
Secondary autothermal reformer, 165
Self heating, 107−108, 113−114, 116
Semi-closed oxy-fuel combustion combined cycle (SCOC-CC), 288, 289f
Separate milling, 191−192
Shale, 10
    formations, 10
    gas, 10−11
    oil, 3
Ships, 53
Shovel or scoop, 81
Silicon dioxide (SiO_2), 43−44
Silos, 109, 113, 114f, 117−118, 123
SiO_2. *See* Silicon dioxide (SiO_2)
Situ gasification, 8
Slagging, 220−221
Slagging-mode operation, 195
Slagging−fouling−corrosion, 231
Small-scale heating systems, 59
Small-scale outdoor storage tests, 132−133
SMR. *See* Steam-methane reforming (SMR)
SO_2 autoreduction. *See* SO_2 reduction
SO_2 reduction, 186, 187f

Index

Sodium oxide (Na$_2$O), 43–44
Softening temperature (ST), 43
Solid fuel types
   coal supply chain main characteristics, 48–53
   fossil fuel feedstocks, 29–44
   fuels derived from coal, 44–48
Solid fuel-flexible power generation, 201
   biomass fuels, 207–210, 209t
   coals, 202–207, 204t
     minerals in, 206t
   degradation mechanisms and modelling, 217–229
   flexible fuel use, 230–231
   power plant types, component operating environments and fuel options, 212–217
   quantification of damage and protective measures, 232–233
   waste-derived fuels, 210–212, 211t
Solid wastes, 243
Solid-recovered fuel (SRF), 99, 186, 211
Solids-handling processes, 100f
Spark-ignited gas engine, 294
SRF. *See* Solid-recovered fuel (SRF)
ST. *See* Softening temperature (ST)
Stationary gas turbines, 273
Stationary reciprocating engines, 293, 302
Steam
   coal, 9
   explosion, 125
   gasification reaction, 247
Steam reforming. *See* Steam-methane reforming (SMR)
Steam-injected gas turbine (STIG), 288, 290f
Steam-methane reforming (SMR), 165–166, 247
STIG. *See* Steam-injected gas turbine (STIG)
Stock rotation, 113–116
Stoichiometric air requirement, 296, 296t
Storage of biomass, 123
Store design, 106
Subbituminous coal, 9, 37
Sulfur (S), 42, 71
Sulphur oxides (SO$_x$), 264
Superheater–reheater corrosion, 224–229
Superheaters–reheaters–waterwalls, 213–217

Surface mining, 48–51
Swamp gas. *See* Biogas
Syngas. *See* Synthesis gas (Syngas)
Synthesis gas (Syngas), 9, 15, 44–45, 145. *See also* Biomass
   conversion to products, 161–168
     carbon chain groups, 162t
     hydrogen–carbon monoxide ratios, 163t
     product properties, 167–168
     technologies, 164–167
   current status and future trends, 168
     environmental aspects, 169–170
     technical aspects, 168–169
   gasification, 145–158
   production methods, 160
     carbonization, 161
     liquefaction, 160–161

T

Tar sand, 11–12
   bitumen, 11–12
   deposits, 12
Thermal treatment, 72
Thermophoresis, 218
Tight shale, 11
Time dependent biomass materials, 113
Titanium dioxide (TiO$_2$), 43–44
Torrefaction, 125, 153–154
Torrefied *Eucalyptus* pellets, 131
Total moisture, 39
TPOX reaction. *See* Noncatalytic partial oxidation reaction (TPOX reaction)
TSE. *See* Turun Seudun Energiantuotanto Oy (TSE)
Turbocharger, 296
Turun Seudun Energiantuotanto Oy (TSE), 191
Typical mining supply chain, 49f

U

UK. *See* United Kingdom (UK)
Ultimate analysis, 41, 83–84, 83f
Ultra-supercritical main boiler (USC main boiler), 194–195
UN. *See* United Nations (UN)
Unconventional energy sources, 11–14. *See also* Conventional energy sources
   biomass, 13–14

Unconventional energy sources (*Continued*)
  oil shale, 12–13
  tar sand bitumen, 11–12
Underground mining, 48, 50f
UNECE. *See* United Nations Economic Commitment for Europe (UNECE)
United Kingdom (UK), 122, 203
United Nations (UN), 210
United Nations Economic Commitment for Europe (UNECE), 34
United States (USA), 203
Unloading/discharge, 122
USA. *See* United States (USA)
USC main boiler. *See* Ultra-supercritical main boiler (USC main boiler)

V

Vapour condensation, 218
Volatile matter, 40, 149
Volumetric energy density
  low and variable effect, 103–105
  variation effect, 106

W

Waste, 18–19
  incineration, 180

Waste-derived fuels, 177, 210–212, 211t
Waste-fired systems, 220, 225
Water (H_2O), 13, 158–159, 263
  gas, 45
Water–gas-shift reaction, 162, 247
Water–gas-shift reactors (WGS reactors), 166–167, 287
Waterwall corrosion, 222–224
Westinghouse advanced gasification technology, 156
Wet basis, 106–107
WGS reactors. *See* Water–gas-shift reactors (WGS reactors)
WI. *See* Wobbe Index (WI)
Wobbe Index (WI), 275, 297
  gaseous fuels, 297t
Wood, 64, 106
  char, 59
  chips, 105, 107, 109–110
  fuels, 59
  pellets, 67, 99, 105, 107–108, 109f, 113–114
Woody
  biomass, 243, 245t–246t, 60f
  by-products and residues, 67